HDL WITH
DIGITAL DESIGN

HDL WITH
DIGITAL DESIGN
VHDL AND VERILOG

Nazeih Botros

MERCURY LEARNING AND INFORMATION

Dulles, Virginia
Boston, Massachusetts
New Delhi

Publisher: David Pallai
Mercury Learning and Information
22841 Quicksilver Drive
Dulles, VA 20166
info@merclearning.com
www.merclearning.com
(800) 232-0223

Nazeih Botros. *HDL with Digital Design: VHDL and Verilog*
ISBN: 978-1-938549-81-6

The publisher recognizes and respects all marks used by companies, manufacturers, and developers as a means to distinguish their products. All brand names and product names mentioned in this book are trademarks or service marks of their respective companies. Any omission or misuse (of any kind) of service marks or trademarks, etc. is not an attempt to infringe on the property of others.

Library of Congress Control Number: 2014950125

151617321 This book is printed on acid-free paper.

Our titles are available for adoption, license, or bulk purchase by institutions, corporations, etc. For additional information, please contact the Customer Service Dept. at 800-232-0223(toll free).

All of our titles are available in digital format at authorcloudware.com and other digital vendors. Companion files (figures and code listings) for this title are available by contacting info@merclearning.com. The sole obligation of Mercury Learning and Information to the purchaser is to replace the disc, based on defective materials or faulty workmanship, but not based on the operation or functionality of the product.

CONTENTS

PREFACE

This book provides the basic knowledge necessary to understand how to design and analyze basic digital logic systems and to know how to simulate these systems using hardware description languages. Systems here include digital logic circuits such as: adders, multiplexers, decoders, multipliers, flip-flops, latches, counters, sequential state machines, cache memories, and basic computers, simplified biological mechanisms that describe the operation of organs such as kidney, mathematical models (e.g., factorial, greatest of N numbers, multiplication algorithms, polynomials), and artificial intelligence (e.g., artificial neural networks). The book covers, in detail, Very High Speed Integrated Circuit Hardware Description Language (VHDL) and Verilog HDL. The book also covers a very important tool in writing the HDL code, the mixed language description where both VHDL and Verilog constructs are implemented in one HDL program. It also covers fundamentals of hardware synthesis. The book classifies the HDL styles of writing into six groups: data flow, behavioral, structural or gate-level, switch-level, mixed-type, and mixed language description.

Book Organization

The following is a brief description of the subjects that are covered in each chapter.

Chapter 1: Covers structure of the HDL module, operators including logical, arithmetic, relational and shift, data types such as scalar, composite and file, and a brief comparison between VHDL and Verilog. The chapter also covers how to simulate and test HDL code using test benches

Chapter 2: Covers: a) Analysis and design of combinational circuits such as adders, subtractors, decoders, multiplexers, comparators and simple multipliers, and sequential circuits such as latches; b) Simulation of the above combinational and sequential circuits using VHDL and Verilog data-flow description. The description includes covering of logical operators, concurrent signal-assignment statements, time delays, and vectors.

Chapter 3: Covers: a) Analysis and design of sequential circuits such as D flip-flop, JK flip-flop, T flip-flop, binary counters, and shift register; b) Understand the concept of some basic genetic and renal systems; c) Implementation of

Booth algorithm; d) Simulation of the systems in (a), (b), and (c) using VHDL and Verilog behavioral description. The description includes covering of the sequential statements `if`, `case`, `loop casex`, `casez`, `when`, `report`, `$display`, `wait`, `loop`, `exit`, `next`, `always`, `repeat`, `forever`, and `initial`.

Chapter 4: Covers: a) Analysis and design of sequential state machines; b) Analysis and design of adders, multiplexers, decoders, comparators, encoders, latches, flip-flops, counters, shift registers, and memory cells; c) Simulation of the systems in (a) and (b) using VHDL and Verilog structural description including the statements: `component`, `use`, `and`, `or`, `not`, `xor`, `nor`, `generate`, `generic`, and `parameter`.

Chapter 5: Covers: a) Analysis and design of primitive gates and simple logics using transistors (switches); b) Simulation of the above logics using HDL switch-level description. The description includes the Verilog statements `nmos`, `pmos`, `cmos`, `supply1`, `supply0`, `tranif0`, `tran`, and `tranif0`.

Chapter 6: Covers: a) Handling of real (fraction) data, Implementation of IEEE 754 Floating point representation and handling of signed numbers; b) Analysis and design of combinational array multiplier; c) Exploring the enzyme-substrate mechanism; d) Simulation of (a) and (b) using VHDL and Verilog `procedure`, `task`, and `function`.

Chapter 7: Covers: a) Implementation of arrays, single and multidimensional; b) Design of a basic computer; c) Simulation of (a) and (b) using VHDL and Verilog mixed description. The description includes VHDL user-defined types and packages.

Chapter 8: Covers: a) Analysis and design of cache memories and simple artificial neural networks; b) Simulation of the above systems in (a); c) File processing, character and string implementation VHDL Assert and Block statements.

Chapter 9: Covers: Mixed language description where both VHDL and Verilog can be implemented in the same program.

Chapter 10: Covers the basics of hardware synthesis.

Who Should Use this Book?

The book is appropriate as a textbook for first or second year electrical engineering, computer engineering, or computer science students; some of the advanced topics in the book can be omitted if desired by the instructor. The book is also appropriate for short courses for digital design engineers.

Suggested courses that could use this book are: digital logic design, computer architecture, HDL programming and synthesis, application-specific integrated circuits (ASICs) design, or digital design projects.

About the Examples covered in this Book

The examples written in this book are comprehensive and numerous. The examples cover a wide span of topics such as digital design logic, artificial neural networks, and simple biological mechanisms. The examples cover the analysis and design of digital logic circuits and the basic microcomputer. The examples cover, in detail, how to write the HDL code to simulate the systems under consideration. Both VHDL and Verilog codes are explained and implemented in the examples. The rules of writing the HDL code are explained in the examples.

There might not be enough time available to cover all the examples. In this case, the instructor can opt to cover only those examples that fit the student's "background."

How to Use this Book

The digital logic design part of the book is designed to cover the basic components in the early chapters (2–5) and then the more complex components in chapters 6–10. The HDL part of the book covers the two major hardware description languages, VHDL and Verilog. The book almost equally focuses on both languages. If readers want to learn one language at a time, they can read the sections with the title of the respective language. Almost all examples in the book are written into two parts, a and b; part a is written in VHDL and part b is written in Verilog. Some examples, however, are written in only one language, when the example is dealing with a very specific language construct that belongs only to one language and has no counterpart in the other language. An example of this exception is the VHDL Assert statement; this statement does not have a clear Verilog counterpart, so it is written only in VHDL.

If the reader wants to learn both languages at the same time; the book is organized to serve as learning tool for both languages. The two languages are not far apart from each other; they have several similarities. I have taught both languages in one course in one semester. I started with one language (VHDL); I covered the VHDL sections in Chapter 1, "Introduction," and Chapter 2, "Data Flow Description." After covering VHDL in Chapters 1 and 2, the student became familiar with the basic rules of HDL language and is ready to learn the other language (Verilog). I covered Verilog material of Chapters 1 and 2. After Chapter 2 until the end of the semester, I have covered both VHDL and Verilog at the same time in the same order as the Chapters of the book. The order of these Chapters after Chapter 2 is: 1) Chapter 3, "Behavioral description"; VHDL and Verilog have several similarities on behavioral statements such as if, case, and loop. 2) Chapter 4, "Structural Description," again both languages have many similarities except the VHDL does not have built-in components as the Verilog does. By including packages, VHDL can use components very similar to that of Verilog. 3) Chapter 5, "Switch-Level Description," -again VHDL does not have built-in constructs for switch-level descriptions, but we can include packages that allow us to write VHDL switch level statements very close to that of Verilog. 4) Chapter 6, "Procedures, Tasks, and Functions," here VHDL and Verilog have many similarities. 5) Chapter 7, "Mixed-Type Description." 6) Chapter 8, "Advanced HDL Description." 7) Chapter 9, "Mixed Language Description"; the student now knows both VHDL and Verilog; in this chapter he will learn how to mix between VHDL and Verilog constructs. 8) Chapter 10, "Synthesis Basics."

Companion Files

Companion files (figures and code listings) for this title are available by contacting info@merclearning.com.

Nazeih Botros
Carbondale, IL
February, 2015

INTRODUCTION

Chapter Objectives

- Understand the basics of hardware description language (HDL)
- Learn how the HDL module is structured
- Learn the use of operators in HDL modules
- Learn the different types of HDL objects
- Understand and analyze the half-adder circuit
- Understand the function of a simulator
- Understand the function of a synthesizer
- Understand the main differences between VHDL and Verilog HDL

1.1 Hardware Description Language

Hardware Description Language (HDL) is an essential computer-aided design (CAD) tool for the modern design and synthesis of digital systems. The recent steady advances in semiconductor technology continue to increase the power and complexity of digital systems. Due to their complexity, such systems cannot be easily realized using discrete integrated circuits (ICs) or even the newer schematic-level simulation. These systems are usually realized using high-density programmable chips, such as application-specific integrated circuits (ASICs) and field-programmable gate arrays (FPGAs), and require sophisticated CAD tools. HDL is an integral part of such tools. HDL offers the designer a very efficient tool for implementing and synthesizing designs on chips.

The designer uses HDL to describe the system in a computer-language code that is similar to several commonly used software languages such as C. Debugging the design is easy because HDL packages implement simulators and test benches. The two widely used hardware description languages are VHDL and Verilog. Because the two languages are implemented in both academia and industry, this book covers both languages.

After writing and testing the HDL code, the user can synthesize the code into digital logic components such as gates and flip-flops that can be downloaded into FPGAs or compatible electronic components. Usually, the CAD package that has HDL will also have a synthesizer. The HDL and synthesizer have made the task of designing complex systems much easier and faster than before. It is worth mentioning here that the currently available synthesizers have some limitations and cannot synthesize all HDL constructs; however, continuous improvement of the synthesizers is being undertaken by the electronic industry.

HDL has gone through continuous improvement since its inception. Verilog was introduced in 1980s and has gone through several iterations and standardization by the Institute of Electrical and Electronic Engineers (IEEE), such as in December 1995 when Verilog HDL became IEEE Standard 1364-1995, in 2001 when IEEE Std. 1364-2001 was introduced, and in 2005 when IEEE 1800-2005 was introduced. VHDL, which stands for very-high-speed integrated circuit (VHSIC) hardware description language, was developed in the early 1980s. In 1987, the IEEE Standard 1076-1987 version of VHDL was introduced, and several upgrades followed. In 1993, VHDL was updated and more futures were added; the result of this update was IEEE Standard 1076-1993. Recently, in 2008, the VHDL IEEE 1076-2008 was introduced.

1.2   Structure of the HDL Module

HDL modules follow the general structure of software languages such as C. The module has a source code that is written in high-level language style. Text editors supplied by the HDL package vendor can be used to write the module, or the code can be written using external text editors and imported to the HDL package by copy and paste. The most recently introduced feature in HDL packages allows automatic generation of HDL code from C-language code. VHDL has a somewhat different structure than Verilog HDL. In this book, Verilog HDL will be simply be referred to as Verilog. In Section 1.2.1, VHDL structure is discussed, and in Section 1.2.2, Verilog structure is discussed.

To illustrate the structure of the HDL module, let's consider a half-adder circuit. A half adder is a combinational circuit, which is a circuit whose output depends only on its input and which adds two input bits and outputs the result as two bits, one bit for the sum and one bit for the carry out. Examples of half addition include: 1 + 0 = 01, 1 + 1 = 10, and 0 + 0 = 00. Table 1.1 shows the truth table of the half adder.

TABLE 1.1 Truth Table for the Half Adder

Input		Output	
a	**b**	**S**	**C**
0	0	0	0
1	0	1	0
0	1	1	0
1	1	0	1

The Boolean function of the output of the adder is obtained from the truth table. The Boolean function of the output is generated using minterms (where the output has a value of 1) or maxterms (where the output has a value of 0). The Boolean function using minterms in the sum of products (SOP) form is

$$S = \bar{a}\, b + a\, \bar{b} = a \oplus b \tag{1.1}$$

$$C = a\, b \tag{1.2}$$

Using the maxterms in the product of sums (POS) forms

$$S = (a + b)(\bar{a} + \bar{b}) = a \oplus b \tag{1.3}$$

$$C = (a + b)(\bar{a} + b)(a + \bar{b}) = ab \tag{1.4}$$

After minimization ($a\bar{a}$ = 0 and $b\bar{b}$ =0), the SOP and the POS yield identical Boolean functions. Figure 1.1a shows the logic symbol of the half adder. Figure 1.1b shows the logic diagram of the half adder.

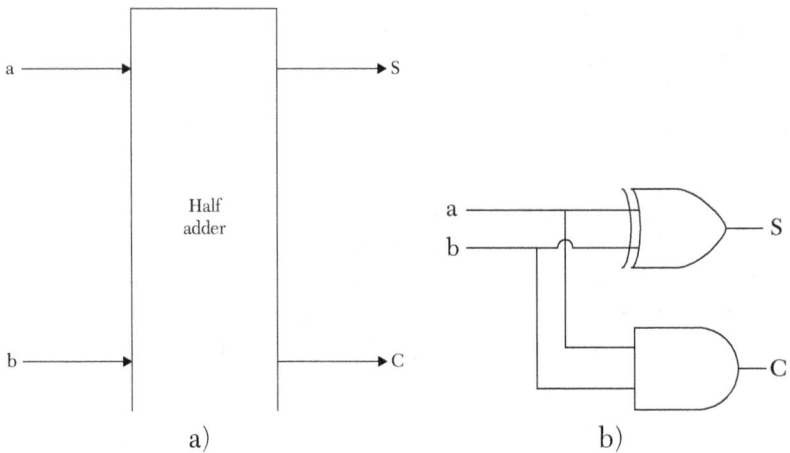

FIGURE 1.1 Half adder. a) Logic symbol. b) Logic diagram.

1.2.1 Structure of the VHDL Module

The VHDL module has two major constructs: entity and architecture. Entity declares the input and output signals of the system to be described and is given a name or identifier by the user. VHDL is case insensitive; for example, the two entity names `Half_ADDER` and `half_adder` are treated as the same name. The name should start with an alphabetical letter and can include the special character underscore (_). Declarations include the name and type of the inputs and outputs of the system. The inputs and outputs here are called input ports and output ports. The name of the port is user selected, and it has the same requirements as the entity's name. The entity that may describe the information depicted in Figure 1.1a is:

```
entity Half_adder is
port(a: in bit; b : in bit; S : out bit;
  C: out bit);
end half_adder;
```

The word `entity` is a predefined word. The name of the entity is `Half_adder`. This name is user selected and does not convey any information about the system; it is just an identifier. The entity could have been given any other name. VHDL does not know that the entity `Half_adder` describes a half adder simply by its name. The entity here has two input ports and two output ports. The term `is` is a predefined word and must be written after the name of the entity. The word `port` is predefined. The names of the input ports are `a` and `b`, and they must be followed by a colon (:). The predefined word `in` instantiates the mode of the port as an input (see Section 1.4.1 for details on port modes). The type of these input signals is `bit` and determines the allowed values that signals a and b can take. Type `bit` allows the signal to take only either logic 0 or logic 1. There are several other types, such as `std_logic`, `real`, and `integer` (see Section 1.6.1). The entity also has two output ports, `s` and `c`; they are declared as outputs with the predefined word `out`, and their type is `bit`. The order in which the input and output ports are written inside the parentheses is irrelevant. The output ports could have been listed before the input ports.

The last line of the entity's code uses the predefined word `end`, and it ends the entity. The name of the entity can follow the word `end`, as in `end Half_adder`, or the name of the entity can be omitted and only `end` is entered.

The semicolon (;) is an important character in HDL. It is used as a separator similar to the carriage return character used in C language. For example, the port statement can be written as:

```
port( a: in bit;
b : in bit;
S : out bit;
C: out bit);
```

The carriage return between the statements does not convey any information; it is the semicolon that signals a new statement. Ports can be declared `in`, `out`, `inout`, `buffer`, or `linkage` (see Section 1.4.1).

The second construct of the VHDL module, the architecture, describes the relationship between the inputs and outputs of the system. Architecture has to be bound to an entity. This relationship can be described using several sources; one of these sources is the Boolean function of the outputs. Other sources for describing the relationship between the output(s) and the input(s) are discussed in Section 1.3. Multiple architectures can be bound to the same entity, but each architecture can be bound to only one entity. Listing 1.1 shows an example of an architecture bound to the entity `Half_adder`. The architecture is declared by the predefined word `architecture`, followed by a user-selected name; this name follows the same name-selecting guidelines as the entity. In Listing 1.1, the name of the architecture is `dtfl_half`. The name is followed by the predefined word `of`, followed by the name of the entity. The predefined word `of` binds the architecture `dtfl_half` to the entity `Half_adder`. Binding here means the information listed in the entity is visible to the architecture.

LISTING 1.1 Example of Entity Architecture

```
entity Half_adder is
  port(a: in bit; b : in bit; S : out bit;
          C: out bit);
  end half_adder;
architecture dtfl_half of Half_adder is
begin
S <= a xor b; -- statement 1
C <= a and b; -- statement 2
--Blank lines are allowed
end dtfl_half;
```

In Listing 1.1, the architecture dtfl_half recognizes the information declared in the entity, such as the name and type of ports a, b, S, and C. After entering the name of the entity, the predefined word is must be entered. The architecture's body starts with the predefined word begin, followed by statements that detail the relationship between the outputs and inputs.

In Listing 1.1, the body of the architecture includes two statements. The two hyphens (--) signal that a comment follows. Statements 1 and 2 constitute the body of the architecture; they are signal assignment statements (see Chapter 2). The two statements describe the relationship between the output ports s and c and the input ports a and b. The xor and and are called logical operators (see Section 1.5.1.1); they simulate EXCLUSIVE-OR and AND logic, respectively. The architecture is concluded by the predefined word end. The name of the architecture can follow, if desired, the predefined word end. Leaving blank line(s) is allowed in the module; also, spaces between two words or at the beginning of the line are allowed.

1.2.2 Structure of the Verilog Module

Verilog module has declaration and body. In the declaration, the name, inputs, and outputs of the module are entered. The body shows the relationship between the inputs and the outputs. Listing 1.2 shows a Verilog description of a half adder based on the Boolean function of the outputs.

Listing 1.2 Example of a Verilog Module

```
module Half_adder(a,b,S,C);
    input a,b;
    output S, C;
    // Blank lines are allowed

    assign S = a ^ b; // statement 1
    assign C= a & b; // statement 2
endmodule
```

The name of the module in Listing 1.2 is a user-selected `Half_adder`. In contrast to VHDL, Verilog is case sensitive. `Half_adder`, `half_adder`, and `half_addEr` are all different names. The name of the module should start with an alphabetical letter and can include the special character underscore `(_)`. The declaration of the module starts with the predefined word module followed by the user-selected name. The names of the inputs and outputs (they are called input and output ports) follow the same guidelines as the module's name. They are written inside parentheses separated by a comma. The parenthesis is followed by a semicolon. In Listing 1.2, a, b, S, and C are the names of the inputs and outputs. The order of writing the input and output ports inside the parentheses is irrelevant. We could have written the module statement as:

```
module half_adder (S, C, a, b);
```

The semicolon (;) plays the same rule as in VHDL module; it is a line separator. Carriage return here does not indicate a new statement, the semicolon does. Following the module statement, the input and output port modes are declared. For example, the statement `input a;` declares signal *a* as an input port. The modes of the ports are discussed in Section 1.4.2. In contrast to VHDL, the type of the input and output port signals need not be declared. The order of writing the inputs and outputs and their declaration is irrelevant. For example, the inputs and outputs in Listing 1.2 can be written as:

```
module half_adder (a,b, S, C);
        output S;
        output C;
        input a;
        input b;
```

Also, more than one input or output could be entered on the same line by using a comma (,) to separate each input or output as:

```
module half_adder (a,b, S, C);
        output S, C;
        input a, b;
```

Statements 1 and 2 in Listing 1.2 are signal assignment statements (see Chapter 2). In statement 1, the symbol ^ represents an EXCULSIVE-OR operation; this symbol is called a logical operator (see Section 1.5.1.2). So, statement 1 describes the relationship between S, a, and b as S = a xor b. In statement 2, the symbol & represents an AND logic; the symbol is called a logical operator. So, statement 2 describes the relationship between C, a, and b as C = a and b. Accordingly, Listing 1.2 simulates a half adder. The double slash (//) is a comment command where a comment can be entered. If the comment takes more than one line, a double slash or pair (/*........*/) can be used. The module is concluded by the predefined word `endmodule`. Leaving blank lines is allowed in the module; also, spaces between two words or at the beginning of the line are allowed.

1.3 Styles (Types) Of Description

Several styles of code writing can be used to describe the system. Selection of the styles depends on the available information on the system. For example, some systems may be easily described by the Boolean function of the output; for other systems, such as biological mechanisms, it will be hard to obtain the Boolean function of the output, but they can be described if the relationship between the changes of the output with the input is known. In the following section, six styles will be discussed: data flow, behavioral, structural, switch level, mixed type, and mixed language.

1.3.1 Data Flow Description

Data flow describes how the system's signals flow from the inputs to the outputs. Usually, the description is done by writing the Boolean function of the outputs. The data-flow statements are concurrent; their execution is controlled by events. The VHDL architecture or Verilog module data-flow description, as defined here, does not include any of the key words that identify behavioral, structural, or switch-level descriptions. Data-flow descriptions are covered in Chapter 2. Data-flow style has been implemented in Section 1.2 where the Boolean function of S and C have been implemented to describe the half adder; see Listing 1.1 (VHDL) and Listing 1.2 (Verilog).

1.3.2 Behavioral Description

A behavioral description models the system as to how the outputs behave with the inputs; usually, a flowchart is used to show this behavior. In the half adder, the S output can be described as "1" if the inputs a and b are not equal, otherwise S = "0," (see Figure 1.2). The output C can be described as acquiring a value of "1" only if each input (a and b) is "1." The HDL behavioral description is the one where the architecture (VHDL) or the module (Verilog) contains the predefined word `process` (VHDL) or `always` or `initial` (Verilog). Behavioral description is usually used when the Boolean function or the digital logic of the system is hard to obtain. Behavioral description is covered in Chapter 3. Listing 1.3 shows a behavioral description of the output S of the half adder.

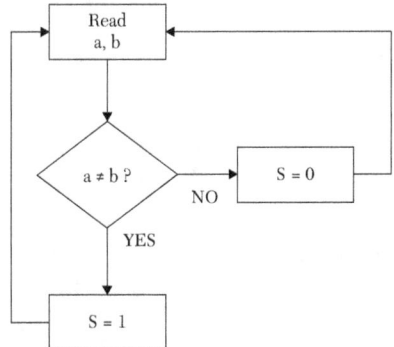

FIGURE 1.2 Behavior of output S with changes in inputs a and b.

LISTING 1.3 Example of Behavioral Description

VHDL1B Description
```
entity Half_adder is
port(a: in bit; b : in bit; S : out bit;
 C: out bit);
end half_adder;

architecture beh_half of Half_adder is
begin
process (a, b)
```

```
begin
if (a /= b) then
S <= '1';
else
S <= '0';
--Blank lines are allowed
end if;
end process;
end beh_half;
```

Verilog Description
```
module Half_adder(a,b,S,C);
    input a,b;
    output S, C;
    reg S,C;
    // Blank lines are allowed
    always @ (a,b)
    begin
    if (a != b)
    S = 1'b1;
    else
    S = 1'b0;
    end
endmodule
```

1.3.3 Structural Description

Structural description models the system as components or gates. This description is identified by the presence of the keyword component in the architecture (VHDL) or gates construct such as and, or, and not in the module (Verilog). Structural description is covered in Chapter 4. For the half adder, Figure 1.1b is used to write the structural code. Listing 1.4 shows a structural description for the half adder.

LISTING 1.4 Example of Structural Description

VHDL Description
```
entity Half_adder is
    port(a: in bit; b : in bit; S : out bit;
        C: out bit);
end half_adder;
architecture struct_exple of Half_adder is
--ADDITIONAL BINDING IS NEEDED TO RUN THIS PROGRAM;
--SEE CHAPTER 4
```

```
component xor2
  --The above statement is a component statement
      port(I1, I2 : in std_logic;
            O1 : out std_logic);
  end component;
component and2
      port(I1, I2 : in std_logic;
      O1 : out std_logic);
  end component;
    begin
  X1: xor2 port map (a,b, S);
  A1: and2 port map (a,b, C);
end struct_exple;
```

Verilog Description
```
module Half_adder1(a,b,S,C);
    input a, b;
    output S,C;
    and a1(C,a,b);
    //The above statement is AND gate
    xor x1(S,a,b);
    //The above statement is EXCLUSIVE-OR gate
endmodule
```

1.3.4 Switch-Level Description

The switch-level description is the lowest level of description. The system is described using switches or transistors. Some of the Verilog predefined words used in the switch level description are nmos, pmos, cmos, tranif0, tran, and tranif1. VHDL does not have built-in switch-level primitives, but a construct package can be built to include such primitives. Details of the switch-level description can be found in Chapter 5. Listing 1.5 shows the switch-level description of an inverter.

LISTING 1.5 An Example of A Switch-Level Description

VHDL Description
```
library IEEE;
use IEEE.STD_LOGIC_1164.ALL;
entity Inverter is
  Port (y : out std_logic; a: in std_logic );
end Inverter;
```

```
architecture Invert_switch of Inverter is
--additional binding is needed to run this program;
--see chapter 5
   component nmos
   --nmos is one of the key words for switch-level.
        port (O1: out std_logic; I1, I2 : in std_logic);
   end component;
   component pmos
   --pmos is one of the key words for switch-level.
        port (O1: out std_logic ;I1, I2 : in std_logic);
end component;
      for all: pmos use entity work. mos (pmos_behavioral);
      for all: nmos use entity work. mos (nmos_behavioral);
      --The above two statements are referring to a package mos
      --See details in Chapter 5
      constant vdd: std_logic := '1';
      constant gnd : std_logic:= '0';
      begin
      p1 : pmos port map (y, vdd, a);
      n1: nmos port map (y, gnd, a);
end Invert_switch;
```

Verilog Description
```
module invert(y,a);
     input a;
     output y;
     supply1 vdd;
     supply0 gnd;
     pmos p1(y, vdd, a);
     nmos n1(y, gnd, a);
          /*The above two statement are using the two primi-
          tives pmos and nmos*/
     endmodule
```

1.3.5 Mixed-Type Description

Mixed-type or mixed-style descriptions are those that use more than one type or style of the above-mentioned descriptions. In fact, most of the descriptions of moderate to large systems are mixed. Some parts of the system may be described using one type and others using other types of description. Mixed-type description is covered in Chapter 7.

1.3.6 Mixed-Language Description

The mixed-language description is a newly added tool to HDL description. The user now can write a module in one language (VHDL or Verilog) and invoke or import a construct (entity or module) written in the other language. Listing 1.6 illustrates the mixed-language description. In this Listing, inside Verilog module `Full_Adder1`, the VHDL entity HA is instantiated (imported). The information given in that entity is now visible to the Verilog module. Mixed-language description is covered in Chapter 9.

LISTING 1.6 Example of Mixed-Language Description

```
module Full_Adder1 ( x,y, cin, sum, carry);
    input x,y,cin;
    output sum, carry;
    wire c0, c1, s0;
     HA H1 (y, cin, s0,c0);
    // Description of HA is written in VHDL in the
    // entity HA
    .................
    endmodule

library IEEE;
use ieee.std_logic_1164.all;
entity HA is
    --For correct binding between this VHDL code and the above
     --Verilog code, the entity has to be named here as HA
    port (a, b : in std_logic; s, c: out std_logic);
end HA;
architecture HA_Dtflw of HA is
    begin
            s <= a xor b;
            c <= a and b;
end HA_Dtflw;
```

1.4 Ports

A simple definition of ports can be stated as a communication means between the system to be described and the environment.

1.4.1 VHDL Ports

In VHDL, ports can take one of the following modes:

- **in:** The port is only an input port. In any assignment statement, the port should appear only on the right-hand side of the statement (i.e., the port is read).

- **out:** The port is only an output port. In any assignment statement, the port should appear only on the left-hand side of the statement (i.e., the port is updated).

- **buffer:** The port can be used as both an input and output but can have only one source (i.e., limited fan out). The port can appear on either the left- or right-hand side of an assignment statement. A buffer port can only be connected to another buffer port or to a signal that also has only one source.

- **inout:** The port can be used as both an input and output.

- **linkage:** Same as **inout** but the port can only correspond to a signal.

1.4.2 Verilog Ports

Verilog ports can take one of the following three modes:

- **input:** The port is only an input port. In any assignment statement, the port should appear only on the right-hand side of the statement (i.e., the port is read).

- **output:** The port is an output port. In contrast to VHDL, the Verilog output port can appear in either side of the assignment statement.

- **inout:** The port can be used as both an input and output. The inout port represents a bidirectional bus.

1.5 Operators

HDL has an extensive list of operators. These operators are used extensively in every chapter of the book. Operators perform a wide variety of functions. These functions can be classified as:

- Logical (see Section 1.5.1), such as AND, OR, and XOR

- Relational (see Section 1.5.2) to express the relation between objects. These operators include equality, inequality, less than, less than or equal, greater than, and greater than or equal.

- Arithmetic (see Section 1.5.3) such as addition, subtraction, multiplication, and division

- Shift (see Section 1.5.4) to move the bits of an object in a certain direction, such as right or left

In the following section, HDL operators are discussed. The reader is advised to briefly study the operators presented here in order to understand their concept. These operators are implemented in almost every chapter of this book. When implemented, the reader can return to this section to read the details of operators used.

1.5.1 Logical Operators

These operators perform logical operations, such as AND, OR, NAND, NOR, NOT, and XOR. The operation can be on two operands or on a single operand. The operand can be single or multiple bits. In Section 1.5.1.1, VHDL logical operators are discussed, and Verilog logical operators are discussed in Section 1.5.1.2.

1.5.1.1 VHDL Logical Operators

Table 1.2 shows a list of VHDL logical operators. These operators should appear only on the right-hand side of statements. The operators are bitwise; they operate on corresponding bits of two signals. For example, consider the statement z: = x xor y. If x is four-bit signal 1011 and y is four-bit signal 1010, then z = 0001.

TABLE 1.2 VHDL Logical Operators

Operator	Equivalent Logic	Operand Type	Result Type
AND		Bit	Bit
OR		Bit	Bit
NAND		Bit	Bit
NOR		Bit	Bit

Operator	Equivalent Logic	Operand Type	Result Type
XOR		Bit	Bit
XNOR		Bit	Bit
NOT		Bit	Bit

1.5.1.2 Verilog Logical Operators

Verilog has extensive logical operators. These operators perform logical operations such as AND, OR, and XOR. Verilog logical operators can be classified into three groups: bitwise, Boolean logical, and reduction. The bitwise operators operate on the corresponding bits of two operands. Consider the statement: $Z = X \& Y$, where the AND operator ($\&$) "ANDs" the corresponding bits of X and Y and stores the result in Z. For example, if X is the four-bit signal 1011, and Y is the four-bit signal 1010, then Z = 1010. Table 1.3 shows bitwise logical operators. For example, the NAND operation on X and Y is written as: $Z = \sim(X \& Y)$.

TABLE 1.3 Verilog Bitwise Logical Operators

Operator	Equivalent Logic	Operand Type	Result Type
&		Bit	Bit
|		Bit	Bit
~ (&)		Bit	Bit
~ (|)		Bit	Bit
^		Bit	Bit

Operator	Equivalent Logic	Operand Type	Result Type
~^		Bit	Bit
~		Bit	Bit

Other types of logical operators include the Boolean logical operators. These operators operate on two operands, and the result is in Boolean: 0 (false) or 1 (true). For example, consider the statement z = x && y where && is the Boolean logical AND operator. If x = 1011 and y = 0001, then z = 1. If x = 1010 and y = 0101, then z = 0. Table 1.4 shows the Boolean logical operators.

TABLE 1.4 Verilog Boolean Logical Operators

Operator	Operation	Number of Operands
&&	AND	Two
| |	OR	Two

The third type of logical operator is the reduction operator. Reduction operators operate on a single operand. The result is in Boolean. For example, in the statement y = &x, where & is the reduction AND operator, and assuming x = 1010, then y = (1 & 0 & 1 & 0) = 0. Table 1.5 shows the reduction logic operators.

TABLE 1.5 Verilog Reduction Logical Operators

Operator	Operation	Number of Operands
&	Reduction AND	One
|	Reduction OR	One
~&	Reduction NAND	One
~ |	Reduction NOR	One
^	Reduction XOR	One
~^	Reduction XNOR	One
!	NEGATION	One

1.5.2 Relational Operators

Relational operators are implemented to compare the values of two objects. The result returned by these operators is in Boolean: false (0) or true (1).

In Section 1.5.2.1, the VHDL relational operators are covered, and in Section 1.5.2.2, the Verilog relational operators are covered.

1.5.2.1 VHDL Relational Operators

VHDL has extensive relational operators. Their main implementations are in the `if` and `case` statements (see Chapter 3). Table 1.6 shows VHDL relational operators.

TABLE 1.6 VHDL Relational Operators

Operator	Description	Operand Type	Result Type
=	Equality	Any type	Boolean
/=	Inequality	Any type	Boolean
<	Less than	Scalar	Boolean
<=	Less than or equal	Scalar	Boolean
>	Greater than	Scalar	Boolean
>=	Greater than or equal	Scalar	Boolean

The following statements demonstrate the implementation of some of the above relational operators.

```
If (A = B) then .....
```

A is compared to B. If A is equal to B, the value of the expression (A = B) is true (1); otherwise, it is false (0).

```
If (A < B) then .....
```

If A is less than B, the value of the expression (A < B) is true (1); otherwise, it is false (0).

1.5.2.2 Verilog Relational Operators

Verilog has a set of relational operators similar to VHDL. Table 1.7 shows Verilog relational operators. As in VHDL, the relational operators return Boolean values: false (0) or true (1).

TABLE 1.7 Verilog Relational Operators

Operator	Description	Result Type
==	Equality	0, 1, x
!=	Inequality	0, 1, x
===	Equality inclusive	0, 1

Operator	Description	Result Type
!==	Inequality inclusive	0, 1
<	Less than	0, 1, x
<=	Less than or equal	0, 1, x
>	Greater than	0, 1, x
>=	Greater than or equal	0, 1, x
?	Conditional operator	0, 1, x

For the equality operator (==) and inequality operator (!=), the result can be of type unknown (x) if any of the operands include "don't care," "unknown (x)," or "high impedance z." These types are covered in Section 1.6.

The following are examples of a Verilog relational operators:

```
if (A == B)  .......
```

If the value of A or B contains one or more "don't care" or z bits, the value of the expression is unknown. Otherwise, if A is equal to B, the value of the expression is true (1). If A is not equal to B, the value of the expression is false (0).

```
if (A === B).....
```

This is a bit-by-bit comparison. A or B can include x or high impedance Z; the result is true (1) if all bits of A match that of B. Otherwise, the result is false (0).

For the conditional operator "?" the format is:

Conditional-expression ? true-expression : false-expression ;

The conditional expression is evaluated; if true, true-expression is executed If false, false-expression is executed. If the result of the conditional-expression is "x," both false and true are executed, and their results are compared bit by bit; if two corresponding bits are the same, the common value of these bits is returned. If they are not equal, an "x" is returned. The conditional operator is discussed in Chapter 2.

1.5.3 Arithmetic Operators

Arithmetic operators can perform a wide variety of operations, such as addition, subtraction, multiplication, and division. In Section 1.5.3.1, VHDL arithmetic operators are covered, and in Section 1.5.3.2, Verilog arithmetic operators are covered.

1.5.3.1 VHDL Arithmetic Operators

VHDL arithmetic operators operate on numeric and physical operand types (see Section 1.6). Physical data types are those that can be measured in units, such as time. An example of an arithmetic operator is the multiplication operator (*); the statement Y: = (A*B) calculates the value of Y as the product of A multiplied by B. Table 1.8 shows the VHDL arithmetic operators and the type of A, B, and Y.

TABLE 1.8 VHDL Arithmetic Operators

Operator	Description	A or B Type	Y Type
+	Addition A + B	A numeric B numeric	Numeric
−	Subtraction A − B	A numeric B numeric	Numeric
*	Multiplication A × B	A integer or real B integer or real	Same as A
*	Multiplication A × B	A physical B integer or real	Same as A
*	Multiplication A × B	A integer or real B physical	Same as B
/	Division A ÷ B	A integer or real B integer or real	Same as A
/	Division A ÷ B	A integer or real B physical	Same as B
/	Division A ÷ B	A physical B integer or real	Same as A
mod	Modulus A mod B	A only integer B only integer	Integer
rem	Remainder A rem B	A only integer B only integer	Integer
abs	absolute abs (A)	A numeric	Positive numeric
&	Concatenation (A & B)	A numeric or array B numeric or array	Same as A
**	Exponent A ** B	A real or integer B only integer	Same as A

1.5.3.2 Verilog Arithmetic Operators

Verilog, in contrast to VHDL, is not extensive type-oriented language. Accordingly, for most operations, only one type of operation is expected for each operator. An example of an arithmetic Verilog operator is the addition operator (+); the statement Y = (A + B) calculates the value of Y as the sum of A and B. Table 1.9 shows the Verilog arithmetic operators.

TABLE 1.9   Verilog Arithmetic Operators

Operator	Description	A or B Type	Y Type
+	Addition A + B	A numeric B numeric	Numeric
–	Subtraction A – B	A numeric B numeric	Numeric
*	Multiplication A × B	A numeric B numeric	Numeric
/	Division A ÷ B	A numeric B numeric	Numeric
%	Modulus A % B	A numeric, not real B numeric, not real	Numeric, not real
**	Exponent A ** B	A numeric B numeric	Numeric
{ , }	Concatenation {A , B}	A numeric or array B numeric or array	Same as A
{N{A}}	Repetition	A numeric or array	Same as A

1.5.3.3 Arithmetic Operator Precedence

The precedence of the arithmetic operators in VHDL or Verilog is the same as in C. The precedence of the major operators is listed below from highest to lowest:

```
**

* / mod (%)

+ -
```

1.5.4   Shift and Rotate Operators

Shift and rotate operators are implemented in many applications, such as in multiplication and division. A shift left represents multiplication by two, and a shift right represents division by two. VHDL shift operators are discussed in Section 1.5.4.1, and Verilog shift operators are discussed in Section 1.5.4.2.

1.5.4.1   VHDL Shift/Rotate Operators

Shift operators are unary operators; they operate on a single operand. To understand the function of these operators, assume that operand A is the four-bit vector 1110. Table 1.10 shows the VHDL shift operators as they apply to operand A.

TABLE 1.10   VHDL Shift Operators

Operation	Description Before Shift	Operand A After Shift	Operand A
A sll 1	Shift A one position left logical	1110	1100
A sll 2	Shift A two positions left logical	1110	10xx
A srl 1	Shift A one position right logical	1110	x111
A srl 2	Shift A two positions right logical	1110	xx11
A sla 1	Shift A one position left arithmetic	1110	110x
A sra 1	Shift A one position right arithmetic	1110	1111
A rol 1	Rotate A one position left	1110	1101
A ror 1	Rotate A one position right	1110	0111

Notice that rotate (ror or rol) keeps all bits of operand A. For example, A ror 1 shifts A one position to the right and inserts the least significant bit (0) in the vacant, most significant position.

1.5.4.2   Verilog Shift Operators

Verilog has the basic shift operators. Shift operators are unary operators; they operate on a single operand. To understand the function of these operators, assume operand A is the four-bit vector 1110. Table 1.11 shows the Verilog shift operators as they apply to operand A.

TABLE 1.11 Verilog Shift Operators

Operation	Description Before Shift	Operand A After Shift	Operand A
A << 1	Shift A one position left logical	1110	1100
A << 2	Shift A two positions left logical	1110	1000
A >> 1	Shift A one position right logical	1110	0111
A >> 2	Shift A two positions right logical	1110	0011
A.>>> 2	Shift A two positions right arithmetic	1110	1111
A.<<< 2	Shift A two positions left arithmetic	1110	1000

1.6   Data Types

Because HDL is implemented to describe the hardware of a system, the data or operands used in the language must have several types to match the need for describing the hardware. For example, if we are describing a signal, we need to specify its type (i.e., the values that the signal can take), such as type bit, which means that the signal can assume only 0 or 1, or type std_logic, in which the signal can assume a value out of nine possible values that include 0, 1, and high impedance. Examples of types include integer, real, vector, bit, and array. In Section 1.6.1, data types for VHDL are discussed, and in Section 1.6.2, data types for Verilog are discussed. The reader is advised to briefly study the data types presented here in order to know their concepts. Data types are implemented in almost every chapter of this book; when implemented, the reader can come back to this section to read the details about a selected data type.

1.6.1 VHDL Data Types

As previously mentioned, VHDL is a type-oriented language; many operations will not be executed if the right type for the operands has not been chosen. The type of any element or object in VHDL determines the allowed values that element can assume. Objects in VHDL can be signal (see Chapter 2), variable (see Chapter 3), or constant (see Chapters 2 and 3).

These objects can assume different types; these types can be classified into five groups depending on the nature of the values the object can assume: scalar, composite, access, file, and other.

1.6.1.1   Scalar Types

The values that a scalar can assume are numeric. Numeric values can be integer, real, physical (such as time), Boolean (0 or 1), or characters when stored as American Standard Code for Information Interchange (ASCII) or compatible code. The following types constitute the scalar types.

Bit *Type*

The only values allowed here are 0 or 1. It is used to describe a signal that takes only 1 (high) or 0 (low). The signal cannot take other values such as high impedance (open). An example of implementing this type is when the type of a port signal is described as:

```
port (I1, I2 : in bit; O1, O2 : out bit);
```

Signals I1, I2, O1, and O2 can assume only 0 or 1. If any of these signals must assume other levels or values, such as high impedance, bit type cannot be used.

Boolean Type

This type can assume two values: false (0) or true (1). Both true and false are predefined words. One of the most frequent applications of the Boolean type is in the if statement (see Chapter 3). Consider the statements:

```
If (y = B) then
     S := '1';
else
     S := '0';
end if;
```

The output of the first line, If (y =B), is Boolean: it is either true or false. If true, then S = 1; if false, S = 0. Boolean can also be specified as the port type:

```
port (I1, I2 : in bit; O1 : out bit; O2 : Boolean);
```

Integer Type

As the name indicates, this type covers all integer values; the values can be negative or positive. The default range is from –2,147,483,648 to

+2,147,483,647. The user can specify a shorter range by using the pre-defined word `range`. The predefined word `natural` can be used instead of `integer` if the values of the object are always positive, including 0. An example of the `integer` type is in the implementation of the exponent operator (see Section 1.5.3.1). The exponent has to be of type `integer`, such as X**2 or X**y, where y is declared as `integer`. The port can also be declared as type `integer`:

```
port (I1 : in natural; I2 : in bit; O1 : out integer; O2 : Boolean);
```

Another predefined type `positive` restricts the values an object can take to be positive and higher than 0.

Real Type

This type accepts fractions, such as .4502, 12.5, and –5.2E–10 where

$E-10 = 10^{-10}$. An example of using real type is:

```
port (I1 : in natural; I2 : in real; O1 : out integer; O2 :
    Boolean);
```

Character Type

This type includes characters that can be used to print a message using, for example, the predefined command `report`, such as:

```
report ("Variable x is greater than Y");
```

Notice that each character in the above message is just printed; no value is assigned to them. The `report` statement is very similar to the `print` statement in C language. Some format can be added to the characters printed by `report`:

```
report ("Variable x is greater than Y.") & CR &
    ("Variable x is > 2.34.");
```

where & is the concatenation operator (see Section 1.5.3.1), and CR is a predefined word for carriage return.

`subtype` and `type`, if used, assign numeric value to each character, as follows:

```
subtype wordChr is character;
type string_chr is array (N downto 0) of wordChr;
```

In addition, `subtype`, `type`, and `array` are predefined words (see arrays and user-defined types in this section and in Chapters 6–8). The two statements above declare an array of $N + 1$ elements, and each element is a character. The characters are associated with ASCII values. For example, character A has the ASCII value of 41 in hex. More discussion on characters can be found in Chapter 8 and Chapter 3.

Physical Type

This type has values that can be measured in units, such as time (e.g., second, millisecond, microsecond) and voltage (e.g., volt, millivolt, micro-volts). An example of type `time` is:

```
constant Delay_inv : time := 1 ns;
```

The above statement states that the constant `Delay_inv` is of type `time`, and its initial value is one nanosecond (1 ns). The word `time` is pre-defined; the units of `time` are as follows:

fs	femtosecond
ps	= 1,000 fs
ns	= 1,000 ps
us	= 1,000 ns
ms	= 1,000 us
sec	= 1,000 ms
min	= 60 sec
hr	= 60 min

User-Defined Types

The user can define a type by using the predefined word `type` as shown below:

```
type op is (add, mul, divide, none);
variable opcode : op := mul;
```

Type `op` is user defined. The variable `opcode` is of type `op` and can therefore be instantiated to: add, mul, divide, or none. More discussion about user-defined types can be found in Chapter 7.

Severity Type

This type is used with the `assert` statement (see Chapter 8). An object with type `severity` can take one of four values: `note`, `warning`, `error`, or `failure`. An example of this type is as follows:

```
assert (Flag_full = false);
report "The stack is full";
severity failure;
```

The `assert` condition is `Flag_full = false.` If `Flag_full` is not false, a message is printed to indicate that the stack is full and simulation is halted.

1.6.1.2 Composite Types

The composite type is a collection of values. There are three composite types: bit vector, arrays (see Chapter 7), and records (see Chapter 8). An array is a collection of values all belonging to a single type; a record is a collection of values with the same or different types.

Bit_vector Type

The `bit_vector` type represents an array of bits; each element of the array is a single bit. The following example illustrates the implementation of type `bit_vector`:

```
Port (I1 : in bit; I2 : in bit_vector (5 downto 0); Sum : out bit);
```

In the above statement, port `I2` is declared as type `bit_vector`; it has six bits. Possible values of `I2` include 110110, 011010, and 000000 or any other six-bit number. More details about `bit_vector` can be found in Chapter 2.

Array Type

This type is declared by using the predefined word `array`. For example, the following statements declare the variable `memory` to be a single-dimensional array of eight elements, and each element is an integer:

```
subtype wordN is integer;
type intg is array (7 downto 0) of wordN;
..........
variable memory : intg;
```

Arrays can be multidimensional. See Chapter 7 for more details on arrays.

Record Type

An object of `record` type is composed of elements of the same or different types. An example of `record` type is shown below:

```
Type forecast is
record
Tempr : integer range -100 to 100;
```

```
Day : real;
Cond : bit;
end record;
```

. . . .

```
variable temp : forecast
```

Variable `temp` is of type `forecast`; type `forecast` includes `record`, and `record` has three different types: `integer`, `real`, and `bit`. More details about records can be found in Chapter 8.

1.6.1.3 Access Types

Values belonging to an `access` type are pointers to objects of other types. For example:

```
type ptr_weathr is access forecast;
```

The type `ptr_weathr` is a pointer to the type `forecast` shown in last example of Section 1.6.1.2.

1.6.1.4 File Types

Objects of type `file` can be read from and written to using built-in functions and procedures that are provided in the standard library. Some of these procedures and functions are `file_open` to open files, `readline` to read a line from the file, `writeline` to write a line into the file, and `file_close` to close the file. More details about file types and operations can be found in Chapter 8.

1.6.1.5 Other Types

There are several other types provided by external libraries. For example, the IEEE library contains a package by the name of std_logic_1164. This package contains an extremely important type: `std_logic`. Type `bit` has only two values: level 0 and level 1. If more values are needed to represent the signal, such as high impedance, `bit` type cannot be used. Instead, type `std_logic`, which can assume nine values including high impedance, can be used.

Std_Logic Type

Std_Logic has nine values, including 1 and 0. Package std_logic_1164 should be attached to the VHDL module. The nine values of std_logic type are shown in Table 1.12.

TABLE 1.12   Values of Std_Logic Type

Value	Definition
U	Uninitialized
X	Unknown
0	Low
1	High
Z	High impedance
W	Weak unknown
L	Weak low
H	Weak high
–	Don't care

Std_logic_vector Type

The type std_logic_vector represents an array. Each element of the array is a single bit of type std_logic. The following example illustrates the implementation of type std_logic_vector:

```
Port (I1 : in bit; I2 : in std_logic_vector (5 downto 0);
Sum : out bit);
```

In the above statement, port I2 is declared as type std_logic_vector; it has six bits. Possible values of I2 include 110110, 011010, or 0Z0Z00. More details about std_logic_vector can be found in Chapter 2.

Signed

Signed is a numeric type. It is declared in the external package numeric_std and represents signed integer data in the form of an array. The leftmost bit is the sign; objects of type signed are represented in 2' complement form. Consider the statement:

```
Variable prod : signed (3 downto 0) := 1010;
```

The above statement declares the variable prod. It is of type signed, has four bits, and its initial value is 1010, or –6 (in decimal). Chapter 3 shows implementations of type signed.

Unsigned

The type `unsigned` represents unsigned integer data in the form of an array of `std_logic` and is a part of the package numeric_std. The following example illustrates type `unsigned`:

```
Variable Qout : unsigned (3 downto 0) := 1010;
```

The above statement declares variable `Qout` as of type `unsigned`, it has four bits, and its initial value is 1010, or 10 (in decimal).

1.6.2   Verilog Data Types

Verilog supports several data types including nets, registers, vectors, integer, real, parameters, and arrays. More details on these types can be found in almost all subsequent chapters.

1.6.2.1   Nets

Nets are declared by the predefined word `wire`. Nets have values that change continuously by the circuits that are driving them. Verilog supports four values for nets, as shown in Table 1.13.

TABLE 1.13   Verilog Net Values

Value	Definition
0	Logic 0 (false)
1	Logic 1 (true)
X	Unknown
Z	High impedance

Examples of net types are as follows:

```
wire sum;
wire S1 = 1'b0;
```

The first statement declares a net by the name `sum`. The second statement declares a net by the name of `S1`; its initial value is `1'b0`, which represents 1 bit with value 0.

1.6.2.2   Register

Register, in contrast to nets, stores values until they are updated. Register, as its name suggests, represents data-storage elements. Register is declared by the predefined word `reg`. Verilog supports four values for register, as shown in Table 1.14.

TABLE 1.14 Verilog Register Values

Value	Definition
0	Logic 0 (false)
1	Logic 1 (true)
X	Unknown
Z	High impedance

An example of register is:

```
reg Sum_total;
```

The above statement declares a register by the name `Sum_total`.

1.6.2.3 Vectors

Vectors are multiple bits. A register or a net can be declared as a vector. Vectors are declared by brackets `[ ]`. Examples of vectors are:

```
wire [3:0] a = 4'b1010;
reg [7:0] total = 8'd12;
```

The first statement declares a net `a`. It has four bits, and its initial value is 1010 (b stands for bit). The second statement declares a register `total`. Its size is eight bits, and its value is decimal 12 (d stands for decimal). Vectors are implemented in almost all subsequent chapters.

1.6.2.4 Integers

Integers are declared by the predefined word `integer`. An example of integer declaration is:

```
integer no_bits;
```

The above statement declares `no_bits` as an integer.

1.6.2.4 Real

Real (floating-point) numbers are declared with the predefined word `real`. Examples of real values are 2.4, 56.3, and 5e12. The value 5e12 is equal to 5×10^{12}. The following statement declares the register weight as real:

```
real weight;
```

1.6.2.5 Parameter

Parameter represents a global constant. It is declared by the predefined word `parameter`. The following is an example of implementing parameters:

```
module compr_genr (X, Y, xgty, xlty, xeqy);
parameter N = 3;
input [N:0] X, Y;
output xgty, xlty, xeqy;
wire [N:0] sum, Yb;
```

To change the size of the inputs x and y, the size of the nets sum, and the size of net Yb to eight bits, the value of N is changed to seven as:

```
parameter N = 7
```

1.6.2.6 Arrays

Verilog, in contrast to VHDL, does not have a predefined word for array. Registers and integers can be written as arrays. Consider the following statements:

```
parameter N = 4;

parameter M = 3;

reg signed [M:0] carry [0:N];
```

The above statements declare an array by the name carry. The array carry has five elements, and each element is four bits. The four bits are in two's complement form. For example, if the value of a certain element is 1001, then it is equivalent to decimal −7. Arrays can be multidimensional. See Chapter 7 for more details on arrays.

1.7   Simulation and Synthesis

The ultimate goal for hardware description is to synthesize the system onto an electronic chip. To synthesize an HDL description, it needs to be simulated and tested. Synthesis basics are covered in Chapter 10. More information about simulators and synthesizers can be found in the manual of the HDL vendors. The steps of simulation and synthesis in general can be summarized as follows:

1. Choose the preferred language to describe the system. The language may be VHDL, Verilog, or mixed-language (both VHDL and Verilog). Mixed-language descriptions are covered in Chapter 9.

2. Choose the style or type of description. Refer to Section 1.6 for selecting a style.

3. Write the code. If writing a VHDL module, be sure to attach all the necessary packages and libraries. At this step, some HDL packages require the user to select the type of synthesis technology and chip type before compilation.

4. Compile the code using the compiler supplied by the HDL package. The compiler checks that the code satisfies the rules of the language and displays any errors. Some compilers suggest how to fix the errors.

5. After successful compilation, the code is tested to see that it correctly describes the system. This test is done by selecting the input and output signals to be tested. For example, if a 2 x 1 multiplexer is being described, the two inputs, the select line, and the output might be selected. The way these signals are selected differs from one simulator to the other; there might be different ways to select signals even within the same simulator. Some simulators are graphical. All signals in the system are displayed in graphical fashion; the user selects the signals and assigns initial values for them. The user then clicks a button to run the simulation, and a simulation screen appears showing the waveform of the selected signals. Some other simulators allow the user to write HDL code, called test bench, for testing the source code.

6. After the simulation verifies that the signals behave as expected, the compiled code can be synthesized. The simulator CAD package usually has a synthesizer. The synthesizer converts the compiled code into a schematic and generates a net list. However, due to limitation in the available synthesizers, some statements may not be synthesized and the user may opt to change these statements if possible. The net list can be downloaded onto a chip, usually field-programmable gate arrays. Chapter 10 illustrates how to convert the HDL code to gate level or RTL, the forms closest to the schematic original that the synthesizer can download onto the chip.

Appendix A shows example of Steps 1–6 including a test bench.

1.8   Brief Comparison of VHDL and Verilog

As previously mentioned, VHDL and Verilog are hardware-description languages that are popular in both industry and academia. Each language, however, has some advantages and disadvantages over the other. These advantages and disadvantages may not be very clear to beginners. The two languages are continuously upgraded, and newer versions are introduced. These newer versions bring the capability of the two languages closer. Verilog is considered better when describing a system at the gate or transistor level due to its use of predefined primitives at this level. VHDL is considered better at the system level; multiple entity/architecture pairs lead to flexibility and ease in writing code for complex systems. Recently, many simulators have acquired the capability to use mixed-language simulations. In mixed-language simulations, a construct of one language can be instantiated into the other. This allows the user to utilize the advantages of both languages (see Chapter 9). In the following sections, the major differences between VHDL and Verilog, as seen by a beginner user, are listed.

Data Types

VHDL: Definitely a type-oriented language, and VHDL types are built in or users can create and define them. User-defined types give the user a tool to write the code effectively; these types also support flexible coding. VHDL can handle objects with multidimensional array types. Another data type that VHDL supports is the physical type; the physical type supports more synthesizable or targeted design code.

Verilog: Compared to VHDL, Verilog data types are very simple and easy to use. All types are defined by the language.

Ease of Learning

VHDL: For beginners, VHDL may seem hard to learn because of its rigid type requirements. Advanced users, however, may find these rigid type requirements easier to handle.

Verilog: Easy to learn, Verilog users just write the module without worrying about what library or package should be attached. Many of the statements in the language are very similar to those in C language.

Libraries and Packages

VHDL: Libraries and packages can be easily attached to the standard VHDL package. Packages can include procedures and functions, and the package can be made available to any module that needs to use it. Packages are used to target a certain design. For example, if the system modeled/designed includes arithmetic functions, a package can be used that includes those functions.

Verilog: Libraries are not as easily implemented as in VHDL, however the basic Verilog package includes several libraries as integer part of the package.

1.9   Summary

In this chapter, several introductory VHDL and Verilog topics have been covered. The structure of the HDL module was discussed. The VHDL module has two major constructs: an `entity` and `architecture`, which are bound to the entity. Verilog has a `module` construct.

Operators, which perform a wide variety of operations, have been covered. Arithmetic operators (see summary in Table 1.15) perform arithmetic operations such as multiplication and division. Relational operators (see summary in Table 1.16) perform comparisons such as greater than and equality. Shift operators (see summary in Table 1.17) perform bit shifts such as a logical shift (a specified number of bit positions) right. Logical operators (see summary in Table 1.18) perform logical operations such as AND.

Data types have also been covered, including `bit`, `std_logic`, `std_logic_vector`, and `array` (for VHDL), and `real`, `integer`, `reg`, and `wire` (for Verilog). The following description styles have been briefly contrasted: data flow, behavioral, structural, switch level, mixed type, and mixed language. Finally, a brief comparison of VHDL and Verilog has been presented.

TABLE 1.15   Summary of Arithmetic Operators for VHDL and Verilog

Operation	Operator VHDL	Verilog
Addition	+	+
Subtraction	–	–
Multiplication	*	*
Division	/	/
Modulus	mod	%
Exponent	**	**
Concatenation	(&)	{ , }

TABLE 1.16 Summary of Relational Operators for VHDL and Verilog

Operation	Operator VHDL	Verilog
Equality	=	==
Inequality	/=	!=
Less than	<	<
Less than or equal	<=	<=
Greater than	>	>
Greater than or equal	>=	>=
Equality inclusive	None	===
Inequality inclusive	None	!==

TABLE 1.17 Summary of Shift Operators for VHDL and Verilog

Operation	Operator VHDL	Verilog
Shift A logical left one position	A sll 1	A << 1
Shift A logical right one position	A srl 1	A >> 1
Shift A arithmetic left one position	A sla 1	A <<< 1
Shift A arithmetic right one position	A sra 1	A >>>1
Rotate A left one position	A rol 1	None
Rotate A right one position	A ror 1	None

TABLE 1.18 Summary of Logical Operators for VHDL and Verilog

Operation	Operator VHDL	Verilog
AND	AND	&
OR	OR	|
NAND	NAND	~ (&)
NOR	NOR	~(|)
XOR	XOR	^
XNOR	XNOR	~^
NOT	NOT	~

1.10 Exercises

1. Determine whether each of the following statements is VHDL, Verilog, or can be both. Justify your answer.

a. Parameter a;

b. assign m=0;

```
c.  port (input1 : bit; output2 : bit; output3 : bit);
d.  module vhdl1(I1, I2, O1, O3);
e.  input D, E;
f.  y = a >>> 3;
g.  process Verlog(a, b, c)
h.  always @ (a, b,c)
i.  end
j.  architecture exc of chapter1 is
k.  endmodule
```

2. If A and B are two unsigned variables, with A = 1100 and B = 1001, find the value of the following expressions:

```
a.  (A AND B)
b.  (A ^ B)
c.  (A XNOR B)
d.  (A & B)
e.  (A && B)
f.  !(A)
g.  ~|(B)
h.  A sll 3
i.  A >> 1
j.  B ror 2
k.  B >>> 2
```

3. Which style(s) would you chose to describe each of the following systems? Explain your answer.

a. A full adder

b. A controller to control the traffic light in five-way intersection

c. A circuit controlling the release of insulin according to the concentration of glucose

d. Two pmos transistors connected in parallel

2

DATA-FLOW DESCRIPTION

Chapter Objectives

- Understand the concept of data-flow description in both VHDL and Verilog
- Understand events and concurrent statements
- Identify the basic statements and components of data-flow description such as logical operators, signal-assignment statements, the assign statement, time delays, and vectors
- Review and understand K-maps, Boolean function, and fundamentals of some digital logic systems such as full adder, full subtractor, 2x1 multiplexer, 2x2 combinational multiplier, two-bit comparator, delay latch, ripple-carry adder, and carry-lookahead adder

2.1   Highlights Of Data-Flow Description

Data flow is one type (style) of hardware description. Other types include behavioral, structural, switch level, mixed type, and mixed language. Listed below are some facts about data-flow description:

- Data-flow description simulates the system to be described by showing how the signal flows from the system inputs to its outputs. For example, the Boolean function of the output or the logical structure of the system shows such signal flow. A data-flow description of a half adder was covered in Section 1.3.1.

▪ Signal-assignment statements are concurrent. At any simulation time, all signal-assignment statements that have an event are executed concurrently (see Section 2.2).

2.2 Signal Declaration And Assignment Statement

Figure 2.1 shows an AND-OR circuit. Signals a, b, c, and d are the inputs, signal y is the output, and signals s1 and s2 are intermediates. The Boolean function of the output y can be written as:

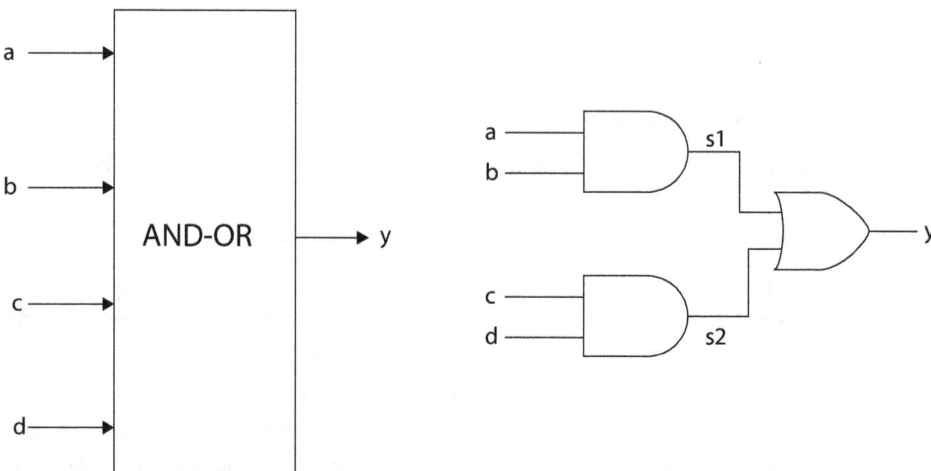

FIGURE 2.1 AND-OR circuit. a) Symbol diagram. b) Logic diagram.

$$y = s1 + s2; \text{ where } s1 = ab \text{ and } s2 = cd \qquad (2.1)$$

The Boolean function of y could be written as:

$$Y = ab + cd \qquad (2.2)$$

Listing 2.1 shows the HDL code of the circuit.

LISTING 2.1 HDL code of Figure 2.1

VHDL Description
```
library IEEE;
use IEEE.STD_LOGIC_1164.ALL;
entity andor is
    port (a,b,c,d: in std_logic; y : out std_logic);
end andor;
```

```
architecture andor_dtfl of andor is
signal s1,s2 : std_logic;
begin
    s1 <= a and b; --statement 1.
    s2 <= c and d; --statement 2.
     y <= s1 or s2; --statement 3.
end andor_dtfl;
```

Verilog description
```
module andor (a,b,c,d, y );
input a,b,c,d;
output y;
wire s1, s2; /* wire statement here is not necessarily
        needed since s1 and s2 are single bit*/
    assign s1 = a & b; //statement 1.
    assign s2 = c & d; //statement 2.
    assign y = s1 | s2; //statement 3.
endmodule
```

Using a CAD package with HDL simulator (see Appendix A), the code in Listing 2.1 can be simulated on the screen of the computer, and a waveform showing a graphical relationship between the input and the output can be obtained. Figure 2.2 shows such a waveform.

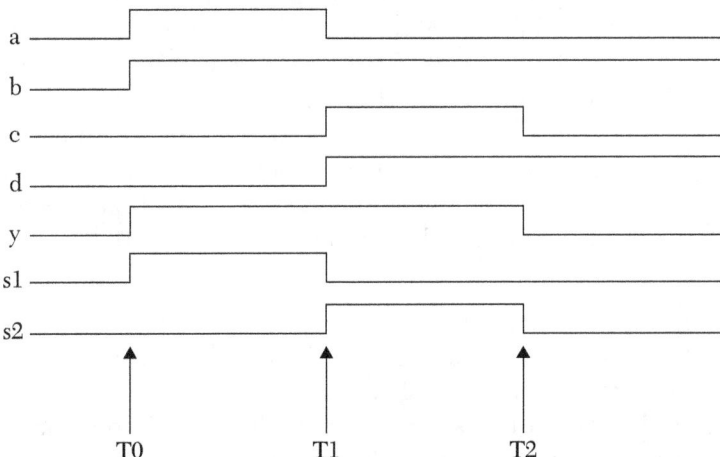

FIGURE 2.2 Simulation waveform for the AND-OR circuit shown in Figure 2.1.

Referring to Listing 2.1, the input and output signals are declared in the entity (module) as ports. In HDL, a signal has to be declared before it can be used (although in Verilog, it is not necessarily needed if the signal is a single bit). Accordingly, signals s1 and s2 have to be declared. In VHDL, s1 and s2 are declared as signals by using the predefined word `signal` in the architecture:

```
signal s1, s2 : bit;
```

In Verilog, s1 and s2 are declared as signals by using the predefined word `wire`:

```
wire s1, s2;
```

By default, all ports in Verilog are assumed to be wires. The value of the wire is continuously changing with changes in the device that is deriving it. For example, s1 is the output of the AND gate in Figure 2.1, and s1 is continuously updated as a or b changes.

A signal-assignment statement is used to assign a value to a signal. The left-hand side of the statement should be declared as a signal. The right-hand side can be a signal, a variable, or a constant. The operator for signal assignment is `<=` in VHDL or the predefined word `assign` in Verilog. In Listing 2.1, statements 1, 2, and 3 are signal-assignment statements.

The execution of the signal-assignment statement in HDL is somehow different in concept from that of software languages such as C. Statements 1–3 need an event to occur on its right-hand side to start execution. If no event occurred on any statement, this statement would not be executed. An event is a change in the value of a signal or variable such as a change from 0 to 1 (from low to high) or from 1 to 0 (from high to low). The statement that receives an event first will be executed first regardless of the order of its placement in the HDL code. If more than one statement have an event at the same time, all of these statements will be executed concurrently (i.e., simultaneously). Accordingly, statement 3, for example, could have been written before statement 1 in Listing 2.1, and the order of execution would not be affected.

The signal-assignment statement is executed in two phases: *calculation* and *assignment*. If an event occurs on the right-hand side of a statement, then this side is calculated at the time of the event; after calculation, the value obtained from the calculation is assigned to the left-hand side, taking into consideration any timing information given in the statement (see Section 2.4 for details of the timing information). Consider Listing 2.1 and Figure 2.2. At T0, an event has occurred in signal a and signal b (both signals changed their value from 0 to 1, which is an event). Accordingly, an event occurred in statement 1; the value of (a and b) is calculated as (1 and 1 = 1).

Because no delay time is specified, the value 1 is assigned immediately to s1, changing s1 from 0 to 1. Changing the value of s1 from 0 to 1 constitutes an event in s1 and in statement 3, which is executed as a result of the event in its right-hand side. The right-hand side of statement 3 is calculated at T0 as (s1 [1] or s2 [0] = 1). The value of 1 is assigned to y; all at T0 because no delay time is specified. At T1, there is event on signals a (1 to 0), c (0 to 1), and d (0 to 1). Statements 1 and 2 will be executed concurrently because an event occurred on their right-hand side. The right-hand side of statement 1 and 2 is calculated at T1 as (0 and 1 = 0) and (1 and 1 = 1); the value of 0 is assigned to s1, and the value of 1 is assigned to s2 at T1. Changing the value of s1 and s2 constitutes an event on s1 and s2, which selects statement 3 for execution at T1; statement 3 is executed (calculation, s1 or s2 = 0 or 1 = 1), and accordingly, 1 is assigned to signal y. At T2, an event occurred on signal c, statement 2 is executed at T2, and the calculation results in 0 and 1 = 0; the value 0 is assigned to s2, changing its value from 1 to 0 and generating an event in s2. Statement 3 is executed because an event (changing the value of s2 from 1 to 0) occurred on the right-hand side. The calculation results in 0 or 0 = 0; the value 0 is assigned to y at T2.

2.2.1 Constant Declaration and Constant Assignment Statements

A constant in HDL is treated as it is in C language; its value is constant within the segment of the program where it is visible. A constant in VHDL can be declared using the predefined word `constant`. In Verilog, a constant can be declared by its type such as `time` or `integer`. For example, the following statements declare period as a constant of type `time`:

```
constant period : time; -- VHDL
time period; // Verilog
```

To assign a value to a constant, use the assignment operator := in VHDL or = in Verilog. For example, to assign a value of 100 nanoseconds to the constant period described above:

```
period := 100 ns; -- VHDL
period = 100; // Verilog
```

In the above Verilog statement, there are no explicit units of time; 100 means 100 simulation screen time units. If the simulation screen time units are defined as nanoseconds (ns), for example, then 100 will mean 100 nanoseconds. The declaration and assignment can be combined in one statement as:

```
Constant period : time := 100 ns; -- VHDL
time period = 100 //Verilog
```

2.2.2 Assigning a Delay Time to the Signal-Assignment Statement

To assign a delay time to a signal-assignment statement, the predefined word `after` in VHDL or `#` in Verilog is used. For example, the following statement assigns a 10 ns delay time to signal `s1`:

```
S1 <= a and b after 10 ns -- VHDL
assign #10 S1 = a & b // Verilog
```

In Verilog, the delay is in simulation screen unit time. Let us assume that there is a delay of 10 ns between the output of each statement 1–3 and its input in Listing 2.1. This is equivalent to saying that operation (and) or (or) takes 10 ns to be completed. Listing 2.2 shows the HDL code for Figure 2.1 with a 10 ns delay for the (and) and (or) operations.

LISTING 2.2 HDL code of Figure 2.1 with 10 ns delay

VHDL description
```
library IEEE;
use IEEE.STD_LOGIC_1164.ALL;
entity andor_dly is
    port (a,b,c,d: in std_logic; y : out std_logic);
end andor_dly;
architecture andor_dtfl of andor_dly is
constant dly : time := 10 ns;
signal s1,s2 : std_logic;
begin
    s1 <= a and b after dly; --statement 1.
    s2 <= c and d after dly; --statement 2.
    y <= s1 or s2 after dly; --statement 3.
 end andor_dtfl;
```

Verilog description
```
module and_orDlyVr( a,b,c,d, y );
input a,b,c,d;
output y;
time dly = 10;
wire s1, s2;
/* wire above is not necessarily needed
since s1 and s2 are single bit*/
    assign # dly s1 = a & b; //statement 1.
    assign # dly s2 = c & d; //statement 2.
    assign # dly y = s1 | s2; //statement 3.
endmodule
```

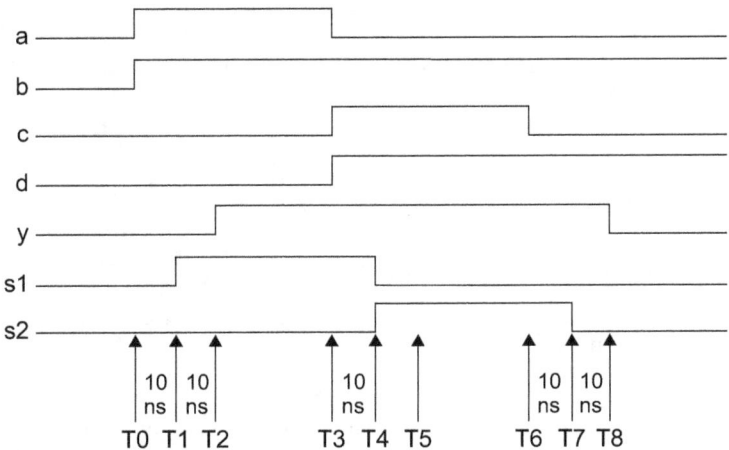

FIGURE 2.3 Simulation waveform of Listing 2.2.

Figure 2.3 shows the simulation waveform of Listing 2.2. Table 2.1 shows analysis of the waveform according to Listing 2.2. At T0, an event occurred on signal a and signal b (both changed from 0 to 1). This event will invoke execution of statement 1. The right-hand side (R.H.S) of statement 1 is calculated at T0 as (1 and 1 = 1). However, this value of 1 will not assigned to s1 at T0; rather, it will be assigned at T0 + 10 ns = T1. The rest of Table 2.1 could be understood by following the same analysis that has been done above at T0.

TABLE 2.1 Analysis of Waveform of Figure 2.3

Event(s) on R.H.S	Time of Event	Statement(s) Affected by Event	R.H.S Calculations	Assignment Value	Time of Assignment the Calculated Value
a (0 to 1) b (0 to 1)	T0	Statement 1	1 and 1 = 1	s1 = 1	T1 (T0 + 10 ns)
s1 (0 to 1)	T1	Statement 3	1 or 0 = 1	Y = 1	T2 (T1 + 10 ns)
a (1 to 0) c (0 to 1) d (0 to 1)	T3	Statements 1 and 2	1 and 0 = 0 1 and 1 = 1	s1 = 0 s2 = 1	T4 (T3 + 10 ns) T4 (T3 + 10 ns)

(Contd.)

Event(s) on R.H.S	Time of Event	Statement(s) Affected by Event	R.H.S Calculations	Assignment Value	Time of Assignment the Calculated Value
s1 (1 to 0) s2 (0 to 1)	T4	Statement 3	0 or 1 = 1	y = 1	T5 (T4 + 10 ns)
c (1 to 0)	T6	Statement 2	0 and 1 = 0	s2 = 0	T7 (T6 + 10 ns)
s2 (1 to 0)	T7	Statement 3	0 or 0 = 0	y = 0	T8 (T7 = 10 ns)

From Table 2.1, the worst total delay time between the input and the output of Figure 2.1, as expected, is 20 ns. It is to be noted that if a signal-assignment statement did not specify a delay time, the assignment to its left-hand side would occur after the default infinitesimally small delay time of D (delta) seconds. This infinitesimally small time cannot be detected on the screen, and the delay time will look as if it is zero. In the following several examples, data-flow descriptions are introduced.

EXAMPLE 2.1   DATA-FLOW DESCRIPTION OF A FULL ADDER

A full adder is a combinational circuit (output depends only on the input) that adds three input bits (a + b + c) and outputs the result as two bits; one bit for the sum and one bit for the carryout. Examples of full addition are: 1 + 0 + 1 = 10 (in decimal 1 + 0 + 1 = 2) and 1 + 1 + 1 = 11 (in decimal 1 + 1 + 1 = 3). Table 2.2 shows the truth table of the full adder.

TABLE 2.2 Truth Table For a Full Adder

Input			Output	
a	b	c	Carryout	Sum
0	0	0	0	0
0	0	1	0	1
0	1	0	0	1
0	1	1	1	0
1	0	0	0	1
1	0	1	1	0
1	1	0	1	0
1	1	1	1	1

The Boolean function of the Sum and Carryout can be obtained from K-maps as shown in Figure 2.4.

bc \ a	00	01	11	10
0		1		1
1	1		1	

Sum = f(a,b,c) = m(1,2,4,7)

bc \ a	00	01	11	10
0			1	
1		1	1	1

Carryout = f(a,b,c) = m(3,5,6,7)

FIGURE 2.4 K-maps for the minterms (m) for the Sum and Carryout.

From Figure 2.4, the Boolean functions can be written as:

$$\text{Sum} = \overline{a}\,\overline{b}c + \overline{a}\,b\,\overline{c} + a\overline{b}\,\overline{c} + abc \qquad (2.3)$$

$$\text{Carryout} = ab + ac + bc \qquad (2.4)$$

The symbol diagram of the full adder is shown in Figure 2.5a. The logic diagram of a full adder based on Equations 2.3 and 2.4 is shown in Figure 2.5b.

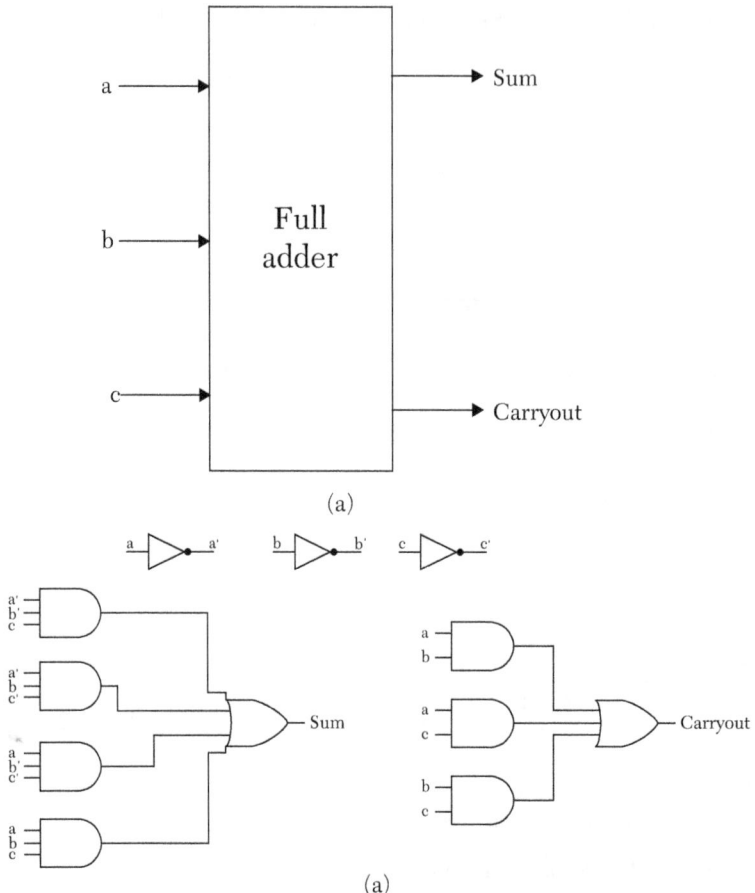

(a)

(a)

FIGURE 2.5 A full Adder. a) Logic symbol. b) Logic diagram.

The full adder can be built from several existing logic components such as two half adders and multiplexers (see Exercise 2.1 at the end of this chapter). Building a full adder from two half adders is based on the following analysis.

The full adder adds a *plus* b *plus* c = carryout sum. If the addition is performed in two steps: a *plus* b = C_1 S, and c plus S = C_2 sum (sum is the sum of the three bits). C_1 and C_2 cannot have a value of 1 at the same time. If C_1 has value of 1, then C_2 has to be 0 and vice versa. For example, to add 1 *plus* 1 *plus* 1, divide the addition in two halves; the first half is 1 *plus* 1 = 10, and the second half is 0 *plus* 1 = 1. The carryout will be (C_1 or C_2); in this example, it is 1 and the sum = 1. Figure 2.6 shows the logic diagram of the full adder built from two half adders.

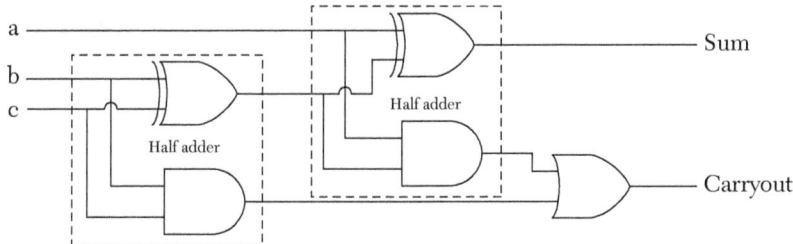

FIGURE 2.6 A full adder built from two half adders.

Listing 2.3 shows the HDL code for the full adder as shown in Figure 2.5. Review Section 1.5.1 to know the VHDL and Verilog logical operators. The code assumes no delay time. The parenthesis in the code, as in C language, gives the highest priority to the expression within the parenthesis and makes the code more readable.

LISTING 2.3 HDL Code of the Full Adder From Figure 2.5

VHDL description
```
library IEEE;
use IEEE.STD_LOGIC_1164.ALL;
entity fulladder is
    Port ( a,b,c : in std_logic;
         sum, Carryout : out std_logic);
end fulladder;
architecture flad_dtfl of fulladder is
begin
    Sum <= (not a and not b and c) or
    (not a and b and not c) or
    (a and not b and not c) or
    (a and b and c);
    Carryout <= (a and b) or (a and c) or (b and c);
end flad_dtfl;


Verilog description
module fulladder(a, b, c);
output Sum, Carryout;
input a, b, c;
    assign Sum = (~ a & ~ b & c)|( ~ a & b & ~c)|
                ( a & ~b & ~c)|( a & b & c) ;

    assign Carryout = (a & b) | (a & c) | (b & c);
endmodule
```

Figure 2.7 shows the waveform of a full adder with no delay time.

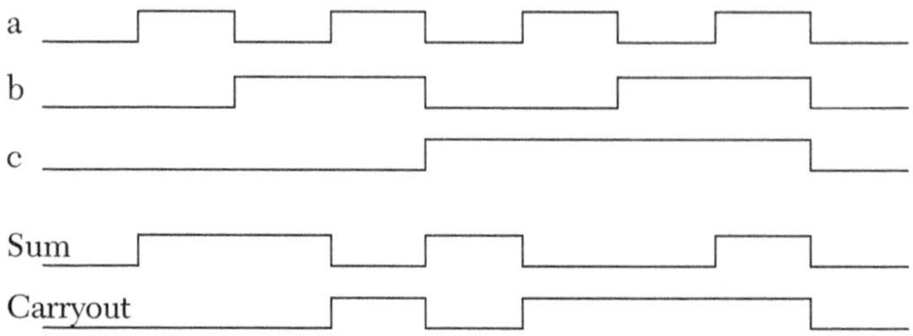

FIGURE 2.7 Simulation waveform of a full adder with no delay time.

EXAMPLE 2.2 FULL SUBTRACTOR

A full subtractor performs the following operation: a - b - c = Borrow Diff. Borrow and Diff are each one-bit output. The Diff is the difference, and Borrow is the borrow. For example, 0 - 1 - 0 = 11. The subtraction is done as follows: 0 - 1 cannot subtract 1 from 0 because 1 is greater than 0, so borrow 1 from the higher-order bit. Accordingly, this 1 has a weight of 2^1, so its value is 2; subtract 2 - 1 = 1. Now, for bit c, 1 - 0 = 1, so the difference is 1, and the borrow is 1. Table 2.3 shows the truth table of a full subtractor.

TABLE 2.3 Truth Table for a Full Subtractor

Input			Output	
a	**b**	**c**	**Borrow**	**Diff**
0	0	0	0	0
0	0	1	1	1
0	1	0	1	1
0	1	1	1	0
1	0	0	0	1
1	0	1	0	0
1	1	0	0	0
1	1	1	1	1

Compare the Diff in Table 2.3 and the Sum in Table 2.2; they are identical, so the Boolean function of the Diff is the same as the sum in Equation 2.3. For the Borrow, draw the K-map as shown in Figure 2.8.

bc a	00	01	11	10
0	0	1	1	1
1	0	0	1	

Borrow = f(a,b,c) = Σ m(1,2,3,7)

FIGURE 2.8 K-map for the borrow of a full subtactor.

From Figure 2.8, the Boolean functions are:

Diff = $\overline{a}\,\overline{b}c + \overline{a}\,b\,\overline{c} + a\overline{b}\,\overline{c} + abc$ .... (2.5)

Borrow = $\overline{a}\,c + \overline{a}\,b + bc$ .... (2.6)

The HDL code of the full subtractor is given as an exercise at the end of this chapter.

EXAMPLE 2.3A   2x1 MULTIPLEXER WITH ACTIVE LOW ENABLE

A 2x1 multiplexer is a combinational circuit; it has two one-bit inputs, a one-bit select line, and a one-bit output. Additional control signals may be added, such as enable. The output of the basic multiplexer depends on the level of the select line. If the select is high (1), the output is equal to one of the two inputs. If the select is low (0), the output is equal to the other input. A truth table for a 2x1 multiplexer with active low enable is shown in Table 2.4.

TABLE 2.4   Truth Table for a 2x1 Multiplexer

Input		Output
SEL	**Gbar**	**Y**
X	H	L
L	L	A
H	L	B

If the enable (Gbar) is high (1), the output is low (0) regardless of the input. When Gbar is low (0), the output is A if SEL is low (0), or the output is B if SEL is high (1). From Table 2.4, the Boolean function of the output Y is:

Y = (S1 and A and *SEL*) or (S1 and B and SEL); S1 is the invert of Gbar

Figure 2.9a shows the logic symbol, and Figure 2.9b shows the gate-level structure of the multiplexer.

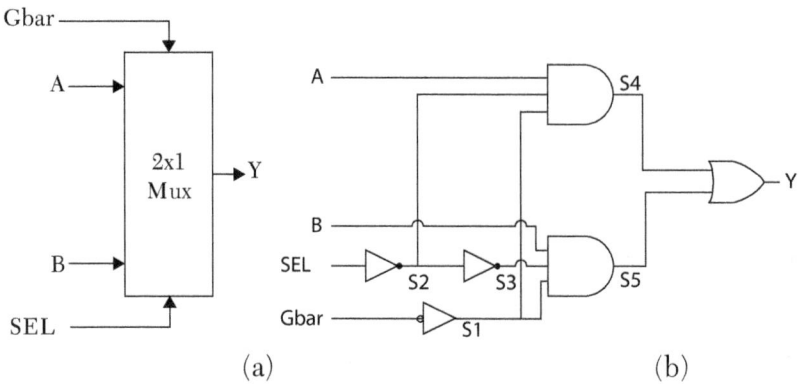

(a)   (b)

FIGURE 2.9 2x1 Multiplexer. a) Logic symbol. b) Logic diagram.

Listing 2.4a shows the HDL code. To generate the code, follow Figure 2.9b. Propagation delay time for all gates is assumed to be 7 ns. Because this is a data-flow description, the order in which the statements are written in the code is irrelevant. For example, statement st6 could have been written at the very beginning instead of statement st1. The logical operators in VHDL and (Verilog) implemented in this Listing are: OR (|), AND (&), and NOT (~).

LISTING 2.4a HDL Code of a 2x1 Multiplexer: VHDL and Verilog

VHDL description
```
library IEEE;
use IEEE.STD_LOGIC_1164.ALL;
entity mux2x1 is
    port (A, B, SEL, Gbar : in std_logic;
    Y : out std_logic);
end mux2x1;
architecture MUX_DF of mux2x1 is
signal S1, S2, S3, S4, S5 : std_logic;
constant dly : time := 7ns;
Begin

-- Assume 7 nanoseconds propagation delay
-- for all and, or, and not operation.
    st1: Y <= S4 or S5 after dly;
```

```
   st2: S4 <= A and S2 and S1 after dly;
   st3: S5 <= B and S3 and S1 after dly;
   st4: S2 <= not SEL after dly;
   st5: S3 <= not S2 after dly;
   st6: S1 <= not Gbar after dly;
 end MUX_DF;
```

Verilog Description
```
module mux2x1 (A, B, SEL, Gbar, Y);
input A, B, SEL, Gbar;
output Y;
wire S1, S2, S3, S4, S5;
time dly = 7;
/* Assume 7 time units delay for all and, or, not operations. The
   delay here is expressedin simulation screen units. */

   assign # dly Y = S4 | S5; //st1
   assign #dly S4 = A & S2 & S1; //st2
   assign #dly S5 = B & S3 & S1; //st3
   assign #dly S2 = ~ SEL; //st4
   assign #dly S3 = ~ S2; //st5
   assign #dly S1 = ~ Gbar; //st6
endmodule
```

Figure 2.10 shows the simulation waveform for the 2x1 multiplexer.

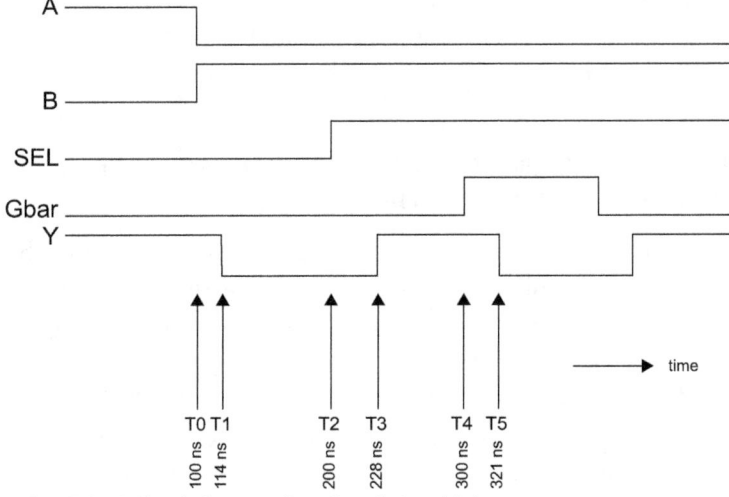

FIGURE 2.10 Simulation waveform for a 2x1 multiplexer.

Analysis of Listing 2.4a

Referring to Listing 2.4a, because the description is a data flow, the order of statements st1 to st6 is irrelevant; statement st5 could have been written before statement st1 without changing the outcome of the HDL program. In Figure 2.10, signal A changes from 1 to 0, and signal B changes from 0 to 1 at T0; these changes constitute an event in signal-assignment statements st2 and st3. Accordingly, statements st2 and st3 are executed simultaneously. As previously mentioned, execution is done in two phases: calculation and assignment. For statement st2, at T0, A = 0, S2 = 1 (the inversion of SEL), and S1 = 1 (the inversion of Gbar); hence, the calculated new value of S4 at T0 is (A AND S1 AND S2) = 0. This is a change in value for S4 from 1 to 0, which is assigned to S4 after 7 ns from time T0 (at 107 ns). For statement st3, at T0, B = 1, S3 = 0, and S1 = 1. The calculated value of S5 is 0, as it was before T0. At T = 107 ns, an event occurs on S4, and this causes execution of statement st1. Y is calculated as (0 or 1) = 1, and this value is assigned to Y after 7 ns, that is, at T1 = 107 + 7 = 114 ns. Alternatively, statements st1 to st5 can be replaced by one statement:

```
-- VHDL:
Y <= not (Gbar) and ((sel and b) or (not sel and A)) after 21 ns;

// Verilog:
assign # 21 Y = ~ (Gbar) & ((SEL & B ) | (~ SEL & A));
```

The above delay time of 21 ns is an estimated average delay time. If either of the above two statements is used, individual delay times cannot be assigned, as was done in Listing 2.4a.

EXAMPLE 2.3B   2x1 MULTIPLEXER WITH ACTIVE LOW ENABLE USING VERILOG CONDITIONAL OPERATOR (?)

The conditional operator ? (see Section 1.5.2.2) can be used to describe a multiplexer or any other similar system that utilizes a selector signal to select between two options. The format of this operator can be written as:

Assign Y = Conditional-expression ? true-expression : false-expression

If the conditional expression is true, the value of the true expression is assigned to Y; if the conditional expression is false, the value of the false expression is assigned to Y. Listing 2.4b illustrates a Verilog code for a 2x1 multiplexer using the conditional operator ? to select the value of the output Y according to the level of the enable Gbar. If Gbar is high (1), that is to say the conditional expression is true, the output Y is assigned to low (0). Otherwise, the output Y is assigned the false expression (SEL & B) | (~ SEL & A). Also, recall from Section 1.5.2.2 that both the true and the false expressions can contain high impedance and don't care; this will allow for describing systems such as multiplexers with tri-state output (see the Exercise section at the end of this chapter).

LISTING 2.4b *HDL Code of a 2x1 Multiplexer Using Verilog Conditional (?)*

```
module Mux2x1_conditional(input A,B,SEL,Gbar, output Y );
    assign Y = (Gbar) ? 1'b0 : (SEL & B ) | (~ SEL & A);
endmodule
```

EXAMPLE 2.4   A 2x4 DECODER

A decoder is a combinational circuit. A 2x4 decoder has two inputs and four outputs. For any input, only one output is active; all others are inactive. For active high output decoders, only one output is high. The output of n-bit input decoder is 2^n bits. Table 2.5 shows the truth table of the 2x4 decoder.

TABLE 2.5 Truth Table for a 2x4 Decoder

Inputs		Outputs			
b	**a**	**D3**	**D2**	**D1**	**D0**
0	0	0	0	0	1
0	1	0	0	1	0
1	0	0	1	0	0
1	1	1	0	0	0

From Table 2.5, the Boolean function of the outputs can be written as:

$$D0 = \overline{a}\,\overline{b}$$

$$D1 = a\,\overline{b}$$

$$D2 = \overline{a}\,b$$

$$D3 = a\,b$$

Figure 2.11 shows the logic symbol and logic diagram of the decoder.

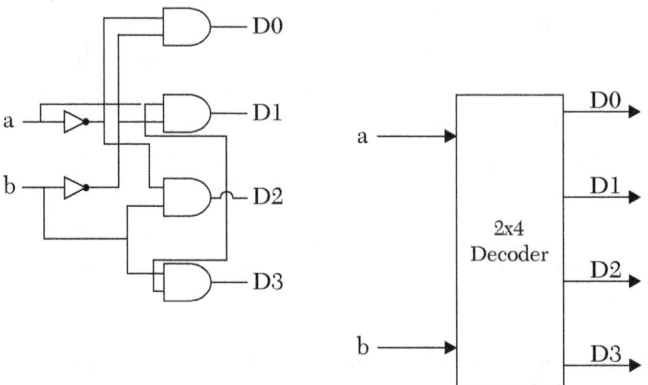

FIGURE 2.11 2x4 Decoder. a) Logic symbol. b) Logic diagram.

Listing 2.4 shows the HDL code of the decoder. Figure 2.12 shows the simulation waveform of the decoder.

LISTING 2.4   HDL Code of a 2x4 Decoder Without Time Delay

VHDL description
```
library IEEE;
use IEEE.STD_LOGIC_1164.ALL;
entity decoder2x4 is
    port ( a, b : in std_logic;
            D : out std_logic_vector (3 downto 0));
end decoder2x4;
architecture decder_dtfl of decoder2x4 is
begin
    D(0) <= not a and not b;
    D(1) <= a and not b;
    D(2) <= not a and b;
    D(3) <= a and b;
end decder_dtfl;
```

Verilog description
```
module decoder2x4( a, b, D);
input a,b;
output [3:0]D;
    assign D[0] = ~a & ~ b;
    assign D[1] = a & ~ b;
    assign D[2] = ~a & b;
    assign D[3] = a & b;
endmodule
```

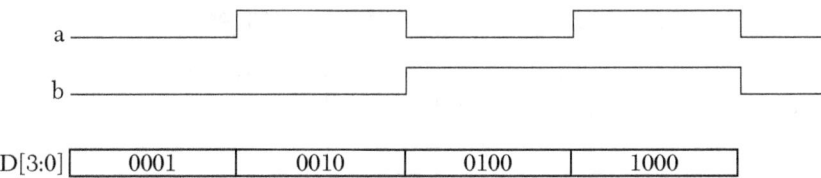

FIGURE 2.12 Simulation Waveform of a 2x4 decoder with no time delay

2.3 Data Type: Vector

The vector data type was briefly covered in Chapter 1. A vector is a data type that declares an array of similar elements, such as declaring an object that has a width of more than one bit. In the previous examples, all signals have been one-bit in width. If signal A has a four-bit width, it can be declared as four different signals, a0, a1, a2, a3, as shown:

```
signal a0, a1, a2, a3 : bit; -- VHDL
wire a0, a1, a2, a3;         // Verilog
```

Or, it can be declared using the vector declaration:

```
signal a : bit_vector (3 downto 0); -- VHDL
wire [3:0] a; // Verilog
```

In VHDL, downto ([3:0] in Verilog) is a predefined operator that describes the width of the vector. If the value of a is 14_d, or $(1110)_2$, then the elements of vector (array) a are:

$$a(3) = 1$$
$$a(2) = 1$$
$$a(1) = 1$$
$$a(0) = 0$$

The following declaration can also be used:

```
signal a : bit_vector (0 to 3); -- VHDL
wire [0:3] a;                    // Verilog
```

where to is a predefined word. In the above declaration, the elements of the vector are:

$$a(0) = 1$$
$$a(1) = 1$$
$$a(2) = 1$$
$$a(3) = 0$$

This means the value of a is considered to be 7_d rather than 14_d.

EXAMPLE 2.5   2x2 UNSIGNED COMBINATIONAL MULTIPLIER

Consider the multiplication of a × b, where a and b are each two-bit numbers. The multiplication is illustrated as follows:

		a(1)	a(0)
		b(1)	b(0)
		b(0) × a(1)	b(0) × a(0)
	b(1) × a(1)	b(1) × a(0)	
P(3)	P(2)	P(1)	P(0)

Because it is only two-bit multiplication, the truth table and K-maps can be easily implemented to find the Boolean function of the product. When the number of bits is large and the K-maps are impractically large, another approach may be taken to design the multiplier (see Chapter 3). The truth table of the 2x2 multiplier is shown in Table 2.6, and Figure 2.13 shows the K-maps of the table.

TABLE 2.6 2x2 Unsigned (Magnitude) Combinational Multiplier

a1	a0	b1	b0		P3	P2	P1	P0
0	0	x	x		0	0	0	0
x	x	0	0		0	0	0	0
0	1	0	1		0	0	0	1
0	1	1	0		0	0	1	0
0	1	1	1		0	0	1	1
1	0	0	1		0	0	`1	0
1	0	1	0		0	1	0	0
1	0	1	1		0	1	1	0
1	1	0	1		0	0	1	1
1	1	1	0		0	1	1	0
1	1	1	1		1	0	0	1

x- indicates don't care

b1b0 a1a0	00	01	11	10
00				
01		1	1	
11		1	1	
10				

P0 = f(a1,a0,b1,b0) = (5,7,13,15)

b1b0 a1a0	00	01	11	10
00				
01			1	1
11		1		1
10		1	1	

P1 = f(a1,a0,b1,b0) = (6,7,9,11,13,14)

b1b0 a1a0	00	01	11	10
00				
01				
11				1
10			1	1

P2 = f(a1,a0,b1,b0) = (10,11,14)

FIGURE 2.13 K-maps of 2x2 Combinational Multiplier.

From the K-maps and Table 2.6, the product can be written as:

P0 = a0 b0

P1= $\overline{a1}$ a0 b1 + a0 $\overline{b0}$ b1 + a1 b0 $\overline{b1}$ + $\overline{a0}$ a1 b0

P2= $\overline{a0}$ a1 b1 + a1 $\overline{b0}$ b1

P3= a0 a1 b0 b1

Figure 2.14a shows the logic symbol, and Figure 2.14b shows the logic diagram of the multiplier. The HDL code is shown in Listing 2.5, and the simulation waveform is shown in Figure 2.15.

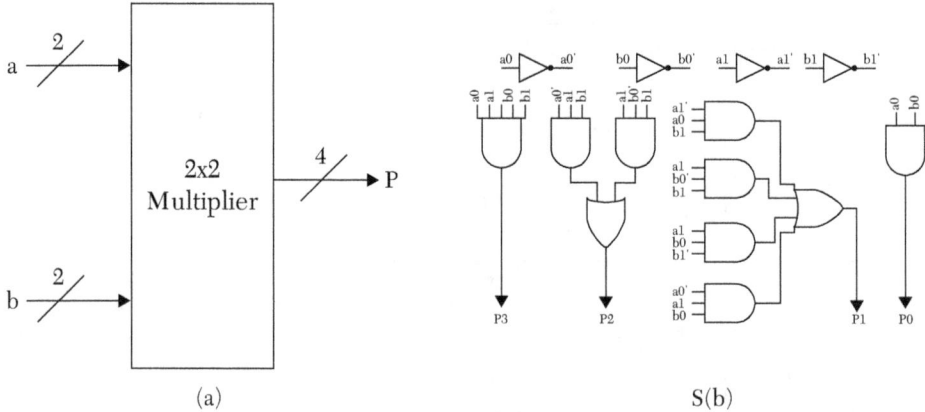

(a)   S(b)

FIGURE 2.14 2x2 combinational multiplier. a) Logic symbol. b) Logic diagram.

LISTING 2.5 HDL Code for a 2x2 Unsigned Combinational Array Multiplier: VHDL and Verilog

VHDL Description
```
--For simplicity, propagation delay times are not
-- considered here.
library IEEE;
use IEEE.STD_LOGIC_1164.ALL;
entity mult_comb is
    port ( a,b: in std_logic_vector (1 downto 0);
           P : out std_logic_vector ( 3 downto 0));
end mult_comb;
architecture mult_dtfl of mult_comb is
begin
    P(0) <= a(0) and b(0);
    P(1) <= (not a(1) and a(0) and b(1)) or
```

```
    (a(0) and not b(0) and b(1)) or
    (a(1) and b(0) and not b(1))or
    (not a(0) and a(1) and b(0));
    P(2) <= (not a(0) and a(1) and b(1)) or
    ( a(1) and not b(0) and b(1));
    P(3) <= a(0) and a(1) and b(0) and b(1);
 end mult_dtfl;
```

Verilog Description
```
module mult_arry (a, b, P);
input [1:0] a, b;
output [3:0] P;
/*For simplicity, propagation delay times are not
considered in this example.*/
    assign P[0] = a[0] & b[0];
    assign P[1] = (~a[1] & a[0]& b [1]) |
                  (a[0] & ~b[0]& b [1])|
                  (a[1] & b[0]& ~b [1]) |
                  (~a[0] & a[1]& b [0]);
    assign P[2] = (~a[0] & a[1]& b [1]) |
                  (a[1] & ~b[0]& b [1]);
    assign P[3] = (a[0] & a[1]& b[0] & b [1]);
```

The simulation output of the multiplier is shown in Figure 2.15.

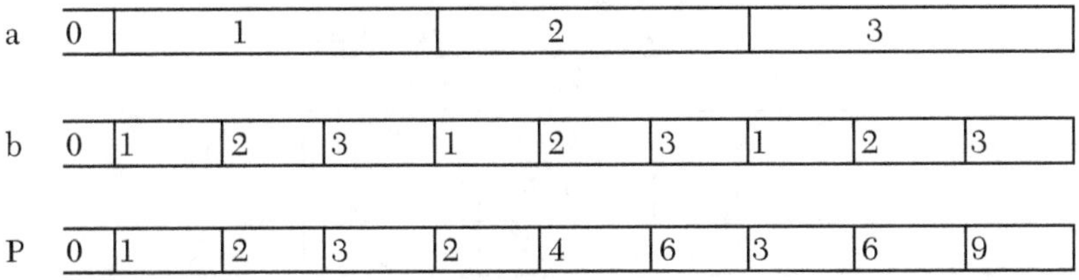

a	0		1			2			3		
b	0	1	2	3	1	2	3	1	2	3	
P	0	1	2	3	2	4	6	3	6	9	

FIGURE 2.15 Simulation output for a two-bit multiplier.

EXAMPLE 2.6   DELAY LATCH

Latches are sequential circuits. The output of a sequential circuit depends on the current state and the input. Figure 2.16 shows the logic symbol of a delay latch (D-latch). At any time (T) the present value of Q is called the current state. At any selected time (T + t_s) the value of Q is called the next state Q^+. The value of the next state depends on the value of the present state and the value of the input (D) (see Table 2.7). In Figure 2.16, the current and next states are the same signal (Q). The current state is the value of Q (0 or 1) before the level of E becomes active. The next state is the value of Q after the enable (E) becomes active. To find the Boolean function of the latch, the excitation table is constructed. Table 2.7 shows the inputs and the corresponding next state. Notice that the current state is considered an input in addition to the input D. Assume an active high enable (E).

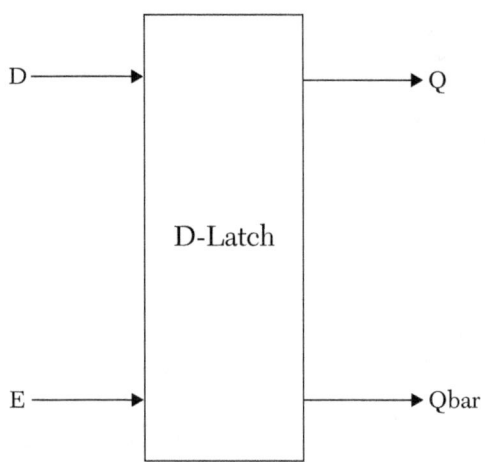

FIGURE 2.16 Logic symbol of D-latch.

TABLE 2.7 Excitation Table of D-Latch with Active High Enable

Inputs			Next State
E	**D**	**Q**	**Q+**
0	x	0	0
0	x	1	1
1	0	x	0
1	1	x	1

Qbar (Qbar$^+$) is always the inverse of Q (Q$^+$). To find the Boolean function, use K-maps to minimize the minterms. The K-map for Q is shown in Figure 2.17.

DQ E	00	01	11	10
0	0	1	1	0
1	0	0	1	

$$Q = f(E,D,Q) = \Sigma(1,3,6,7)$$

FIGURE 2.17 K-map for Q.

From Figure 2.17, Q and Qbar are found:

$$Q = \bar{E}Q + ED$$
$$Qbar = \bar{Q}$$

Figure 2.18 shows the logic diagram of D-latch; the diagram is drawn to be identical to that of the chip 74LS75.

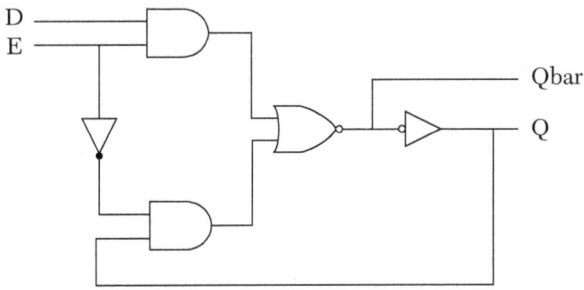

FIGURE 2.18 Gate-level diagram of a D-latch.

Listing 2.6 shows the HDL description of the D-latch. A delay time of 9 ns is assumed between the input and Qbar and 1 ns between Q and Qbar. Note the use of the port-mode buffer in VHDL for the signal Qbar (see section 1.4). The buffer mode is assigned to a port if the port signal appears as read (on the right-hand side of a signal-assignment statement) and as updated (on the left-hand side of a signal-assignment statement). In Listing 2.6, although Q can be declared as only output, it is written with Qbar as a buffer just to improve reading the code; if a signal can be declared as output, it can also be declared as a buffer. In Verilog, Qbar does not necessarily have to be declared as inout because the Qbar is not a bidirectional external bus (see Section 1.4).

LISTING 2.6   HDL Code for a D-Latch: VHDL and Verilog

VHDL Description

```
library IEEE;
use IEEE.STD_LOGIC_1164.ALL;

entity D_Latch is
    port (D, E : in std_logic;
    Q, Qbar : buffer std_logic);

 --Q and Qbar are declared as buffer because they act
 --as both input and output, they appear on the
 --right and left hand side of a signal assignment.
 --inout or linkage could have been used instead of buffer.

end D_Latch;

architecture DL_DtFl of D_Latch is
constant Delay_EorD : Time := 9 ns;
constant Delay_inv : Time := 1 ns;
begin
--Assume 9-ns propagation delay time between
--E or D and Qbar; and 1 ns between Qbar and Q.

    Qbar <= (D and E) nor (not E and Q) after Delay_EorD;
    Q <= not Qbar after Delay_inv;
 end DL_DtFl;
```

Verilog Description

```
module D_latch (D, E, Q, Qbar);
 input D, E;
 output Q, Qbar;

 /* Verilog treats the ports as internal ports,
    so Q and Qbar are not considered here as
    both input and output. If the port is
    connected externally as bidirectional,
    then it should be declared as inout. */

time Delay_EorD = 9;
time Delay_inv = 1;
    assign #Delay_EorD Qbar = ~((E & D) | (~E & Q));
    assign #Delay_inv Q = ~ Qbar;
endmodule
```

Figure 2.19 shows the simulation waveform of the D-latch.

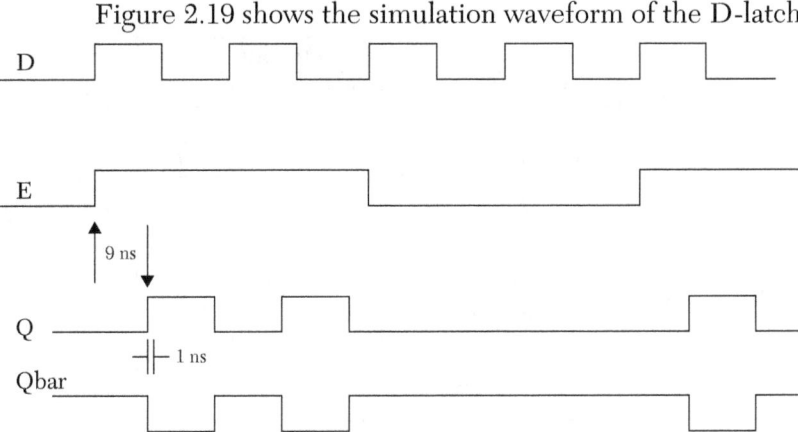

FIGURE 2.19 Simulation waveform of a D-latch with active high enable.

EXAMPLE 2.7   TWO-BIT MAGNITUDE COMPARATOR

A two-bit comparator is a combinational circuit that compares two words (numbers); each word has two bits. Figure 2.20 shows the logic symbol of the comparator. In Figure 2.20, the two words are X and Y. The output of the comparator indicates the result of the comparison: X > Y, X = Y, or X < Y. Because the number of input bits is small (a total of four input bits), a truth table of the comparator can be used to find the Boolean function. Table 2.8 shows the truth table of the 2x2 comparator.

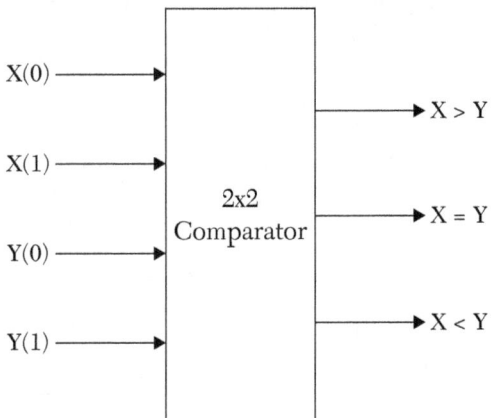

FIGURE 2.20 Logic symbol of a 2x2 magnitude comparator.

TABLE 2.8 Truth Table for a 2x2 Comparator

Input				Output		
X(1)	X(0)	Y(1)	Y(0)	X > Y	X < Y	X = Y
0	0	0	0	0	0	1
0	0	0	1	0	1	0
0	0	1	0	0	1	0
0	0	1	1	0	1	0
0	1	0	0	1	0	0
0	1	0	1	0	0	1
0	1	1	0	0	1	0
0	1	1	1	0	1	0
1	0	0	0	1	0	0
1	0	0	1	1	0	0
1	0	1	0	0	0	1
1	0	1	1	0	1	0
1	1	0	0	1	0	0
1	1	0	1	1	0	0
1	1	1	0	1	0	0
1	1	1	1	0	0	1

If the number of bits increases, the table becomes huge, and other approaches should be used, such as implementation of n-full adders to construct $n \times n$ comparators (see Chapter 4). After constructing the truth table, K-maps are used (see Figure 2.21) to obtain the minimized Boolean function of the output of the comparator. Listing 2.7 shows the HDL description. The simulation waveform is shown in Figure 2.22.

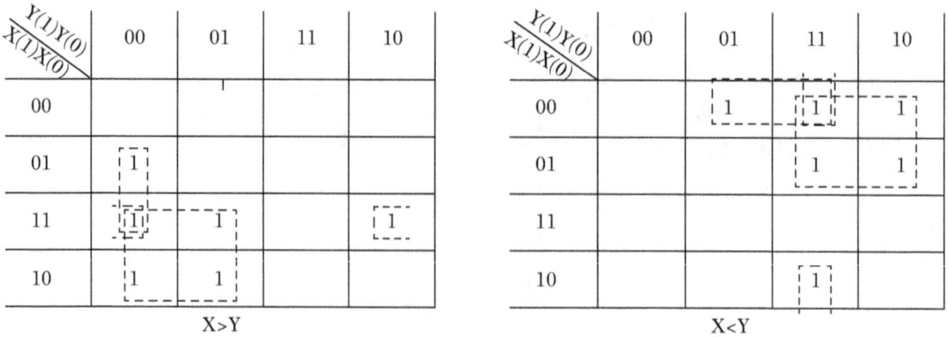

FIGURE 2.21 K-maps for Table 2.4.

$$(X > Y) = X(1)\overline{Y(1)} + X(0)\overline{Y(1)}\,\overline{Y(0)} + X(0)X(1)\overline{Y(0)}$$

$$(X < Y) = \overline{X(1)}Y(1) + \overline{X(0)}\,\overline{X(1)}Y(0) + \overline{X(0)}Y(0)Y(1)$$

$$(X = Y) = \overline{(X > Y) + (X < Y)}$$

LISTING 2.7   HDL Code of a 2x2 Magnitude Comparator

VHDL Description

```
library IEEE;
use IEEE.STD_LOGIC_1164.ALL;

entity COMPR_2 is
     port (x, y : in std_logic_vector(1 downto 0);
     xgty, xlty : buffer std_logic;
     xeqy : out std_logic);

end COMPR_2;

architecture COMPR_DFL of COMPR_2 is
begin
     xgty <= (x(1) and not y(1)) or (x(0) and not y(1)
             and not y(0)) or
             x(0) and x(1) and not y(0));

     xlty <= (y(1) and not x(1)) or ( not x(0) and y(0)
             and y(1)) or (not x(0) and not x(1) and y(0));
     xeqy <= xgty nor xlty;

end COMPR_DFL;
```

Verilog Description

```
module compr_2 (x, y, xgty, xlty, xeqy);
input [1:0] x, y;
output xgty, xlty, xeqy;
     assign xgty = (x[1] & ~ y[1]) | (x[0] & ~ y[1]
             & ~ y[0]) | (x[0] &
                 x[1] & ~ y[0]);
     assign xlty = (y[1] & ~ x[1] ) | (~ x[0] &
                 y[0] & y[1]) |(~ x[0] &
                 ~ x[1] & y[0]);
     assign xeqy = ~ (xgty | xlty);

 endmodule
```

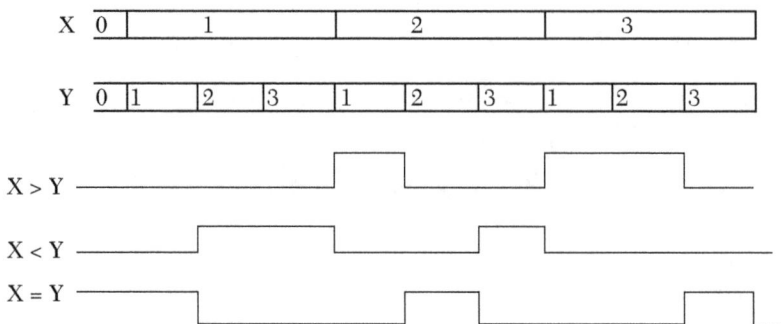

FIGURE 2.22 Simulation waveform of a 2x2 comparator.

CASE STUDY 2.1

In this case study, a three-bit adder is described. The adder is designed using two approaches: ripple carry and carry lookahead. The description is simulated, and timing characteristics of the two adders are compared. Figure 2.23 shows a block diagram of a three-bit ripple-carry adder.

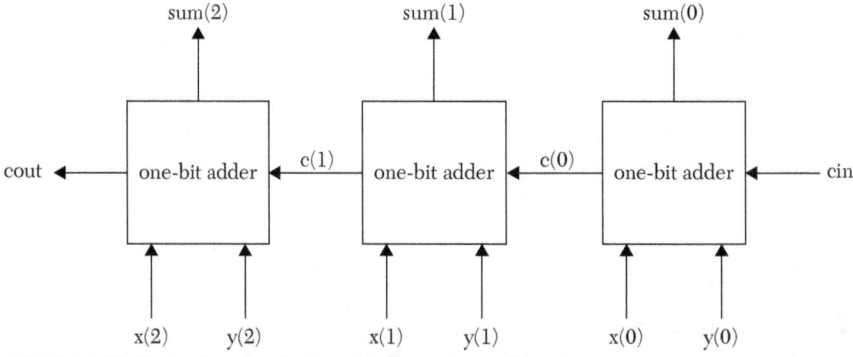

FIGURE 2.23 Block diagram of a three-bit ripple-carry adder.

The Boolean functions of a three-bit ripple-carry adder can be written as (see Example 2.1):

$$\text{sum}(i) = x(i) \text{ XOR } y(i) \text{ XOR } c(i-1), 0 \leq i \leq 2 \tag{2.7}$$

$$c(i) = x(i)y(i) + x(i)c(i-1) + y(i)c(i-1), 0 \leq i \leq 2 \tag{2.8}$$

$$\text{cout} = c(2), c(-1) = \text{cin} \tag{2.9}$$

Each one-bit adder in Figure 2.23 is described by Equations 2.7 and 2.8. To produce the sum and the carryout, each one-bit adder has to wait until the preceding one-bit adder generates its carryout (c[0], c[1], or cout).

The maximum signal-propagation delay of the adder described above is 3d, where d is the delay of a one-bit adder; for an n-bit adder, this delay is n × d.

Figure 2.24 shows a block diagram of a three-bit carry-lookahead adder. The major difference between this adder and the ripple-carry adder is how the carryout of each one-bit full adder is generated and propagated. In ripple carry, each one-bit adder has to wait until the preceding adder unit generates its carryout; in carry lookahead, each one-bit adder generates its carryout at the same time. This simultaneous generation of carries leads to shorter signal-propagation delays. The maximum delay in lookahead adders is 4 × gd, where gd is the average gate delay. This delay is independent of the number of one-bit adders.

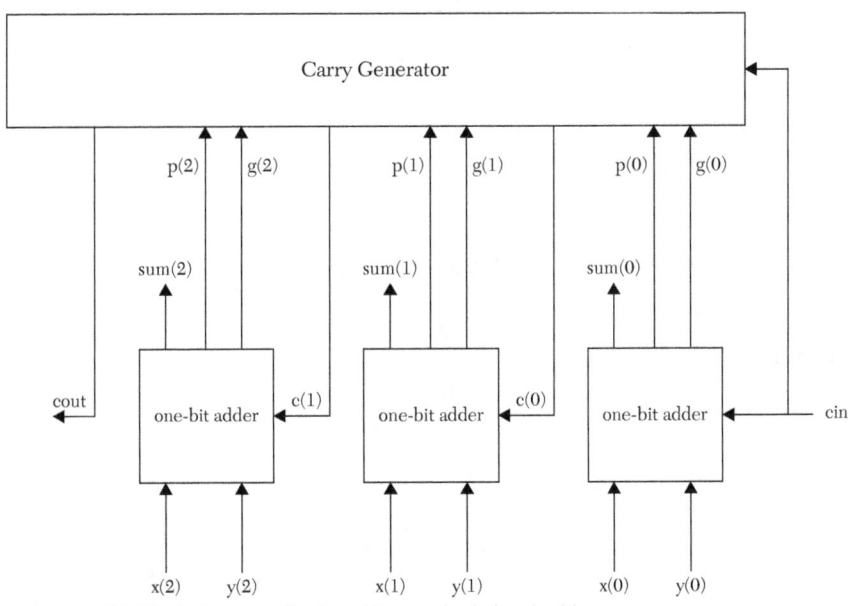

FIGURE 2.24 Block diagram of a three-bit carry-lookahead adder.

The Boolean functions of the carry-lookahead adder are:

$$\text{sum}(i) = x(i) \text{ XOR } y(i) \text{ XOR } c(i-1), 0 \leq i \leq 2 \qquad (2.10)$$

$$g(i) = x(i) \, y(i) \qquad (2.11)$$

$$p(i) = x(i) + y(i) \qquad (2.12)$$

$$c(0) = g(0) + p(0)\text{cin}, \, c(1) = g(1) + p(1)g(0) + p(1)p(0)\text{cin} \qquad (2.13)$$

$$\text{cout} = c(2) = g(2) + p(2)g(1) + p(2)p(1)g(0) + p(2)p(1)p(0)\text{cin} \quad (2.14)$$

Listings 2.8 and 2.9 show the HDL code for the ripple-carry adder and the carry-lookahead adders, respectively. A 4.0-ns delay is assumed for all gate types. A constant of type time `delay_gt` is declared, and 4 ns is assigned to it:

```
constant delay_gt : time := 4 ns; -- VHDL
time delay_gt = 4;                 // Verilog
```

LISTING 2.8   Three-Bit Ripple-Carry Adder Case Study

VHDL Description
```
library IEEE;
use IEEE.STD_LOGIC_1164.ALL;
entity adders_RL is
    port (x, y : in std_logic_vector (2 downto 0);
    cin : in std_logic;
    sum : out std_logic_vector (2 downto 0);
    cout : out std_logic);
end adders_RL;


--I. RIPPLE-CARRY ADDER


architecture RCarry_DtFl of adders_RL is
--Assume 4.0-ns propagation delay for all gates.

signal c0, c1 : std_logic;
constant delay_gt : time := 4 ns;

begin
    sum(0) <= (x(0) xor y(0)) xor cin after 2*delay_gt;

--Treat the above statement as two 2-input XOR.

    sum(1) <= (x(1) xor y(1)) xor c0 after 2*delay_gt;

--Treat the above statement as two 2-input XOR.
    sum(2) <= (x(2) xor y(2)) xor c1 after 2*delay_gt;
--Treat the above statement as two 2-input XOR.
c0 <= (x(0) and y(0)) or (x(0) and cin) or
      (y(0) and cin) after 2*delay_gt;
c1 <= (x(1) and y(1)) or (x(1) and c0) or
      (y(1) and c0) after 2*delay_gt;
cout <= (x(2) and y(2)) or (x(2) and c1) or
```

```
          (y(2) and c1)after 2*delay_gt;
end RCarry_DtFl;
```

Verilog Description
```
module adr_rcla (x, y, cin, sum, cout);
input [2:0] x, y;
input cin;
output [2:0] sum;
output cout;
// I. RIPPLE CARRY ADDER
wire c0, c1;
time delay_gt = 4;
//Assume 4.0-ns propagation delay for all gates.

  assign #(2*delay_gt) sum[0] = (x[0] ^ y[0]) ^ cin;
//Treat the above statement as two 2-input XOR.

  assign #(2*delay_gt) sum[1] = (x[1] ^ y[1]) ^ c0;
//Treat the above statement as two 2-input XOR.

  assign #(2*delay_gt) sum[2] = (x[2] ^ y[2]) ^ c1;
//Treat the above statement as two 2-input XOR.

  assign #(2*delay_gt) c0 = (x[0] & y[0]) |
          (x[0] & cin) | (y[0] & cin);
  assign #(2*delay_gt) c1 = (x[1] & y[1]) |
          (x[1] & c0) | (y[1] & c0);

  assign #(2*delay_gt) cout = (x[2] & y[2]) |
          (x[2] & c1) | (y[2] & c1);
endmodule
```

LISTING 2.9   Three-Bit Carry-Lookahead Adder Case Study

VHDL Description
```
--II. CARRY-LOOKAHEAD ADDER
architecture lkh_DtFl of adders_RL is
--Assume 4.0-ns propagation delay for all gates
--including a 3-input xor.

signal c0, c1 : std_logic;
signal p, g : std_logic_vector (2 downto 0);
constant delay_gt : time := 4 ns;
```

```
begin

    g(0) <= x(0) and y(0) after delay_gt;
    g(1) <= x(1) and y(1) after delay_gt;
    g(2) <= x(2) and y(2) after delay_gt;
    p(0) <= x(0) or y(0) after delay_gt;
    p(1) <= x(1) or y(1) after delay_gt;
    p(2) <= x(2) or y(2) after delay_gt;
    c0 <= g(0) or (p(0) and cin) after 2*delay_gt;

    c1 <= g(1) or (p(1) and g(0)) or (p(1) and p(0)
          and cin) after 2*delay_gt;
cout <= g(2) or (p(2) and g(1)) or (p(2) and p(1)
     and g(0)) or(p(2) and p(1) and
      p(0) and cin) after 2*delay_gt;

sum(0) <= (p(0) xor g(0)) xor cin after delay_gt;
sum(1) <= (p(1) xor g(1)) xor c0 after delay_gt;
sum(2) <= (p(2) xor g(2)) xor c1 after delay_gt;
end lkh_DtFl;
```

Verilog Description

```
// II. CARRY-LOOKAHEAD ADDER
module lkahd_adder (x, y, cin, sum, cout);
input [2:0] x, y;
input cin;
output [2:0] sum;
output cout;
/*Assume 4.0-ns propagation delay for all gates
including a 3-input xor.*/

wire c0, c1;
wire [2:0] p, g;
time delay_gt = 4;
    assign #delay_gt g[0] = x[0] & y[0];
    assign #delay_gt g[1] = x[1] & y[1];
    assign #delay_gt g[2] = x[2] & y[2];
    assign #delay_gt p[0] = x[0] | y[0];
    assign #delay_gt p[1] = x[1] | y[1];
    assign #delay_gt p[2] = x[2] | y[2];
    assign #(2*delay_gt) c0 = g[0] | (p[0] & cin);
```

```
assign #(2*delay_gt) c1 = g[1] | (p[1] & g[0]) |
         (p[1] & p[0] & cin);

assign #(2*delay_gt) cout = g[2] | (p[2] & g[1]) |
         (p[2] & p[1] & g[0]) | (p[2] & p[1] &
         p[0] & cin);

assign #delay_gt sum[0] = (p[0] ^ g[0]) ^ cin;
assign #delay_gt sum[1] = (p[1] ^ g[1]) ^ c0;
assign #delay_gt sum[2] = (p[2] ^ g[2]) ^ c1;
endmodule
```

Figure 2.25 shows the waveform for both ripple-carry and carry-lookahead adders without taking gate delay into consideration. Because there is no delay, the two adders have identical waveforms. From the waveform, it can be concluded that both adders are functioning correctly. Figures 2.26a and 2.26b show the waveforms for ripple-carry and carry-lookahead after taking the gate delay into consideration, respectively.

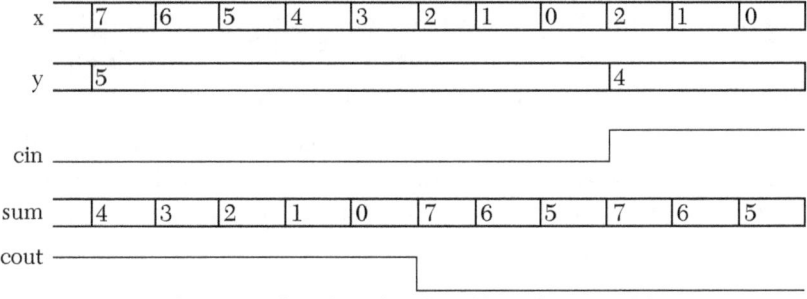

FIGURE 2.25 Simulation waveform for a three-bit adder with no gate delay.

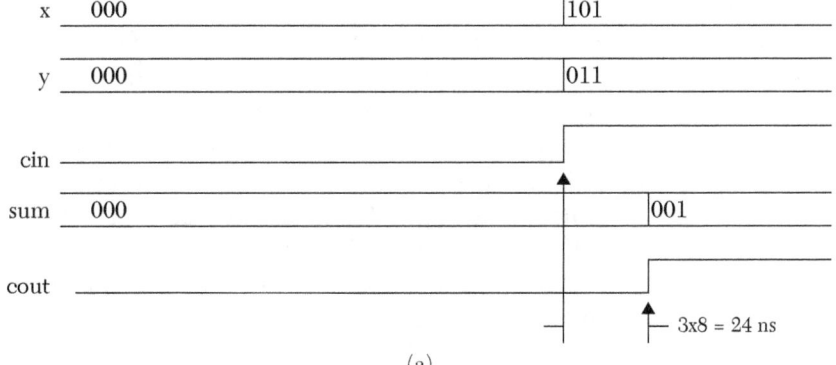

(a)

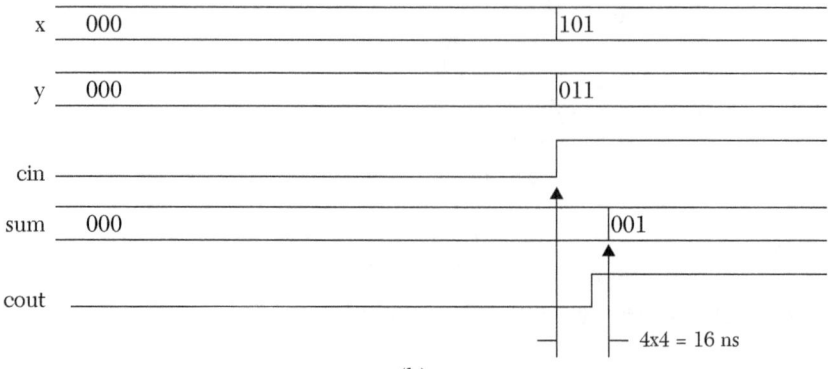

(b)

FIGURE 2.26 Simulation waveforms for three-bit adders with a 4-ns gate delay. a) Ripple-carry adder. b) Carry-lookahead adder.

To calculate the worst delay, values are selected for the inputs x, y, and cin to obtain the maximum possible delay; this is done by selecting those values that cause a change in all the carryout signals. The values x = y = cin = 0 are selected to generate a zero signal on all the outputs, and then the values x = 5, y = 3, and cin = 1. In Figure 2.26a, the total worst delay is 24 ns. Because there are three one-bit adders, and each has a worst delay of 8 ns (two XOR gates), the total worst delay is 8 × 3 = 24 ns, which is equal to the number of one-bit adders times the delay of one one-bit adder.

In Figure 2.26b, the total worst delay is 16 ns, which is four times the delay of a single gate (4 ns). If the number of input bits of the lookahead adder is increased, the total worst delay is still the same 16 ns. More adders will be discussed in Chapter 4.

2.4 Common Programming Errors

This section discusses common programming errors. These errors are classified as either syntax or semantic errors. Syntax errors are those that result from not following the rules of the language. For example, consider the sentence: "Jim am a policeman." The sentence has a syntax error. According to the rules of English language, the word "is" should replace "am." The sentence, after correcting the syntax error, may still have a semantic error if Jim is not a policeman. A semantic error is an error in the meaning of the statement, rather than an error in the mechanics of the statement. The example above applies to HDL; there can be syntax and semantic errors. Syntax errors terminate compilation of the program. Semantic errors may not terminate the program, but the outcome of the program may not be as expected.

2.4.1 Common VHDL Programming Errors

This section briefly discusses some common syntax and semantic errors when writing VHDL programs. Table 2.9 shows a code written in VHDL for two entities and the errors (if any) in that code.

TABLE 2.9 Errors in VHDL Code

Code	Error
`entity mult_comb`	The word `is` is missing
`port (a; b : in std_logic_vector(1 downto 0));`	Semicolon is inserted instead of comma (`a, b`)
`P : std_logic_vector (3 downto 0)`	The direction of the port `P` is missing (`out`)
`architecture MULT_DF of mult_cmb is`	The name of the entity is misspelled: it should be (`mult_comb`)
`P(0) = a(0) and b(0);`	The signal-assignment statement operator is wrong ("`<=`" should replace "`=`")
`P(3) <= a(0) and a(1) b(0) and b(1);`	The "and" operator is missing in `a(1)b(0);`
`P(0) <= a(0) and b(2);`	The index of "b" is out of range: it should be 0 or 1
`end MUL_DF;`	The name of the architecture is misspelled: it should be `MULT_DF`
`P(0) <= a(0) and b0;`	`b0` is not the same as `b(0)`
--No Library listed on first line of code	

`entity errors is`
`port (t, t1: in std_logic ;`
`b,c: out std_logic);`
`end errors;`
`architecture Behavioral of errors is`
`begin`
`b <= t;`
`c <=b;`
`Behavioral;`
	`IEEE.STD_LOGIC_1164.`
	`ALL Library` has to be entered to use std-logic
	
	`b` should be declared as buffer since it is appearing on both right- and left-hand end

2.4.2 Common Verilog Programming Errors

Here, some common syntax and semantic errors in writing Verilog programs are discussed. One of the most common errors for beginners is in not adhering to Verilog's case-sensitive nature. Table 2.10 lists Verilog code and errors (if any).

TABLE 2.10 Possible Errors in Modified Listing 2.4 (Verilog)

Modified Code	Error
`module mult_comb (a, b, P)`	The semicolon (`;`) is missing at the end of the statement
`input [1:0] A, b;`	"`A`" is not defined: it should be lowercase
`output (3:0) P;`	Brackets [3:0] should be used instead of parentheses
`P[0] = a[0] and b[0];`	The word "`assign`" is missing
`assign P[0] = a[0] and b[0];`	The word "`and`" cannot be used here: in Verilog, the logical operator "`&`" should be used
Assign p[0] = S[0] | a[0];	Because S[0] is vector, it has to be declared: if it is scalar (such as S0), it may not need to be declared
`endmodule;`	No semicolon at the end of "`endmodule`"

2.5 Summary

This chapter discussed data-flow descriptions based mainly on writing the Boolean function(s) of the system. The Boolean function is coded as signal-assignment statements. In VHDL, the signal-assignment operator `<=` is implemented to assign a value to a signal; in Verilog, the signal-assignment operator is `assign`. Logical operators such as and (`&`), or (`|`), and xor (`^`) have been implemented to describe the Boolean function in VHDL (Verilog) code. The following table summarizes the commands that have been used in this Chapter. Table 2.11 lists data-flow commands/components in VHDL and their counterparts (if any) in Verilog.

TABLE 2.11 VHDL Versus Verilog Data-Flow Components

VHDL Command/Components	Verilog Counterpart
entity	module
<=	assign
and, or, xor, not	&, |, ^, ~
signal	wire
after	#
in, out, inout	input, output, inout
(2 downto 0)	[2:0]
(0 to 2)	[0:2]

2.6   Exercises

1. Construct a full adder from two 4x1 multiplexers. One multiplexer is to generate the sum, and the other generates the carryout. Write a dataflow description (in both VHDL and Verilog) of the full adder. Use a 5-ns delay for any gate including XOR. Draw the truth table of this adder and derive the Boolean function after minimization. Simulate and verify the circuit.

2. Write a data-flow description (in both VHDL and Verilog) of a system that has three one-bit inputs, a(1), a(2), and a(3), and one one-bit output b. The least significant bit is a(1). The output b is 1 only when {a(1)a(2) a(3)} = 1, 2, 4, or 7 (all in decimal); otherwise, b is 0. Derive a minimized Boolean function of the system and write the data-flow description. Simulate the system and verify that it works as designed. What is the function of this system?

3. Given the following Verilog description code, fill the values of s1 and s2 into the table. T = time in nanoseconds. Do not use a computer to solve this problem.

```
module problem (a, b, s1, s2);
input a, b;
output s1, s2;
    assign #10 s1 = a ^ b;
    assign #10 s2 = a | s1;
```

```
endmodule
```

	T=100	T=150	T=165	T=200	T=250	T=300
a	1	0	0	1	0	1
b	1	1	1	0	0	1
s1	0					
s2	0					

Explain how you obtained the values for s1 and s2 at time T = 165 ns. Translate the Verilog code to VHDL.

4. Referring to Case Study 2.1, increase the number of bits from three to four. Derive the Boolean functions of both the ripple-carry and the carry-lookahead adders. Simulate the adders and calculate the worst delay between the input and output using Verilog description. Contrast your results with Figure 2.16 and explain.

5. The following VHDL code describes an SR-latch. Translate the code to Verilog.

```
entity SR is
    port (S, R : in bit; Q : buffer bit; Qb : out bit);
end SR;
architecture SR_DtFL of SR is
begin
    Q <= S or (not R and Q);
    Qb <= not Q;
end SR_DtFL;
```

6. Describe a system that divides D/V to give a quotient, Q, and Remainder, R. The dividend, D, is three bits; the divisor, V, is two bits. If V = 0, set a flag Z to 1. Write the truth table of the system and obtain the Boolean functions of Q, R, and Z. Use VHDL and Verilog to describe the system.

7. Change the multiplier in Example 2.5 to multiply X*Y where X is three bits and Y is two bits. Find the Boolean function of the output and describe the system using VHDL and Verilog.

8. Write the VHDL and Verilog code describing the full subtractor shown in Example 2.2.

9. Use the conditional operator in Example 2.3b to describe a 2x1 multiplexer with active high enable. If the enable is inactive (low), the output

CHAPTER 3

BEHAVIORAL DESCRIPTION

Chapter Objectives

- Understand the concept of sequential statements and how they differ from concurrent statements
- Identify the basic statements and components of behavioral descriptions such as `process`, variable-assignment statements `if`, `case`, `casex`, `casez`, `when`, `report`, `$display`, `wait`, `loop`, `exit`, `next`, `always`, `repeat`, `forever`, **and** `initial`
- Review and understand the basics of digital logic systems such as D flip-flop, JK flip-flop, T flip-flop, binary counters, and shift register
- Understand the concept of some basic genetic and renal systems
- Both VHDL and Verilog descriptions are discussed

3.1   Behavioral Description Highlights

In Chapter 2, data-flow simulations were implemented to describe digital systems with known digital structures such as adders, multiplexers, and latches. The behavioral description is a powerful tool to describe systems for which digital logic structures are not known or are hard to generate. Examples of such systems are complex arithmetic units, computer control units, and biological mechanisms that describe the physiological action of certain organs such as the kidney or heart.

Facts

■ The behavioral description describes the system by showing how outputs behave with the changes in inputs.

■ In this description, details of the logic diagram of the system are not needed; what is needed is how the output behaves in response to a change in the input.

■ In VHDL, the major behavioral-description statement is `process`. In Verilog, the major behavioral-description statements are `always` and `initial`.

■ For VHDL, the statements inside the process are sequential. In Verilog, all statements are concurrent (see "Analysis of VHDL Code" in Example 3.5).

3.2 Structure of the HDL Behavioral Description

Listing 3.1 shows a simple example of HDL code describing a system (`half_add`) using behavioral description. Usually sequential statements such as IF or Case are used to describe the change of the output; however, in this section, Boolean functions are used to describe the change. This is done here to explain how the HDL executes signal-assignment statements written inside `process` (VHDL) or inside `always` or `initial` (Verilog). The code in Listing 3.1 mainly consists of signal-assignment statements.

Referring to the VHDL code, the entity `half_add` has two input ports, `I1` and `I2`, and two output ports, `O1` and `O2`. The ports are of type `bit`; this type is recognized by the VHDL package without the need to attach a library. If the type is `std_logic`, for example, the IEEE library must be attached. The name of the architecture is `behave_ex`; it is bound to the entity `half_add` by the predefined word `of`. `Process` is the VHDL behavioral-description keyword. Every VHDL behavioral description has to include a process. The statement `process (I1, I2)` is a concurrent statement, so its execution is determined by the occurrence of an event. `I1` and `I2` constitute a sensitivity list of the process. The process is executed (activated) only if an event occurs on any element of the sensitivity list; otherwise, the process remains inactive. If the process has no sensitivity list, the process is executed continuously. The process in Listing 3.1 includes two signal-assignment statements: statement 1 and statement 2.

All statements inside the body of a process are executed sequentially. Recall from Section 2.2 that the execution of a signal-assignment statement has two phases: calculation and assignment. The sequential execution here means *sequential calculation*, which means the calculation of a statement will not wait until the preceding statement is assigned; it will only wait until the calculation is done. To illustrate this sequential execution, refer to Figure 3.1. Assume that in Listing 3.1, at $T = T_0$, I1 changes from 0 to 1, while I2 stays at 1. This change constitutes an event on I1, which in turn activates the process. Statement 1 is calculated as O1 = (I1 XOR I2) = (1 XOR 0) = 1. Then, the value of O2 is calculated, still at T0, as (I1 and I2)= (1 and 0)= 0. After calculation, the value of 1 is assigned to O1 after the delay of 10 ns at T0 +10 ns; the value of 0 is assigned to O2 after the delay of 10ns at T0 + 10ns. For the above example, both data-flow and behavioral descriptions yield the same output for the two signal-assignment statements. This is not the case when a signal appears on both the right-hand side of the statement and the left-hand side of another statement, which will be seen later.

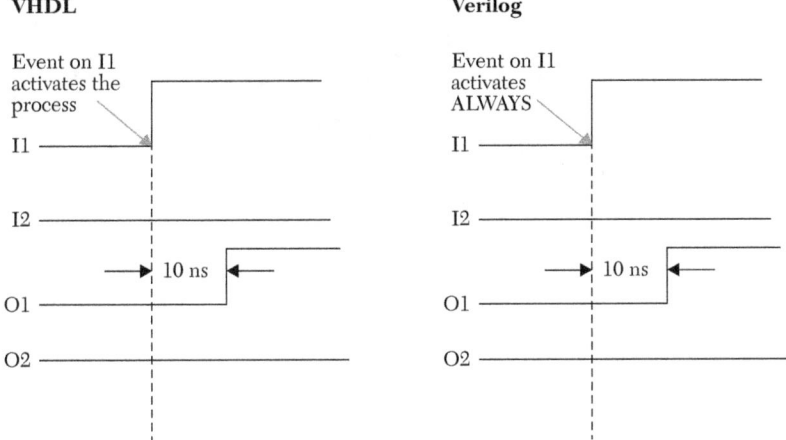

VHDL

Event on I1 activates the process

Verilog

Event on I1 activates ALWAYS

1. Calculate for O1 (1 xor 0) = 1
2. Calculate for O2 (1 and 0) = 0
3. Assign O1 = 1 after 10 ns
4. Assign O2 = 0 after 10 ns

1. Calculate: O1 (1 xor 0) = 1, assign 1 to O1 after 10 ns
2. Calculate: O2 (1 and 0) = 0, assign 0 to O2 after 10 ns

FIGURE 3.1 Execution of signal-assignment statements inside process (VHDL) or inside always (Verilog).

Referring to the Verilog code in Listing 3.1, always is the Verilog behavioral statement. In contrast to VHDL, all Verilog statements inside always are treated as concurrent, the same as in the data-flow description (see

Section 2.2). Also, here any signal that is declared as an output or appears at the left-hand side of a signal-assignment statement should be declared as a register (reg) if it appears inside always. In Listing 3.1, O1 and O2 are declared outputs, so they should also be declared as reg.

LISTING 3.1   Example of an HDL Behavioral Description

VHDL Description
```
entity half_add is
port (I1, I2 : in bit; O1, O2 : out bit);
-- Since we are using type bit, no need for attaching a
-- Library.
-- If we use std_logic, we should attach the IEEE
-- Library.

end half_add;
architecture behave_ex of half_add is
begin
process (I1, I2)
    begin
        O1 <= I1 xor I2 after 10 ns; -- statement 1
        O2 <= I1 and I2 after 10 ns; -- statement 2
-- The above two statements are signal-assignment
-- statements with 10 nanoseconds delays.
--
--Other behavioral (sequential) statements can be added
-- here
    end process;
end behave_ex;
```

Verilog Description
```
module half_add (I1, I2, O1, O2);
input I1, I2;
output O1, O2;
reg O1, O2;
/* Since O1 and O2 are outputs and they are
   written inside "always," they should be
   declared as reg */

always @(I1, I2)
   begin
        #10 O1 = I1 ^ I2; // statement 1.
        #10 O2 = I1 & I2; // statement 2.
```

```
/*The above two statements are
signal-assignment statements with 10 simulation screen units
delay*/
/*Other behavioral (sequential) statements can be added here*/
   end
endmodule
```

3.3   The VHDL Variable-Assignment Statement

The use of variables inside processes is a common practice in VHDL behavioral description. Consider the following two signal-assignment statements inside a process, where S1, S2, and t1 are signals:

```
Sign1 : process(t1)
begin
st1 : S1 <= t1;
st2 : S2 <= not S1;
end process;
```

In VHDL, a statement can be labeled, and the label should be followed by a colon. In the above code, Sign1, st1, and st2 are labels. VHDL code in this example does not use these labels for compilation or simulation; they are optional. Labels are used here to refer to a certain statement by its label. For example, to explain the statement S1 <= t1, it can be referred to by statement st1.

In the above code, signal S1 appears on both the left-hand side of statement st1 and on the right-hand side of statement st2. Assume at simulation time T_0, t1 = 0 and S1 = 0, and at simulation time T_1, t1 changes from 0 to 1 (see Figure 3.2). This change constitutes an event, and the process labeled Sign1 is activated. For statement st1, S1 is calculated as 1. S1 does not acquire this new value of 1 at T1, but rather at T1 + D. For statement st2, S2 at T1 is calculated using the old value of S1 (0). Alternately, variable-assignment statements can be used instead of the above signal- assignment statement as follows:

```
Varb : process(t1)
variable temp1, temp2 : bit; -- This is a variable
                            -- declaration statement
begin
  st3 : temp1 := t1; -- This is a variable assignment
                    -- statement
```

```
st4 : temp2 := not temp1; -- This is a variable
                          -- assignment statement
st5 : S1 <= temp1;
st6 : S2 <= temp2;
end process;
```

Signl: process(t1)
begin
st1: S1 <= t1;
st2: S2 <= not S1;
end process;

Varb: process(t1)
    variable temp1, temp2: bit;
    begin
    st3: temp1: = t1;
    st4: temp2: = not temp1;
    st5: S1 <= temp1;
    st6: S2 <= temp2;
    end process;

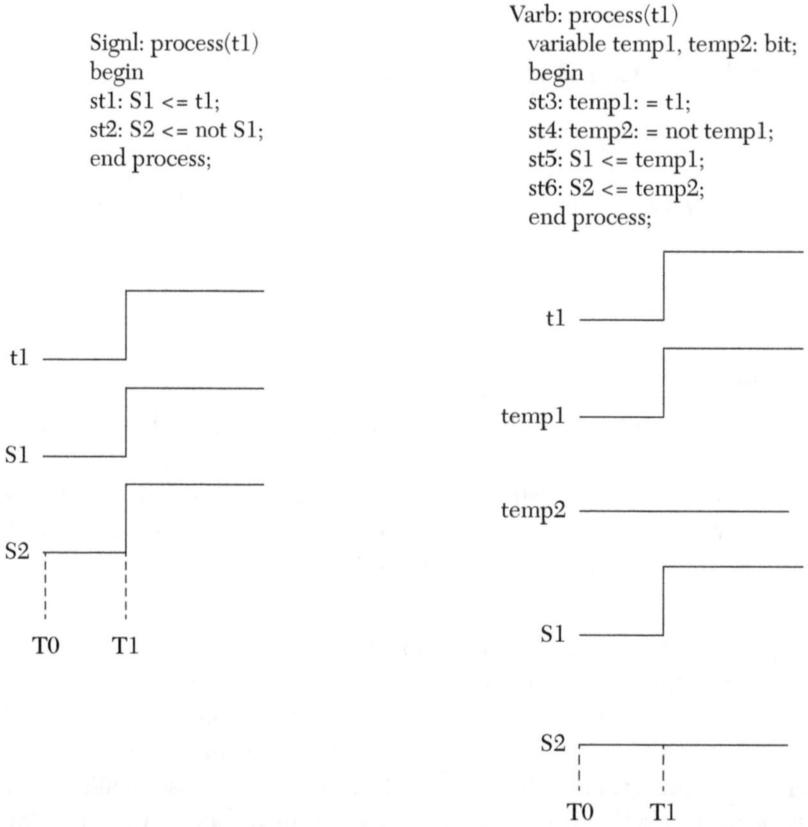

FIGURE 3.2 Signal versus variable in VHDL.

Variable-assignment statements, as in C language, are calculated and assigned immediately with no delay time between calculation and assignment. The assignment operator is :=. If t1 acquires a new value of 1 at T_1, then momentarily temp1 = 1 and temp2 = 0. For statements st5 and st6, S1 acquires the value of temp1 (1) at T1 + D, and S2 acquires the value of temp2 (0) at T1 + D. Because D is infinitesimally small, S1 and S2 appear on the simulation screen as if they acquire their new values at T_1.

3.4 Sequential Statements

There are several statements associated with behavioral descriptions. These statements have to appear inside `process` in VHDL or inside `always` or `initial` in Verilog. The following sections discuss some of these statements.

3.4.1 IF Statement

`IF` is a sequential statement that appears inside `process` in VHDL or inside `always` or `initial` in Verilog. It has several formats, some of which are as follows:

```
VHDL IF-Else Formats
if (Boolean Expression) then
statement 1;
statement 2;
statement 3;
.......
   else
statement a;
statement b;
statement c;
.......
end if;
```

```
Verilog IF-Else Formats
if (Boolean Expression)
begin
   statement 1; /* if only one statement, begin and end
                  can be omitted */
   statement 2;
   statement 3;
.......
end
   else
begin
   statement a; /* if only one statement, begin and end
                  can be omitted */
   statement b;
   statement c;
.......
end
```

The execution of IF statement is controlled by the Boolean expression. If the Boolean expression is true, then statements 1, 2, and 3 are executed. If the expression is false, statements a, b, and c are executed.

EXAMPLE 3.1 BOOLEAN EXPRESSION AND EXECUTION OF IF

VHDL
```
if (clk = '1') then
temp := s1;
else
temp := s2;
end if;
```

Verilog
```
if (clk == 1'b1)
// 1'b1 means 1-bit binary number of value 1.
temp = s1;
else
temp = s2;
```

In Example 3.1, if `clk` is high (1), the value of `s1` is assigned to the variable `temp`. Otherwise, `s2` is assigned to the variable `temp`. The `else` statement can be eliminated, and in this case, the IF statement simulates a latch, as shown in Example 3.2.

EXAMPLE 3.2 EXECUTION OF IF AS A LATCH

VHDL

```
if clk = '1' then
    temp := s1;
end if;
```

Verilog

```
if (clk == 1)
begin
   temp = s1;
end
```

If `clk` is high, the value of `s1` is assigned to `temp`. If `clk` is not high, `temp` retains its current value, thus simulating a latch. Another format for the IF statement is Else-IF.

EXAMPLE 3.3   EXECUTION OF IF AS ELSE-IF

VHDL

```
if (Boolean Expression1) then
statement1; statement2;...
elsif (Boolean expression2) then
statement i; statement ii;...
else
statement a; statement b;...
end if;
```

Verilog
```
if (Boolean Expression1)
begin
    statement1; statement 2;.....
end
else if (Boolean expression2)
begin
    statementi; statementii;.....
end
else
begin
    statementa; statement b;....
    end
```

EXAMPLE 3.4   IMPLEMENTING ELSE-IF

VHDL

```
if signal1 ='1' then
temp := s1;
elsif signal2 = '1' then
temp := s2;
else
temp := s3;
end if;
```

Verilog
```
if (signal1 == 1'b1)
temp = s1;
else if (signal2 == 1'b1)
```

```
temp = s2;
else
temp = s3;
```

After execution of the above IF statement, temp acquires the values shown in Table 3.1.

TABLE 3.1 Output Signals (temp) for Else-IF Statements in Example 3.4

signal1	signal2	temp =
0	0	s3
1	0	s1
0	1	s2
1	1	s1

The Boolean expression may specify other relational operations such as inequality or greater than or less than (see Chapter 1 for details on relational operators).

To illustrate the difference between signal- and variable-assignment statements in VHDL code, the behavioral description of a D-latch is written in Example 3.5. A process is written based on signal-assignment statements, and another process is written based on variable-assignment statements. A comparison of the simulation waveforms of the two processes will highlight the differences between the two assignment statements.

EXAMPLE 3.5 BEHAVIORAL DESCRIPTION OF A LATCH USING VARIABLE AND SIGNAL ASSIGNMENTS

The functionality of a D-latch can be explained as follows: if the enable (E) is active, the output of the latch (Q) follows the input (d); otherwise, the outputs remain unchanged. Also, Qb, the invert output, is always the invert of Q. Figure 3.3a shows the logic symbol of a D-latch. A flowchart that illustrates this functionality is shown in Figure 3.3b. Listing 3.2 shows the VHDL code of the D-latch using variable-assignment statements.

LISTING 3.2 VHDL Code for Behavioral Description of D-Latch Using Variable-Assignment Statements

```
entity DLTCH_var is
    port (d, E : in bit; Q, Qb : out bit);
-- Since we are using type bit, no need for attaching a
```

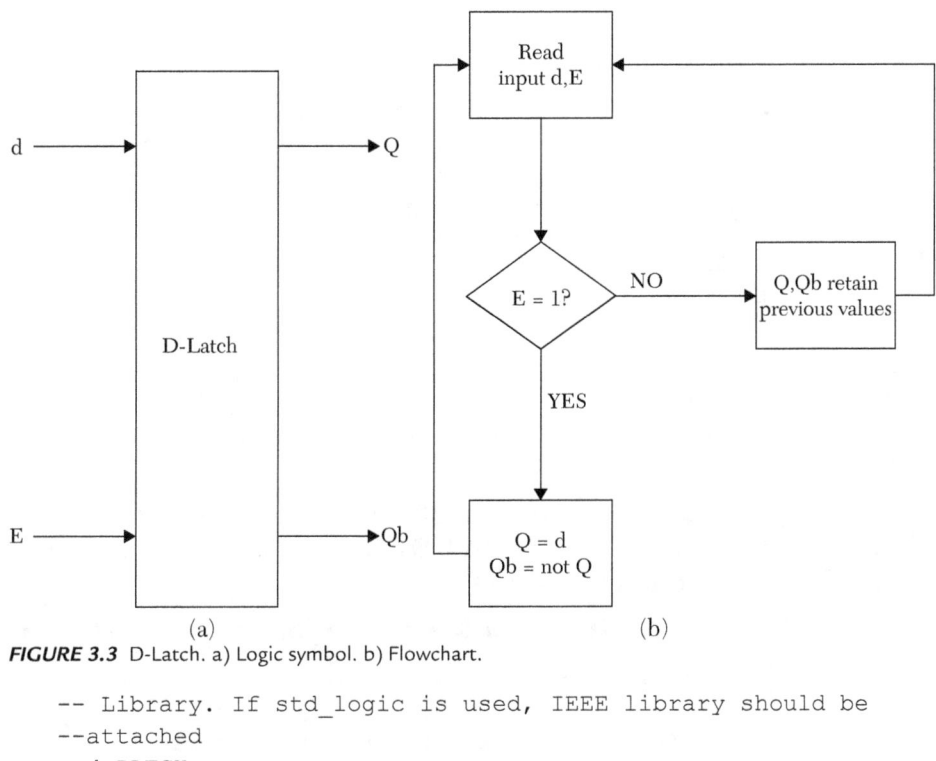

FIGURE 3.3 D-Latch. a) Logic symbol. b) Flowchart.

```
-- Library. If std_logic is used, IEEE library should be
--attached
end DLTCH_var;
architecture DLCH_VAR of DLTCH_var is
begin
VAR : process (d, E)
variable temp1, temp2 : bit;
begin
    if E = '1' then
    temp1 := d; -- This is a variable assignment statement.
    temp2 := not temp1; -- This is a variable assignment
                        -- statement.
end if;
Qb <= temp2; -- Value of temp2 is passed to Qb
Q <= temp1; -- Value of temp1 is passed to Q
end process VAR;
end DLCH_VAR;
```

Figure 3.4 shows the waveform for Listing 3.2. Clearly, from the waveform, the code correctly describes a D-latch where Q follows d when E is high; otherwise, d retains its previous value. Also, Qb is the invert of Q at all times.

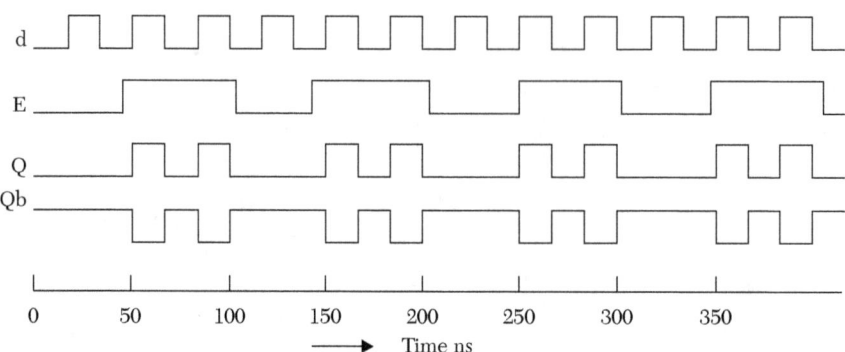

FIGURE 3.4 Simulation waveform of a D-Latch using variable-assignment statements. The waveform correctly describes a D-latch.

Next, the same VHDL code from Listing 3.2 is rewritten using signal-assignment statements. Listing 3.3 shows the VHDL behavioral code for a D-Latch using signal-assignment statements.

LISTING 3.3 VHDL Code for Behavioral Description of a D-Latch Using Signal-Assignment Statements

```
entity Dltch_sig is
port (d, E : in bit; Q : buffer bit; Qb : out bit);
--Q is declared as a buffer because it is an
--input/output signal; it appears on both the left
-- and right hand sides of assignment statements.
end Dltch_sig;
architecture DL_sig of Dltch_sig is
begin
process (d, E)
    begin
    if E = '1' then
        Q <= d; -- signal assignment
        Qb <= not Q; -- signal assignment
    end if;
end process;
end DL_sig;
```

Figure 3.5 shows the simulation waveform of Listing 3.3. The figure shows Q is following Qb, which is an error because Qb should be the invert of Q. This error is due to the sequential execution of the signal-assignment statements in the behavioral description (see details below).

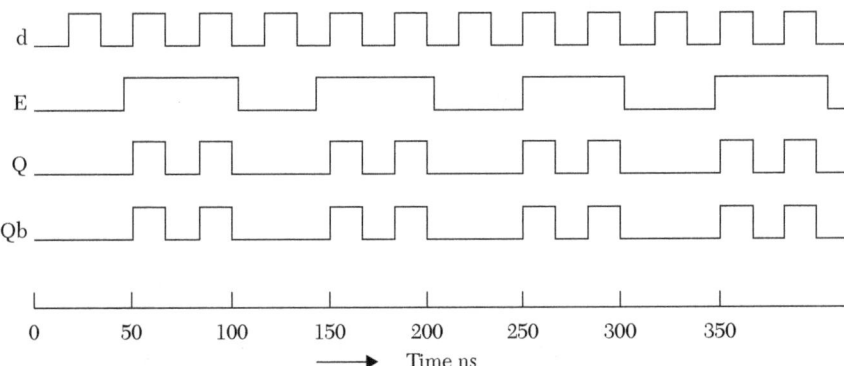

FIGURE 3.5 Simulation waveform of a D-Latch using signal-assignment statements. Qb is following Q instead of being the invert of Q.

3.4.1.1 Analysis of VHDL Code in Listings 3.2 and 3.3

The variable-assignment statements in Listing 3.2 are `temp1 := d` and `temp2 := not temp1`. Referring to Figure 3.4, at simulation time T = 0 ns, initial values are: E = 0, d = 0, Q = 0, and Qb = 0. At T = 50 ns, signal E changes from 0 to 1. Because temp1 and temp2 are variables, they instantaneously acquire their new values 1 and 0, respectively. These correct values are passed to Q and Qb.

Listing 3.3 shows two signal-assignment statements inside the body of the process, `Q <= d` and `Qb <= not q`. Initial values at T ≤ 50 ns are: E = 0, d = 0, Q = 0, and Qb = 0. Recall that execution of a signal-assignment statement inside a process is done in two phases (calculation and assignment). At T = 50 ns, E changes from 0 to 1, and d is 1 at T = 50 ns. Q is calculated as Q = d = 1. Q does not acquire this new value of 1 at T = 50 ns but at T = 50 + Δ. At T = 50 ns, Qb is calculated as 1 (using the old value of Q because Q has not yet acquired its new value of 1). After calculation, a value of 1 is assigned to Q, and the same (wrong) value of 1 is assigned to Qb.

One of the major differences between VHDL and Verilog is that Verilog treats all signal-assignment statements as concurrent, whether they are written as data flow or inside the body of `always`. Listing 3.4 shows the Verilog code for a D-latch; the code generates the same waveform as in Figure 3.4.

LISTING 3.4 Verilog Code for Behavioral Description of a D-Latch

```
module D_latch (d, E, Q, Qb);
input    d, E;
```

```
output Q, Qb;
reg Q, Qb;
always @ (d, E)
begin
    if (E == 1)
        begin
        Q = d;
        Qb = ~ Q;
        end
end
endmodule
```

EXAMPLE 3.6 BEHAVIORAL DESCRIPTION OF A 2x1 MULTIPLEXER WITH TRI-STATE OUTPUT

To describe the behavior of the output of a multiplexer with the change in the input, a flowchart is developed. Figure 3.6a shows the logic symbol of the multiplexer, and Figure 3.6b shows diagram a flowchart describing the functionality of the multiplexer. The flowchart shows how the output behaves with the input. The output is high impedance if the enable (Gbar) is high. When the enable is low, the output is equal to input B if select is high; otherwise, the output is equal to A. The logic diagram of the multiplexer is not needed to write the HDL behavioral description. Although the flowchart here represents a 2x1 multiplexer, it can represent any other applications that have the same behavior; these applications may come from a variety of fields such as electrical engineering, computer engineering, science, business, biomedical engineering, and many other fields. In this example, for simplicity, the propagation delays between the input and the output are not considered.

Listing 3.5 shows the HDL description of the multiplexer using the IF-Else statement, and Listing 3.6 shows the HDL description with the Else-IF statement. The VHDL code uses variable-assignment statements to declare the variable temp; this variable is treated as if it is the output. After calculation of its value, the variable is assigned to the output Y. VHDL executes variable-assignment statements, as does C language; no delay time is involved in the execution. The signal-assignment statements Y <= 'z'; in VHDL and Y = 1'bz; in Verilog assign high impedance to the single-bit Y. If Y is a three-bit signal, then the two statements in VHDL and Verilog are Y <= "zzz"; and Y = 3'bzzz;, respectively. Figure 3.7 shows the simulation waveform of the multiplexer.

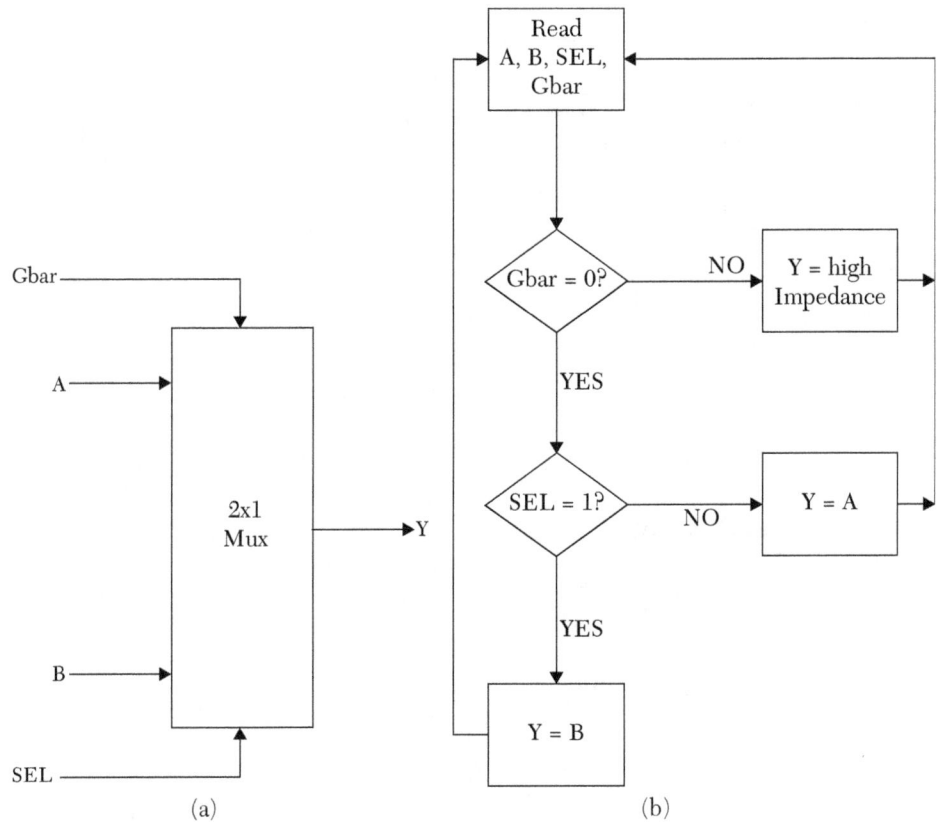

FIGURE 3.6 2x1 Multiplexer. a) Logic symbol. b) Flow chart.

LISTING 3.5 HDL Description of a 2x1 Multiplexer Using `IF-Else`

VHDL Description

```
library IEEE;
use IEEE.STD_LOGIC_1164.ALL;
entity MUX_if is
port (A, B, SEL, Gbar : in std_logic;
          Y : out std_logic);
end MUX_if;
architecture MUX_bh of MUX_if is
begin
process (A, B, SEL, Gbar)
-- A, B, SEL, and Gbar are the sensitivity list of the process.
   variable temp : std_logic;
-- Above statement is declaring temp as a variable; it
-- will be calculated as if it is the output of the
```

```
-- multiplexer.
begin
    if Gbar = '0' then
        if SEL = '1' then
            temp := B;
            else
            temp := A;
        end if;
--Now assign the variable temp to the output
    Y <= temp;
    else
    Y <= 'Z';
    end if;
end process;
end MUX_bh;
```

Verilog Description
```
module mux2x1 (A, B, SEL, Gbar, Y);
input A, B, SEL, Gbar;
output Y;
reg Y;
always @ (SEL, A, B, Gbar)
begin
    if (Gbar == 1)
    Y = 1'bz;
    else
    begin
        if (SEL)
        Y = B;

        else
        Y = A;
    end
end
endmodule
```

LISTING 3.6 HDL Description of a 2x1 Multiplexer Using `Else-IF`

VHDL Description
```
library IEEE;
use IEEE.STD_LOGIC_1164.ALL;
entity MUXBH is
```

```vhdl
      port (A, B, SEL, Gbar : in std_logic;
      Y : out std_logic);
end MUXBH;
architecture MUX_bh of MUXBH is
begin
process (SEL, A, B, Gbar)
variable temp : std_logic;
   begin
       if (Gbar = '0') and (SEL = '1') then
       temp := B;
       elsif (Gbar = '0') and (SEL = '0')then
       temp := A;
       else
       temp := 'Z'; -- Z is high impedance.
   end if;
   Y <= temp;
end process;
end MUX_bh;
```

Verilog Description
```verilog
module MUXBH (A, B, SEL, Gbar, Y);
input A, B, SEL, Gbar;
output Y;
reg Y; /* since Y is an output and appears inside
       always, Y has to be declared as reg( register) */

always @ (SEL, A, B, Gbar)
begin
    if (Gbar == 0 & SEL == 1)
    begin
        Y = B;
    end
    else if (Gbar == 0 & SEL == 0)
    Y = A;
    else
    Y = 1'bz; //Y is assigned to high impedance
end
endmodule
```

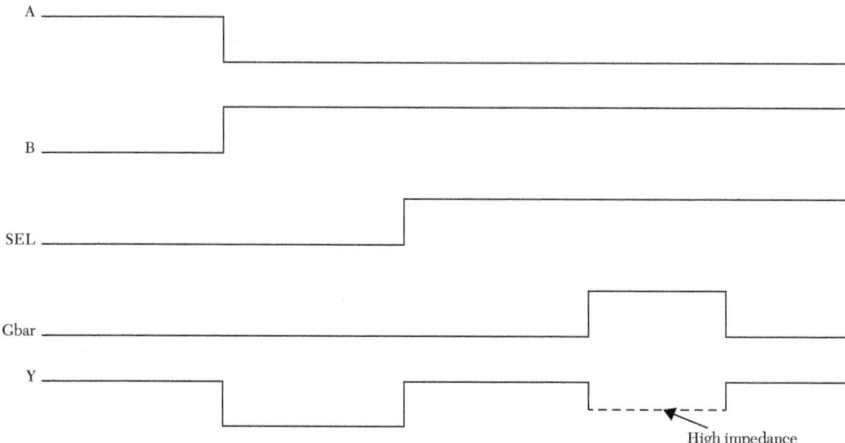

FIGURE 3.7 Simulation waveform of a 2x1 multiplexer.

3.4.2 The case Statement

The case statement is a sequential control statement. It has the following format:

VHDL Case Format
```
case (control-expression) is
when test value or expression1 => statements1;
when test value or expression2 => statements2;
when test value or expression3 => statements3;
when others => statements4;
end case;
```

Verilog Case Format
```
case (control-expression)
test value1 : begin statements1; end
test value2 : begin statements2; end
test value3 : begin statements3; end
default : begin default statements end
endcase
```

If, for example, test value1 is true (i.e., it is equal to the value of the control expression), statements1 is executed. The case statement must include all possible conditions (values) of the control-expression. The statement when others (VHDL) or default (Verilog) can be used to guarantee that all conditions are covered. The case resembles IF except the correct condition in case is determined directly, not serially as in IF statements. The begin and end are not needed in Verilog if only a single statement is specified for a certain test value. The `case` statement can be used to describe data listed into tables.

EXAMPLE 3.7 THE CASE STATEMENT

VHDL
```
case sel is
when "00" => temp := I1;
when "01" => temp := I2;
when "10" => temp := I3;
when others => temp := I4;
end case;
```

Verilog
```
case sel
2'b00 : temp = I1;
2'b01 : temp = I2;
2'b10 : temp = I3;
default : temp = I4;
endcase
```

In Example 3.7, the control is `sel`. If `sel` = 00, then `temp` = `I1`, if `sel` = 01, then `temp` = `I2`, if `sel` = 10, then `temp` = `I3`, if `sel` = 11 (others or default), then `temp` = `I4`. All four test values have the same priority; it means that if `sel` = 10, for example, then the third (VHDL) statement (`temp :=` `I3`) is executed directly without checking the first and second expressions (00 and 01).

EXAMPLE 3.8 BEHAVIORAL DESCRIPTION OF A POSITIVE EDGE-TRIGGERED JK FLIP-FLOP USING THE CASE STATEMENT

Edge-triggered flip-flops are sequential circuits. Flip-flops are triggered by the edge of the clock, in contrast to latches where the level of the clock (enable) is the trigger. Positive (negative) edge-triggered flip-flops sample the input only at the positive (negative) edges of the clock; any change in the input that does not occur at the edges is not sampled by the output. Figures 3.8a and 3.8b show the logic symbol and the state diagrams of a positive edge-triggered JK flip-flop, respectively.

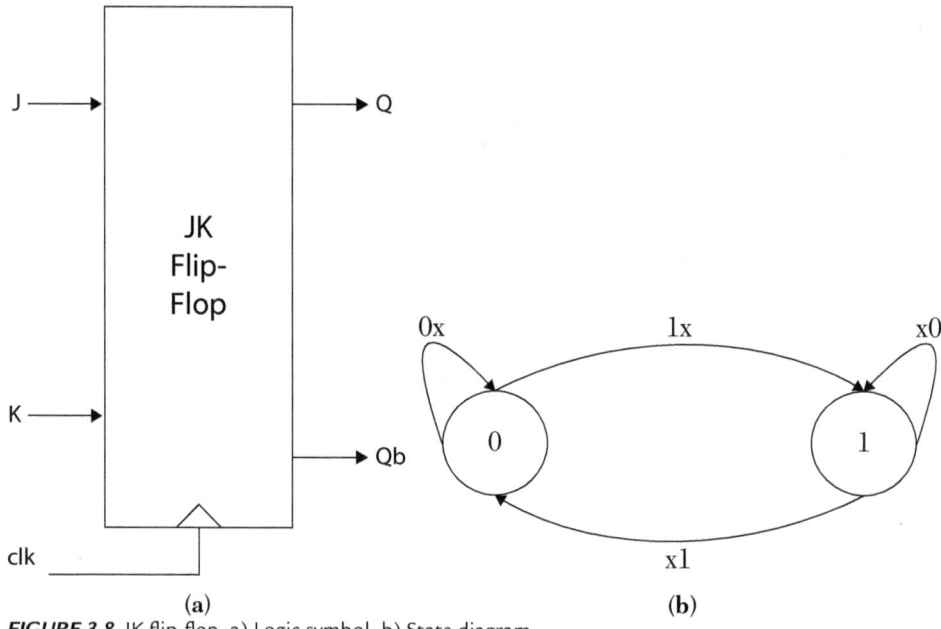

(a) **(b)**
FIGURE 3.8 JK flip-flop. a) Logic symbol. b) State diagram.

Table 3.2 shows the excitation table of the JK flip-flop. It conveys the same information as the state diagram. The state diagram (Figure 3.8b) shows the possible states (two in this case: q can take 0 or 1), state 0 and state 1. The transition between these states has to occur only at the positive edges of the clock. If the current state is 0 (q = 0), then the next state is 0(1) if JK = 0x(1x), where x is "don't care." If the current state is 1 (q = 1), then the next state is 1(0) if JK = x0(x1). Table 3.2 shows the same results as the state diagram. For example, a transition from 0 to 1, according to the excitation table, can occur if JK = 10 or JK = 11, which is JK = 1x.

TABLE 3.2 Excitation Table of a Positive Edge-Triggered JK Flip-Flop

J	K	clk	q (next state)
0	0		No change (hold), next = current
1	0		1
0	1		0
1	1		Toggle (next state) = invert of (current state)
x	x	no +ve edge	No change (hold), next = current

Listing 3.7 shows the HDL code for a positive edge-triggered JK flip-flop using the case statement. In the Listing, `rising_edge` (VHDL) and

posedge (Verilog) are predefined words called *attributes*. They represent the positive edge of the clock (clk). If the positive edge is present, the attribute yields to true. For VHDL, the clk has to be in std_logic to use this attribute. Other attributes are covered in Chapters 4, 6, and 7. Any of the four case statements can be replaced with others (VHDL) or default (Verilog). For example:

```
when "00" => temp1 := temp1; -- VHDL
2'd3 : q =~ q;                // Verilog
```

can be replaced by:

```
when others => temp1 := not temp1; -- VHDL
default : q =~ q;                   // Verilog
```

Because others here refers to 00, this replacement does not change the output of the simulation as long as J and K values are either 0 or 1. The waveform of the flip-flop is shown in Figure 3.9.

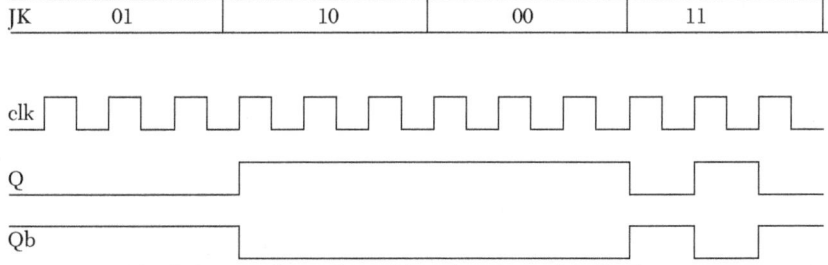

FIGURE 3.9 Simulation waveform of a positive edge-triggered JK flip-flop.

LISTING 3.7 HDL Code for a Positive Edge-Triggered JK Flip-Flop Using the case Statement

VHDL Description
```
library ieee;
use ieee.std_logic_1164.all;
entity JK_FF is
port(JK : in bit_vector (1 downto 0);
clk : in std_logic; q, qb : out bit);
end JK_FF;
architecture JK_BEH of JK_FF is
begin
P1 : process (clk)
variable temp1, temp2 : bit;
begin
if rising_edge (clk) then
```

```
case JK is
when "01" => temp1 := '0';
when "10" => temp1 := '1';
when "00" => temp1 := temp1;
when "11" => temp1 := not temp1;
end case;
q <= temp1;
temp2 := not temp1;
qb <= temp2;
end if;
end process P1;
end JK_BEH;
```

Verilog Description
```
module JK_FF (JK, clk, q, qb);
input [1:0] JK;
input clk;
output q, qb;
reg q, qb;
always @ (posedge clk)
begin
    case (JK)
    2'd0 : q = q;
    2'd1 : q = 0;
    2'd2 : q = 1;
    2'd3 : q =~ q;
    endcase
 qb =~ q;
 end

 endmodule
```

EXAMPLE 3.9 BEHAVIORAL DESCRIPTION OF A THREE-BIT BINARY COUNTER WITH ACTIVE HIGH SYNCHRONOUS CLEAR

Counters are sequential circuits. For count-up counters (or simply up counters), the next state is the increment of the present state. For example, if the present state is 101, then the next state is 110. For down-count counters (or simply down counters), the next state is the decrement of the present state. For example, if the present state is 101, then the next state is 100. A three-bit binary up counter counts from 0 to 7 (Mod 8). Decade counters count from 0 to 9 (Mod10). Synchronous clear means that `clear` resets the counter when the clock is active; in contrast, asynchronous clear resets the counter instantaneously. The counter can be depicted by a flowchart showing its function (see Figure 3.10). Although the flowchart here represents a counter, it could have represented any other system with the same behavior. The excitation table for the three-bit binary counter is as shown in Table 3.3. The logic symbol is shown in Figure 3.10a.

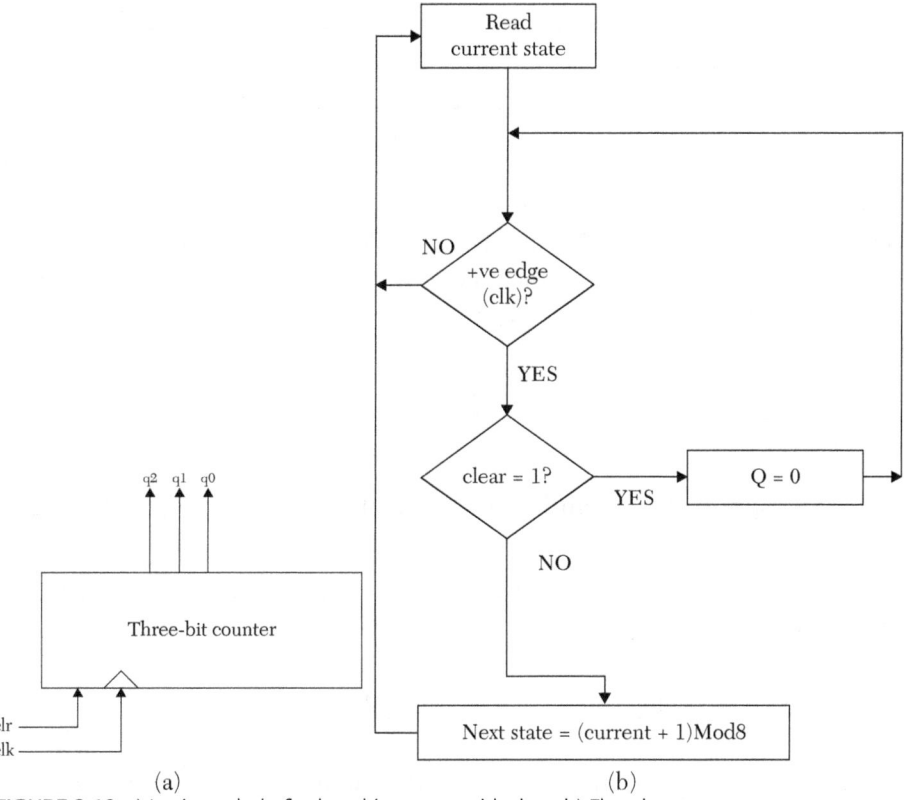

(a)   (b)

FIGURE 3.10 a) Logic symbol of a three-bit counter with clear. b) Flowchart.

TABLE 3.3 Excitation Table of a Three-Bit Binary Counter with Synchronous Active High Clear

| | Input | | Output |
clk	clr	Current State	Next State
↑	H	xxx	000
↑	L	000	001
↑	L	001	010

clk	Input clr	Current State	Output Next State
↑	L	010	011
↑	L	011	100
↑	L	100	101
↑	L	101	110
↑	L	110	111
↑	L	111	000
L	x		hold

The most efficient approach to describe the above counter is to use the fact that the next state is the increment of the present for upward counting. The goal here, however, is to use the case statement. Table 3.3 is treated as a look-up table. Listing 3.8 shows the HDL code for the counter. To assign initial values, such as 101, to the count at the start of simulation in Verilog, the procedural initial is used as follows:

```
initial
begin
q = 3'b101;
end
```

The begin and end can be omitted if there is a single initial statement.

In VHDL, the initial value is assigned to the variable temp after the statement process, as shown:

```
ctr : process (clk)
variable temp : std_logic_vector (2 downto 0) := "101";
begin
```

Any value assigned to a variable written between process and its begin is acquired only once at the beginning of the simulation; subsequent execution of the process will not reassign that value to the variable unless a new simulation is executed. Figure 3.11 shows the simulation waveform of the counter.

LISTING 3.8 HDL Code for a Three-Bit Binary Counter Using the case Statement

```
VHDL Description
library IEEE;
```

```
use IEEE.STD_LOGIC_1164.ALL;
entity CT_CASE is
port (clk, clr : in std_logic;
     q : buffer std_logic_vector (2 downto 0));
end CT_CASE;
architecture ctr_case of CT_CASE is
begin
ctr : process(clk)
variable temp : std_logic_vector (2 downto 0) := "101";
--101 is the initial value, so the counter starts from
-- 110
begin
    if rising_edge (clk) then
    if clr = '0' then
        case temp is
            when "000" => temp := "001";
            when "001" => temp := "010";
            when "010" => temp := "011";
            when "011" => temp := "100";
            when "100" => temp := "101";
            when "101" => temp := "110";
            when "110" => temp := "111";
            when "111" => temp := "000";
            when others => temp := "000";
        end case;
    else
    temp := "000";
end if;
end if;
q <= temp;
end process ctr;

end ctr_case;
```

Verilog Description

```
module CT_CASE (clk, clr, q);
input clk, clr;

output [2:0] q;
reg [2:0] q;
initial
/* The above initial statement is to force
the counter to start from initial count q=110 */
```

```
q = 3'b101;
always @ (posedge clk)
begin
if (clr == 0)
begin
    case (q)
        3'd0 : q = 3'd1;
        3'd1 : q = 3'd2;
        3'd2 : q = 3'd3;
        3'd3 : q = 3'd4;
        3'd4 : q = 3'd5;
        3'd5 : q = 3'd6;
        3'd6 : q = 3'd7;
        3'd7 : q = 3'd0;
    endcase
end
else
q = 3'b000;
end
endmodule
```

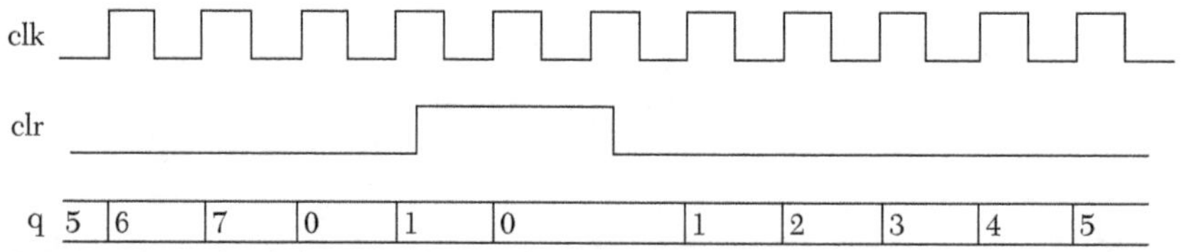

FIGURE 3.11 Simulation waveform of a three-bit positive edge-triggered counter with active high synchronous clear.

EXAMPLE 3.10A   MODELING THE GENOTYPE AND PHENOTYPE OF HUMAN BLOOD USING BIT_VECTOR

In this example, some biomedical engineering applications are considered. The example is about determining the blood type of a child given the blood type of the parents. First, consider some biological definitions to help in understanding the example:

■ **Cells**: The simplest basic structural units that make up all living things.

■ **Chromosomes:** Rod-like structures that appear in the nucleus of the cell, they contain the genes responsible for heredity. Humans have a total 46 different chromosomes in most cells: 23 paternal (from the father) and 23 maternal (from the mother). Sex cells (sperm and ova) each contain half the total number of chromosomes (i.e., 23).

▪ **Deoxyribonucleic acid (DNA):** A polymer of deoxyribonucleotides in the form of a double helix. It is the genetic molecule of life and codes the sequence of amino acids in proteins. Only identical twins have identical DNA. Otherwise, DNA differs from one person to another.

▪ **Gametes:** Sex cells that contain half of the number of chromosomes. In humans, these cells comprise the genetic makeup of eggs and sperm. Each gamete cell contains 23 chromosomes. When a male mates with a female, the two sex cells (egg and sperm) combine to form a single cell called a *zygote*. Gametes for blood types have a single allele: A, B, or O.

▪ **Gene:** A heritable unit in a chromosome, it is a series of nucleotide bases on the DNA molecule that codes for polypeptides (chains of amino acids). Humans have about 30,000 genes.

• **Allele:** An alternate form of a gene.

• **Codominant alleles:** Both alleles are expressed equally. The alleles for blood types A and B are codominant. If combined from a male and a female, the children will be blood type AB.

• **Dominant allele:** An allele that, if combined with other recessive alleles, suppresses their expressions. In blood types, alleles A and B are dominant.

• **Recessive allele:** An allele that, if combined with other dominant alleles, is suppressed. For example, the brown-eye allele is dominant to the blue-eye allele. If a male with blue eyes mates with a female with brown eyes, their children (assuming complete dominance of the brown-eye allele) will have brown eyes. For blood types, the O allele is recessive to A and B.

▪ **Genotype:** The type of alleles in the cell. In the blood example, geneotype is the concatenation of the parental and maternal alleles such as AO, AB, OO.

▪ **Heterozygous in a gene:** Two different alleles are inherited. For blood types, heterozygous alleles can be AB, AO, or BO.

▪ **Homozygous genes:** These cells contain the same alleles of the gene. A person who is homozygous for the brown-eye gene has inherited two alleles for brown eyes, one from their mother and one from their father.

A person who is homozygous for blood type A has two A alleles, one parental and one maternal.

- **Phenotype:** The expression that results from allele combinations. For example, the phenotype of the genotype AO is blood type A because A is dominant and O is recessive. The phenotype of genotype AB is blood type AB because A and B are codominant.

To find all possible genotypes and phenotypes of human blood, a table is constructed to show all possible blood alleles (A, B, O) from male and female gametes. Then, determine the offspring's genotype. From the genotype, the phenotype is determined according to the type of allele (recessive, dominant, or codominant). Table 3.4a shows all possible genotypes, and Table 3.4b shows all possible phenotypes for the offspring.

TABLE 3.4 Genotypes and Phenotypes of Human Blood

A. Genotypes			
♂	A	B	O
♀			
A	AA	AB	AO
B	AB	BB	BO
O	AO	BO	OO
B. Phenotype			
♂	A	B	O
♀			
A	A	AB	A
B	AB	B	B
O	A	B	O

Tables 3.4a and 3.4b are look-up tables, and the case statement can be used to describe the table. Listing 3.9 shows the code for describing the genotypes and phenotypes using case. As shown in the Listing, the alleles are decoded into two bits and entered in the entity as type bit_vector; the output it is decoded in three bits and entered in the entity as a three-bit vector. The two statements

```
geno := allelm & allelf; -- VHDL
    geno   =    {allelm   ,    allelf};   //   Verilog
```

concatenate allelm and allelf into one vector, geno, using the concatenation operator & for VHDL or { , } for Verilog (see Section 1.5.3). For example, if `allelm` = 10, and `allelf` = 11, after concatenation, `geno` = 1011.

LISTING 3.9 HDL Code for Genotypes and Phenotypes Using the `case` Statement: VHDL and Verilog

This program takes the blood genotypes (alleles) of a male and a female and generates the   possible blood phenotypes of their offspring. The statement `report` (VHDL) or `display` (Verilog) is used to print the phenotype on the screen of the simulator. The male allele is `allelm`, and `allelf` is the female allele. Both `allelm` and `allelf` are decoded as 00 for genotype A, 01 for B, or 10 for O. Phenotype A is decoded as 000, B as 001, AB as 010, O as 011, and an illegal allele entry as 111. Figure 3.12 shows the simulation waveform for genotypes and phenotypes of human blood.

VHDL Description

```
library ieee;
use ieee.std_logic_1164.all;
entity Bld_type is
    port (allelm, allelf : in bit_vector (1 downto 0);
        pheno : out bit_vector (2 downto 0));
end Bld_type;
architecture GEN_BLOOD of Bld_type is
begin
Bld : process (allelm, allelf)
variable geno : bit_vector(3 downto 0);
begin
    geno := allelm & allelf;

-- The operator (&) concatenates the two 2-bit vectors
-- allelf and allelm into one 4-bit vector geno.

    case geno is
    when "0000" => pheno <= "000";
    report "phenotype is A ";
--report statement is close to printf in C language.
--The statement here prints on the screen whatever
--written between the quotations.
    when "0001" => pheno <= "010";
    report "phenotype is AB ";
    when "0010" => pheno <= "000";
    report "phenotype is A ";
```

```
        when "0100" => pheno <= "010";
        report "phenotype is AB ";
        when ("0101") => pheno <= "001";
        report "phenotype is B ";
        when ("0110") => pheno <= "001";
        report "phenotype is B ";
        when "1000" => pheno <= "000";
        report "phenotype is A ";
        when ("1001") => pheno <= "001";
        report "phenotype is B ";
        when "1010" => pheno <= "011";
        report "phenotype is O ";
        when others =>pheno <= "111";
        report "illegal allele entry ";
end case;
end process;
end GEN_BLOOD;
```

Verilog Description
```
module bld_type (allelm, allelf, pheno);
input [1:0] allelm, allelf;
output [2:0] pheno;
reg [2:0] pheno;
reg [3:0] geno;
always @ (allelm, allelf)
begin

geno = {allelm , allelf};
/* { , } concatenates the two 2-bit vectors
allelm and allelf into one 4-bit vector geno */
case (geno)
4'd0 : begin pheno = 3'd0;
$display ("phenotype is A "); end
4'd1 : begin pheno = 3'd2;
$display ("phenotype is AB "); end
/* $display statement is close to printf in C language.
The statement here prints on the screen whatever
written between the quotations.*/

4'd2 : begin pheno = 3'd0;
$display ("phenotype is A "); end
```

```
4'd4 : begin pheno = 3'd2;
$display ("phenotype is AB "); end
4'd5 : begin pheno = 3'd1;
$display ("phenotype is B "); end
4'd6 : begin pheno = 3'd10;
$display ("phenotype is B "); end
4'd8 : begin pheno = 3'd0;
$display ("phenotype is A "); end
4'd9 : begin pheno = 3'd1;
$display ("phenotype is B "); end
4'd10 : begin pheno = 3'd3;
$display ("phenotype is O "); end
default: begin pheno = 3'd7;
$display ("illegal allele entry "); end
endcase
end
endmodule
```

allelm	11	10	01	00	11	10	01	00	11	10	01	00

allelf	10				01				00			

geno	1110	1010	0110	0010	1101	1001	0101	0001	1100	1000	0100	0000

pheno	111	011	001	000	111	001	001	010	111	000	010	000

FIGURE 3.12 Simulation waveforms for genotypes and phenotypes of human blood. The phenotype is also printed (not shown here) on the main screen of the simulator.

EXAMPLE 3.10B   MODELING THE GENOTYPE AND PHENOTYPE OF HUMAN BLOOD USING CHARACTER TYPE

In Listing 3.9, the inputs `allelm` and `allelf` and the output pheno had to be decoded into bits so they can be entered as `bit_vector`. Reading the code in decoded bits is not easy because the reader has to memorize what code was given to each signal. Using charcter type (see Section 1.6.1.1) is more convenient in this case because reading the alleles as A, B, and O is more convenient than reading them as 00, 01, and 10.

For VHDL, the *string* type is used to declare a signal in characters; it resembles `bit_vector`, but the elements are ASCII characters rather than bits. If the signal is six charcters in length, for example, the string is declared as *string (1 to 6)*. The double quotaion mark is used to assign the value of the signal in ASCII such as "ABCDEF."

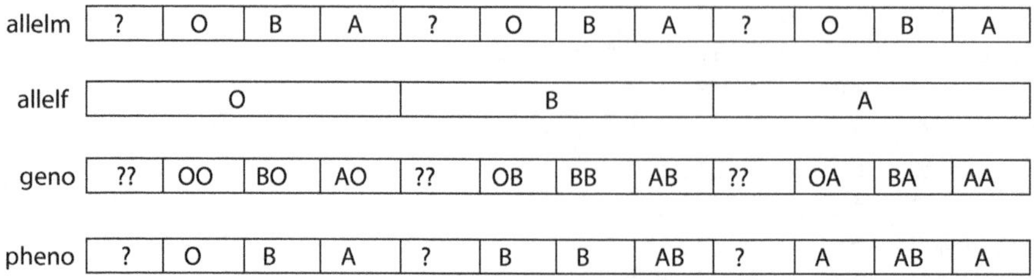

allelm	?	O	B	A	?	O	B	A	?	O	B	A

allelf	O				B				A			

geno	??	OO	BO	AO	??	OB	BB	AB	??	OA	BA	AA

pheno	?	O	B	A	?	B	B	AB	?	A	AB	A

FIGURE 3.13 Simulation waveforms for genotypes and phenotypes of human blood using character type.

For Verilog, each ASCII character is represented by eight bits (two hex digits). In Listing 3.10, `allelm` is represented as one character (two hex digits); the output pheno is represented to two charcters (four hex digits of a total of sixteen bits). The character assignment, same as in VHDL, is done between double quotations. Figure 3.13 shows the simulation waveform of Listing 3.10.

LISTING 3.10 HDL Code for Genotypes and Phenotypes Using the `case` Statement and Character Type

VHDL Description

```
library IEEE;

use IEEE.STD_LOGIC_1164.ALL;

entity bld_charctr is
port ( allelem, allelef : in string(1 to 1) ;
pheno : out string (1 to 2));
end bld_charctr;
architecture Bld_beh of bld_charctr is

begin

process (allelem, allelef)

variable geno: string (1 to 2);
begin
geno := (allelem & allelef);
case (geno ) is
when "AA" => pheno <= "A ";
when "AB" => pheno <= "AB";
when "AO" => pheno <= "A ";
```

```
when "BA" => pheno <= "AB";
when "BB" => pheno <= "B ";
when "BO" => pheno <= "B ";

when "OA" => pheno <= "A ";
when "OB" => pheno <= "B ";
when "OO" => pheno <= "O ";

when others => pheno <= "??";
end case;

end process;
end Bld_beh;
```

Verilog Description

```
module Bld_typeCharctr(allelm, allelf, pheno);

input [8:1] allelm, allelf;
output [2*8:1] pheno;

reg [2*8:1] pheno; /*Since phenol is two characters;
      two ASCII characters are allocated to it.*/
reg [2*8:1] geno;
always @ (allelm, allelf)
begin

geno = {allelm , allelf};

case (geno)
"AA": pheno = "A ";
"AB": pheno = "AB";
"AO": pheno = "A ";
"BB": pheno = "B ";
"BA": pheno = "AB";
"BO": pheno = "B ";
"OA": pheno = "A ";
"OB": pheno = "B ";
"OO": pheno = "O ";
default : pheno = "??"; //?? means invalid entry

endcase
end
endmodule
```

3.4.2.1   Verilog `casex` and `casez`

Section 3.2.3 covered the `case` statement for both VHDL and Verilog. Verilog has another two variations of `case`: `casex` and `casez`. `casex` ignores the "don't care" values of the control expression, and `casez` ignores the high impedance in the control expression. For example, in the code

```
casex (a)
4'bxxx1:  b = 4'd1;
4'bxx10:  b = 4'd2;
.................. .
 endcase;
```

all occurrences of x are ignored; b = 1 if and only if the least significant bit of a (bit order 0) is 1, regardless of the value of the higher order bits of a, and b = 2 if the bits of order 0 and 1 are 10, regardless of the value of all other bits. For the Verilog variation `casez`, all high-impedance values (z) in control expressions are ignored. For example:

```
casez (a)
4'bzzz1 :  b = 4'd1;
4'bzz10 :  b = 4'd2;
.................. .
endcase;
```

b = 1 if and only if the least significant bit (bit of order 0) of a = 1, and b = 2 if bit 0 of a = 0 and bit 1 of a = 1.

EXAMPLE 3.11   VERILOG DESCRIPTION OF A PRIORITY ENCODER USING CASEX

A priority encoder encodes the inputs according to a priority set by the user, such as when the input represents interrupt requests. If two or more interrupt requests are issued at the same time by the devices needing service, and the central processing unit (CPU) can only serve one device at a time, then one of these requests should be given priority over the others and be served first. A priority encoder can handle this task. The input to the encoder is the interrupt requests, and the output of the encoder can be memory addresses where the service routine is located or an address leading to the actual address of the routines. Table 3.5 shows the truth table of a four-bit encoder; bit 0 of input a has the highest priority. Listing 3.11 shows the Verilog description for a four-bit priority encoder. Figure 3.14 shows the simulation waveform of Listing 3.11.

TABLE 3.5 Truth Table for Four-Bit Encoder

Input	Output
a	b
xxx1	1
xx10	2
x100	4
1000	8
Others	0

LISTING 3.11 *Verilog Description for a Four-Bit Priority Encoder Using* **casex**

```
module Encoder_4 (Int_req, Rout_addrs);
input [3:0] Int_req;
output [3:0] Rout_addrs;
reg [3:0] Rout_addrs;

always @ (Int_req)
begin
casex (Int_req)
4'bxxx1 : Rout_addrs=4'd1;
4'bxx10 : Rout_addrs=4'd2;
4'bx100 : Rout_addrs=4'd4;
4'b1000 : Rout_addrs= 4'd8;
default : Rout_addrs=4'd0;

endcase
end
endmodule
```

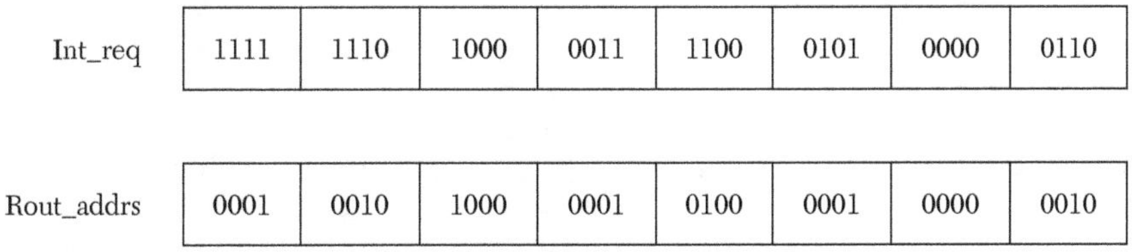

Int_req	1111	1110	1000	0011	1100	0101	0000	0110

Rout_addrs	0001	0010	1000	0001	0100	0001	0000	0010

FIGURE 3.14 Simulation waveform of a four-bit priority encoder.

3.4.3 The `wait-for` Statement

The `wait` statement has several formats; in this section, only *wait for* a time period is discussed. For example:

VHDl : `wait for 10 ns;`

Verilog `# 10;`

The `wait` statement can be implemented to generate clocks, as it is usually common in bench marks. Listing 3.12 shows an example of using the `wait-for` statement to generate three different clocks: a with a period of 20 ns, b with a period of 40 ns, and c with a period of 80 ns. Note that if a `process` (VHDL) or `always` (Verilog) does not have a sensitivity list, this `process` or `always` will run indefinitely. Figure 3.15 shows the waveform of Listing 2.12.

LISTING 3.12 Implementation of the `wait-for` Statement to Generate Clocks

VHDL
```
Library IEEE;
use IEEE.STD_LOGIC_1164.ALL;
entity waittestVHDL is
port ( a,b,c : out std_logic);
end waittestVHDL;

architecture Behavioral of waittestVHDL is

begin
  p1 :process
  variable a1: std_logic := '0';
    begin
      a <= a1;
      wait for 10 ns;
      a1 := not a1;

    end process;
p2 :process
variable b1: std_logic := '0';
  begin
    b <= b1;
    wait for 20 ns;
    b1 := not b1;
end process;
```

```
p3 :process
variable c1: std_logic := '0';
  begin
    c <= c1;
    wait for 40 ns;
    c1 := not c1;
end process;
END;
```

Verilog
```
module waitstatement(a,b,c);
output a,b,c;
reg a,b,c;

initial
begin
// Initialize Inputs
    a = 0;
    b = 0;
    c = 0;
     end
always
   begin
   #10 ;
   a = ~ a;
   end

always
   begin
   #20 ;
   b = ~ b;
    end
always
   begin
   #40 ;
   c = ~ c;
    end

endmodule
```

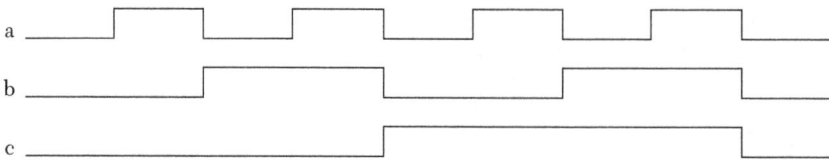

FIGURE 3.15 Simulation waveform of Listing 2.12.

3.4.4 The Loop Statement

Loop is a sequential statement that has to appear inside process in VHDL or inside always or initial in Verilog. Loop is used to repeat the execution of statements written inside its body. The number of repetitions is controlled by the range of an index parameter. The loop allows the code to be compressed; instead of writing a block of code as individual statements, it can be written as one general statement that, if repeated, reproduces all statements in the block. There are several ways to construct a loop. Some of those ways are discussed here.

3.4.4.1 For-Loop

The HDL general format for a For-Loop is:

```
for <lower index value> <upper index value> <step>
statements1; statement2; statement3; ....
end loop
```

If the value of index is between lower and upper, all statements written inside the body of the loop are executed. For each cycle, the index is modified at the end loop according to the step. If the value of index is not between the lower and upper values, the loop is terminated.

> **EXAMPLE 3.12   FOR-LOOP: VHDL AND VERILOG**

VHDL For-Loop
```
for i in 0 to 2 loop
if temp(i) = '1' then
result := result + 2**i;
end if;
end loop;
statement1; statement2; ....
```

Verilog For-Loop
```
for (i = 0; i <= 2; i = i + 1)
begin
```

```
    if (temp[i] == 1'b1)
        begin
            result = result + 2**i;
        end
    end
statement1; statement2; ....
```

The index is `i`, the lower value is 0, the upper value is 2, and the step is 1. All statements between the `for` statement and `end loop` (VHDL) or `end` (Verilog) are executed until the index `i` goes out of range. At the very beginning of the loop, `i` takes the value of 0, and the statements `if` and `result` are executed as:

```
if temp(0) = '1' then
result := result + 2**0;
```

When the program encounters the end of the loop, it increments `i` by 1. If `i` is less than or equal to 2, the loop is repeated; otherwise, the program exits the loop and executes statement1, statement2, and so on. In VHDL, index `i` does not have to be declared, but in Verilog, it has to be declared. If the `loop` statement is stated without range, the loop will run indefinitely.

3.4.4.2 `While-Loop`

The general format of the `While-Loop` is:

```
while (condition)
Statement1;
Statement2;
............
end
```

As long as the condition is true, all statements written before the end of the loop are executed. Otherwise, the program exits the loop.

EXAMPLE 3.13   **WHILE-LOOP: VHDL AND VERILOG**

VHDL `While-Loop`
```
while (i < x)loop
    i := i + 1;
    z := i * z;
end loop;
```

Verilog While-Loop
```
while (i < x)
    begin
        i = i + 1;
        z = i * z;
    end
```

In the above example, the condition is `(i < x)`. As long as `i` is less than `x`, `i` is incremented, and the product `i * z` (`i` multiplied by `z`) is calculated and assigned to `z`.

3.4.4.3 Verilog repeat

In Verilog, the sequential statement `repeat` causes the execution of statements between its `begin` and `end` to be repeated a fixed number of times; no condition is allowed in `repeat`.

EXAMPLE 3.14   VERILOG REPEAT

```
repeat (32)
begin
  #100 i = i + 1;
end
```

In the above example, `i` is incremented 32 times with a delay of 100 screen time units. This describes a five-bit binary counter with a clock period of 100 screen time units.

3.4.4.4 Verilog forever

The statement `forever` in Verilog repeats the loop endlessly. One common use for `forever` is to generate clocks in code-oriented test benches. The following code describes a clock with a period of 20 screen time units:

```
initial
begin
    Clk = 1'b0;
    forever #20 clk = ~clk;
end
```

3.4.4.5 VHDL next and exit

In VHDL, `next` and `exit` are two sequential statements associated with `loop`; `exit` causes the program to exit the loop, and `next` causes the program to jump to the end of the loop, skipping all statements written between `next` and `end loop`. The index is incremented, and if its value is still within the loop's range, the loop is repeated. Otherwise, the program exits the loop.

EXAMPLE 3.15   VHDL NEXT-EXIT

```
for i in 0 to 2 loop
......
.....
next When z = '1';
statements1;
end loop;
statements2;
```

In the above example, at the very beginning of the loop's execution, `i` takes the value 0; at the statement `next When z = '1'`, the program checks the value of `z`. If `z = 1`, then `statements1` is skipped and `i` is incremented to 1. The loop is then repeated with `i = 1`. If `z` is not equal to 1, then `statements1` is executed, `i` is incremented to 1, and the loop is repeated.

EXAMPLE 3.16   BEHAVIORAL DESCRIPTION OF A FOUR-BIT POSITIVE EDGE-TRIGGERED SYNCHRONOUS UP COUNTER

In this example, the `Loop` statement is used to convert values between binary and integer and use this conversion to describe a binary up counter. The HDL package is assumed to not contain predefined functions that will increment a binary input or convert values between binary and integer. In addition, the current and next state are expressed in binary rather than integer. Describing a counter using the above binary-to-integer conversion is not the most efficient way; the main goal here is to demonstrate the implementation of the `Loop` statement.

The next state of a binary counter is generated by incrementing the current state. Because, in this example, a binary value cannot be incremented directly by the HDL code (as was assumed), it is first converted to an integer. HDL packages can easily increment integers. We increment the integer and convert it back to binary. To convert an integer to binary, the predefined operator MOD in VHDL or % in Verilog (see Section 1.5.3.1.) is used. For example: (X MOD 2) equals 1 if X is 1 (odd) or equals 0 if X is 0 (even, divisible by 2). By successively dividing the integer by 2 and recording the remainder from the outcome of the MOD2, the integer is converted to binary. To convert a binary to integer, multiply each bit by its weight and accumulate the products: $1011_2 = (1 \times 1) + (1 \times 2) + (0 \times 4) + (1 \times 8) = 11_{10}$. If the bit is equal to 0, it can be ignored.

Listing 3.13 shows the HDL code of the counter. The simulation waveform is the same as that shown in Figure 3.11, except the count here is from 0 to 15 rather than from 0 to 7 as in the figure.

LISTING 3.13 HDL Code for a Four-Bit Counter With Synchronous Clear: VHDL and Verilog

VHDL Description
```
library ieee;
use ieee.std_logic_1164.all;
entity CNTR_LOP is
port (clk, clr : in std_logic; q :
    buffer std_logic_vector (3 downto 0));
end CNTR_LOP;
architecture CTR_LOP of CNTR_LOP is
begin
ct : process(clk)
variable temp :
                std_logic_vector (3 downto 0) := "0000";
variable result : integer := 0;
begin
if rising_edge (clk) then
    if (clr = '0') then
        result := 0;
-- change binary to integer
        lop1 : for i in 0 to 3 loop
            if temp(i) = '1' then
            result := result + 2**i;
            end if;
        end loop;
-- increment result to describe a counter
        result := result + 1;
-- change integer to binary
        for j in 0 to 3 loop
        if (result MOD 2 = 1) then
            temp (j) := '1';
```

```
            else temp (j) := '0';
        end if;
-- integer division by 2
        result := result/2;
        end loop;
    else temp := "0000";
    end if;
q <= temp;
end if;
end process ct;
end CTR_LOP;
```

Verilog Description

```verilog
module CNTR_LOP (clk, clr, q);
input clk, clr;
output [3:0] q;
reg [3:0] q;
integer i, j, result;
initial
begin
q = 4'b0000; //initialize the count to 0
end
always @ (posedge clk)
begin
    if (clr == 0)
    begin
        result = 0;
        //change binary to integer
        for (i = 0; i < 4; i = i + 1)
            begin
                if (q[i] == 1)
                result = result + 2**i;
            end
        result = result + 1;
        for (j = 0; j < 4; j = j + 1)
        begin
            if (result %2 == 1)
            q[j] = 1;
            else
            q[j] = 0;
            result = result/2;
        end
    end
```

```
        else q = 4'b0000;
end
endmodule
```

A more efficient approach to describe a binary counter is to directly increment the current state. As mentioned before, the approach implemented in Listing 3.13 is not the most efficient way to describe a counter. To write an efficient code for a four-bit counter, direct increment of the current state is used. The following Verilog code describes a four-bit binary counter using direct increment of the current state:

```
module countr_direct (clk, Z);
input clk;
output [3:0] Z;
reg [3:0] Z;
initial
Z = 4'b0000;

/*This initialization is needed if we want to start counting
from 0000 */

always @ (posedge clk)
Z = Z + 1;
endmodule
```

EXAMPLE 3.17   BEHAVIORAL DESCRIPTION OF A FOUR-BIT COUNTER WITH SYNCHRONOUS HOLD USING THE LOOP STATEMENT

To write the code for the counter, binary-integer conversion is used. As mentioned in Example 3.16, this approach is not the most efficient way to describe a counter, but it will be implemented here to demonstrate the use of Loop and the Exit statements. The hold signal in a counter, when active, retains the value of the output and keeps it unchanged until the hold is inactivated. The flowchart of the counter is shown in Figure 3.16. In VHDL, an exit statement is used to exit the loop when the hold is active. Verilog, however, does not have an explicit exit statement, but the loop can be exited by assigning the index a value higher than its upper value. Listing 3.14 shows the HDL code for the counter. Figure 3.17 shows the simulation waveform of the counter.

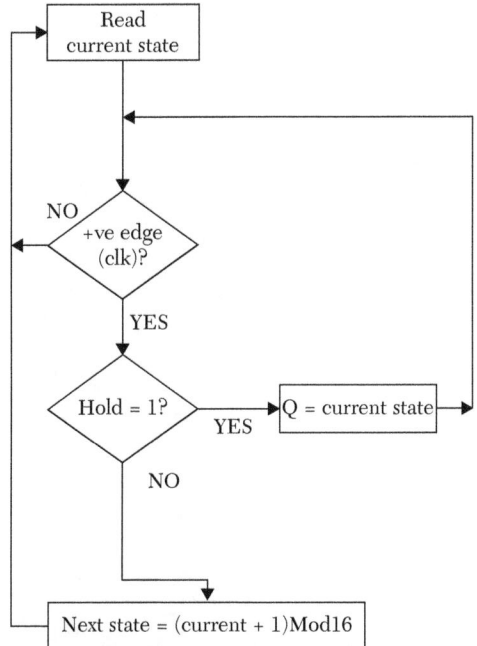

FIGURE 3.16 Flowchart of a four-bit counter with active high hold.

LISTING 3.14 HDL Code for a Four-Bit Counter with Synchronous Hold: VHDL and Verilog

VHDL Description
```
library ieee;
use ieee.std_logic_1164.all;
entity CNTR_Hold is
port (clk, hold : in std_logic;
q : buffer std_logic_vector (3 downto 0));

end CNTR_Hold;
architecture CNTR_Hld of CNTR_Hold is
begin
ct : process (clk)
variable temp : std_logic_vector
         (3 downto 0) := "0000";
-- temp is initialized to 0 so count starts at 0
variable result : integer := 0;
begin
if rising_edge (clk) then
    result := 0;
```

```
-- change binary to integer
   lop1 : for i in 0 to 3 loop
   if temp(i) = '1' then
      result := result + 2**i;
      end if;
   end loop;
-- increment result to describe a counter
   result := result + 1;
   -- change integer to binary
   lop2 : for i in 0 to 3 loop
-- exit the loop if hold = 1
   exit when hold = '1';
-- "when" is a predefined word
   if (result MOD 2 = 1) then
      temp (i) := '1';
   else
      temp (i) := '0';
   end if;
--Successive division by 2
   result := result/2;
   end loop;
   q <= temp;
end if;
end process ct;
end CNTR_Hld;
```

Verilog 4-Bit Counter with Synchronous Hold Description
```
module CT_HOLD (clk, hold, q);
input clk, hold;
output [3:0] q;
reg [3:0] q;
integer i, result;
initial
begin
q = 4'b0000; //initialize the count to 0
end
always @ (posedge clk)
begin
result = 0;

//change binary to integer
```

```
for (i = 0; i <= 3; i = i + 1)
begin
if (q[i] == 1)
result = result + 2**i;
end
result = result + 1;
for (i = 0; i <= 3; i = i + 1)
begin
if (hold == 1)
i = 4; //4 is out of range, exit.
else
    begin
        if (result %2 == 1)
        q[i] = 1;
        else
        q[i] = 0;
        result = result/2;
    end
end
end
endmodule
```

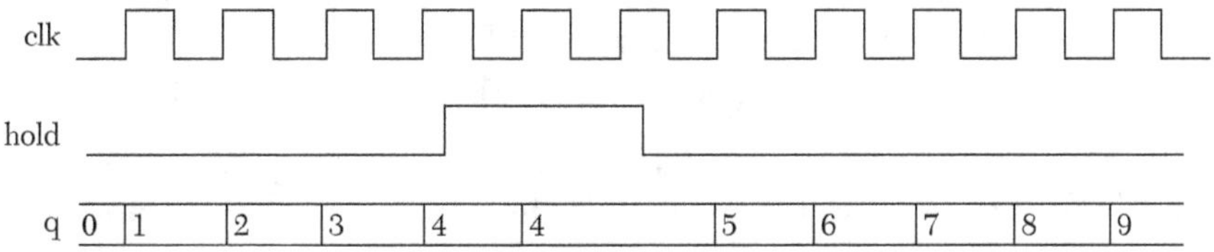

FIGURE 3.17 Simulation waveform of a four-bit binary counter with synchronous hold.

EXAMPLE 3.18   SHIFT REGISTERS DESCRIPTION USING THE LOOP STATEMENT

The main function of a general-purpose register is to store data. The data can be retrieved, or it can be stored indefinitely. The data in the register can be manuplated by several actions such as shift. The data can be shifted right or left logically (Figure 3.18), where zeros are used to fill the vacant bits after shifting; in this shift, some data can be lost. The data can also be shifted arithmatically (Figure 3.18), where if shifted right, the sign of the data (the most significant bit) is preserved. The data in the register can also be rotated left or right (Figure 3.18); here no data are lost. Shift operation is widely used in many areas of digital design such as arithmetic units and serial communications. Shift registers may have an external input bit that replaces the vacant bit after shift. Other registers may have load and bidirectional shifts; these registers are called universal shift registers and are covered in Chapter 4.

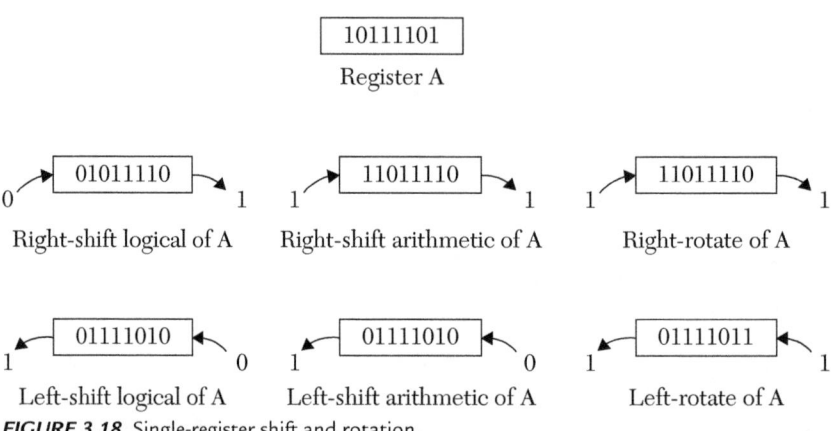

FIGURE 3.18 Single-register shift and rotation.

Listing 3.15 shows a HDL code for describing a logical shift, as shown in Figure 3.18, using the Loop statement. The code shifts register q n bits right or left logically. The number of bits to be shifted is determined by user-selected parameter N. The code resembles the preserved statement sll and slr in VHDL and (<< and >>) in Verilog. See Section 1.5.4.

$display statement in Listing 3.15 is one of Verilog's system tasks that displays values of objects on the console of the simulator. The statement

```
$display (" i= %d", i);
```

will display a printout of the text between the quotation marks (i =) excluding the %d, which determines that the object should be displayed in decimal. The i after the comma is the object to be displayed. The $display is a tool that can be used to display objects that are not listed as an output. Several other formats can be selected for display such as:

%b for binary

%o for octal

$d for decimal

%h for hexadecimal

%t for time

%e or %f or %g for real

%c for character

%s for string

%v for binary and strength

LISTING 3.15 HDL Code for Logical Shifting of a Register Using the Loop Statement

VHDL Description

```vhdl
library IEEE;
use IEEE.STD_LOGIC_1164.ALL;
entity shift_register is
    port(start : in std_logic; shft: in std_logic;
        N: in natural;
    q : out std_logic_vector(7 downto 0))
end shift_register;
--N is number of shifts selected by the user
architecture shift_righLift of shift_register is

begin
st: process (start)
variable vq : std_logic_vector (7 downto 0)
                := "11001110";
--initial values for the vector is selected to be
  -- 1100110
begin
if (start ='1') then
lop2: for j in 1 to N loop
lop1: for i in 0 to 6 loop
if shft ='0' then
--shft = 0 is logical right shift; =1 logical left
-- shift

 vq(i) := vq(i+1);
 vq(7) :='0';
 else
 vq(7-i ) := vq(6-i);
 vq(0) := '0';
 end if;
end loop lop1;
end loop lop2;
end if;
q <= vq;
```

```
end process st;

end shift_righLift;
```

Verilog Description

```verilog
module shft_regVerilog(start,shft, N,q);
input start,shft;
input [7:1] N;
//N is number of requested shifts
output [7:0]q;
reg [7:0]q;
integer i,j;

initial
q = 8'b01100110;
/*initial values for the vector is selected to be
1100110 */

always @ (posedge start)
begin
lop2: for (j= 1; j <= N; j = j +1)
begin
lop1: for (i= 0; i <= 6; i = i +1)
begin
if (shft == 1'b0 )
/*shft = 0 is logical right shift; =1 logical left
Shift */
begin
$display (" shft = %d", shft);/*This is a system task
         to display The value of shift on the console's
         screen of the simulator*/
$display (" i= %d", i);
$display ("q[i] = %b", q[i]);
$display ("q[i+1] = %b", q[i+1]);
  q[i] = q[i+1];
  q[7] =1'b0; $display (" q = %b", q);end

else
begin q[7-i] = q[6-i];
q[0] = 1'b0; end
$display (" shft = %d", shft);
```

```
end
end
end

endmodule
```

EXAMPLE 3.19 **CALCULATING THE FACTORIAL USING BEHAVIORAL DESCRIPTION WITH WHILE-LOOP**

In this example, a HDL behavioral description is written to find the factorial of a positive number N. The factorial of N is (N!) = Nx(N-1)x(N-2)x(N-3)x ....x1. For example, 4!=4×34×24×1=24. In VHDL, N and the output z are declared as natural; this restricts the values that N and z can assume to positive integers. If N and z are declared as `std_logic`, the multiplication operator (∗) cannot be used directly; they must be converted to integers before multiplication or an external library should be attached. In VHDL, be sure to include all the necessary libraries. If the appropriate libraries are not included in the code, the simulator will not accept the declaration and will report it as undefined.

In Verilog, the default declaration of inputs and outputs allows for the direct use of arithmetic operators such as multiplication. Listing 3.16 shows the HDL code for calculating the factorial.

LISTING 3.16 HDL Code for Calculating the Factorial of Positive Integers: VHDL and Verilog

VHDL Description
```
library IEEE;
use IEEE.STD_LOGIC_1164.ALL;
--The above library statements can be omitted;
--however no error if it is not omitted.
--The basic VHDL has type "natural."
entity factr is
port(N : in natural; z : out natural);
end factr;
architecture factorl of factr is
begin
process (N)
variable y, i : natural;
begin
    y := 1;
```

```
    i := 0;
    while (i < N) loop
    i := i + 1;
    y := y * i;
    end loop;
    z <= y;
end process;
end factorl;
```

Verilog Description
```
module factr (N, z);
input [5:0] N;
output [15:0] z;
reg [15:0] z;
/* Since z is an output, and it will appear inside
"always," then Z has to be declared "reg" */

integer i;
always @ (N)
begin
    z = 16'd1;
    i = 0;
    while (i < N)
    begin
        i = i + 1;
        z = i * z;
    end
end
endmodule
```

CASE STUDY 3.1 BOOTH ALGORITHM

The Booth algorithm is used to multiply two signed numbers. The signed numbers are in twos-complement format. The function of the algorithm is to determine the beginning and end of a string of 1s in the multiplier and perform multiplicand addition-accumulation at the end of the string or perform subtraction-accumulation at the beginning of the string. A string consists of one or more consecutive 1s. For example, 01110 has one string, 1011 has two strings (1 and 11). Any signed number can be written in terms of its bit order at the beginning and end of the string. For example, the number 0111011 has the following bit order:

Bit order   6 5 4 3 2 1 0

                0 1 1 1 0 1 1

The number above has two strings. One string has two 1s, begins at bit 0, and ends at bit 1. The other string has three 1s, begins at bit 3, and ends at bit 5. The value of any binary number is equal to $(2^{end1+1} - 2^{begin1})$ + $(2^{end2+1} - 2^{begin2})$+ ....., where begin1 and begin2 are the bit orders of the beginning of string1 and string2, respectively, and end1 and end2 are the bit orders of the end of string1 and string2, respectively. So, 0111011 = $(2^2 - 2^0)$ +$(2^6 - 2^3)$ = 3 + 56 = 59. For the multiplication Z = multiplier (X) × multiplicand (Y), we can write:

$$Z = \{(2^{end1+1} - 2^{begin1}) + (2^{end2+1} - 2^{begin2})+...\}Y$$

$$Y = \{(2^{end1+1}Y - 2^{begin1}Y) + (2^{end2+1}Y - 2^{begin2}Y)+...\} \quad (3.1)$$

Multiplication of Y by positive power(s) of 2 is a shift left of Y. For example, Y × 2^3 is a three-left shift of Y. From Equation 3.1, it can be seen that the calculation of the product Z consists of addition at the end of the string, subtraction at the beginning of the string, and a number of shifts equal to the number of the bits of the multiplicand or the multiplier; here we assume multiplier and multiplicand have the same number of bits). To guarantee no overflow, Z is selected to

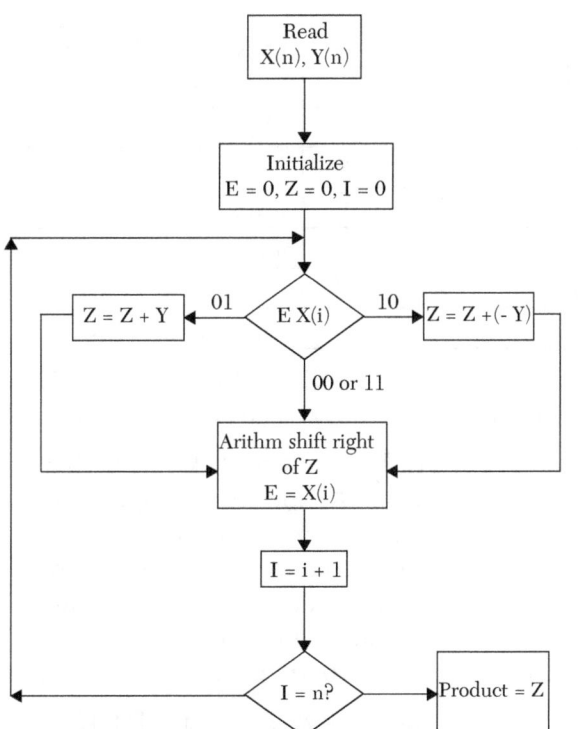

FIGURE 3.19 Flowchart of the Booth multiplication algorithm.

be double the width of X or Y. For example, if X is four bits, then Z is eight bits. The beginning of a string is the transition from 0 to 1, while the end is the transition from 1 to 0. To detect the transition, the one-bit register (E) is used to hold 0 initially. By comparing E with the bits of X, the beginning and end of the string can be detected. The flowchart of the algorithm is shown in Figure 3.19.

To illustrate the algorithm, consider multiplication of two four-bit numbers: –5 (1011) multiplied by 7 (0111). To avoid any possibility of overflow in the product, we assign eight bits to the product. The steps of the Booth algorithm are shown in Table 3.6.

TABLE 3.6 Example of the Booth Algorithm

X = 1011, Y = 0111, –Y = 1001

Step	X(i)E	Action	E	Z
Initial			0	00000000
1, i = 0	10	subtract Y		1001
				10010000
		arithm. shift Z, E = x(i)	1	11001000
2, i = 1	11	arithm. shift Z, E = x(i)	1	11100100
3, i = 2	01	add Y		0111
				01010100
		arithm shift Z, E = x(i)	0	00101010
4, i = 3	10	subtract Y		1001
last step				10111010
		arithm shift Z, E = x(i)	1	11011101

The answer is Z = 11011101 = –35. Note that Z – Y = Z + (–Y), so subtraction of Y from Z is an addition of the twos-complement of Y to Z.

The HDL code for a 4x4-bit Booth algorithm is shown in Listing 3.17. The multiplier (X) and the multiplicand (Y) have to be declared as signed numbers. To do this declaration, the predefined word signed is used. In VHDL, be sure that the appropriate libraries are attached to the code. The statement sum (7 downto 4) represents four bits of sum starting from bit order seven and ending at bit order four. For example, if sum is the eight-bit number 11001010, then sum (7 downto 4) is 1100.

The statement Y := –Y in VHDL (Y = –Y in Verilog) changes Y to its twos complement. If Y = 1101, then –Y = 0011. The statement sum := sum `srl` in VHDL (`z = z >> 1` in Verilog) is the logical shift right of `sum(Z)` one position. For example, if `sum` or `z = 11010100`, then after right shift, `sum(Z) = 01101010`. In Listing 3.17, `sum` and `z` are signed numbers; this means that the most significant bit is the sign bit. If this bit is 0, the number is positive, and if it is 1, the number is negative. Notice that after the logical shift, the sign may change, as in our example where `sum(Z)` changes from 11010100 (a negative number) to 01101010 (a positive number) after a one-position right shift. Another type of shift is arithmetic, where the sign is preserved. An arithmetic right shift of 11010100 yields 11101010. The shift in the Booth algorithm is arithmetic; the following two statements perform arithmetic shift:

VHDL	**Verilog**
`sum := sum srl 1;`	`Z = Z >>> 1;`
`sum (7):= sum(6);`	

The first statement performs logical shift, and the second performs sign preservation. VHDL code has a predefined arithmetic shift operator, `sra;`. For example, `sum := sum sra 2` executes a right shift of two positions and preserves the sign. To use this shift, be sure that the appropriate libraries and simulator are used. The simulation waveform of the Booth algorithm is shown in Figure 3.20.

LISTING 3.17 4x4-Bit Booth Algorithm: VHDL and Verilog

VHDL Description

```
library ieee;
use ieee.std_logic_1164.all;
use ieee.numeric_std.all;
entity booth is
  port (X, Y : in signed (3 downto 0);
  Z : buffer signed (7 downto 0));
end booth;
architecture booth_4 of booth is
begin
```

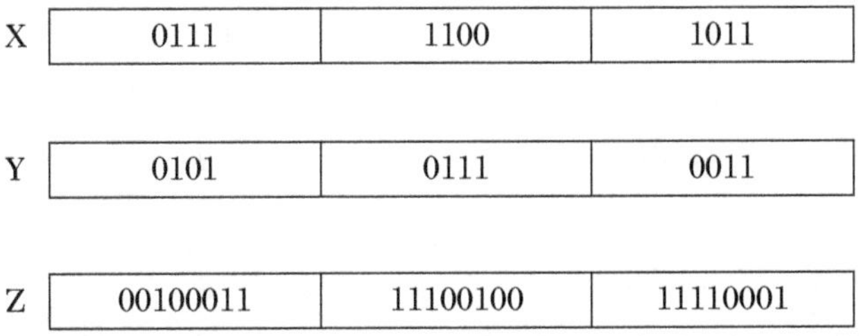

X	0111	1100	1011

Y	0101	0111	0011

Z	00100011	11100100	11110001

FIGURE 3.20 Simulation waveform of a Booth multiplication algorithm.

```
process (X, Y)
variable temp : signed (1 downto 0);
variable sum : signed (7 downto 0);
variable E1 : unsigned (0 downto 0);
variable Y1 : signed (3 downto 0);
begin
sum := "00000000"; E1 := "0";
for i in 0 to 3 loop
temp := X(i) & E1(0);
Y1 := - Y;
case temp is
    when "10" => sum (7 downto 4) :=
    sum (7 downto 4) + Y1;
    when "01" => sum (7 downto 4) :=
    sum (7 downto 4) + Y;
    when others => null;
end case;
sum := sum srl 1; --This is a logical
--shift of one position to the right
sum (7) := sum(6);

--The above two statements perform arithmetic
--shift where the sign of the
--number is preserved after the shift.

E1(0) := x(i);
end loop;
    if (y = "1000") then

--If Y = 1000; then according to our code,
--Y1 = 1000 (-8 not 8 because Y1 is 4 bits only).
--The statement sum = -sum adjusts the answer.

        sum := - sum;
    end if;
z <= sum;
end process;
end booth_4;
```

Verilog Description
```
module booth (X, Y, Z);
input signed [3:0] X, Y;
```

```
output signed [7:0] Z;
reg signed [7:0] Z;
reg [1:0] temp;
integer i;
reg E1;
reg [3:0] Y1;
always @ (X, Y)
begin
Z = 8'd0;
E1 = 1'd0;
for (i = 0; i < 4; i = i + 1)
begin
temp = {X[i], E1};

//The above statement is catenation

Y1 = - Y;

//Y1 is the 2' complement of Y

case (temp)
2'd2 : Z [7 : 4] = Z [7 : 4] + Y1;
2'd1 : Z [7 : 4] = Z [7 : 4] + Y;
default : begin end
endcase
Z = Z >>> 1;
/*The above statement is arithmetic shift of one position to
the right*/

E1 = X[i];
   end
if (Y == 4'd8)

/*If Y = 1000; then according to our code,
Y1 = 1000 (-8 not 8, because Y1 is 4 bits only).
The statement sum = - sum adjusts the answer.*/
   begin
       Z = - Z;
   end
 end
endmodule
```

CASE STUDY 3.2 BEHAVIORAL DESCRIPTION OF A SIMPLIFIED RENAL ANTIDIURETIC HORMONE MECHANISM

In this case study, the action of antidiuretic hormone (ADH) on water excreted by the kidney is discussed. One function of the kidney is to regulate the amount of water excreted by the body as urine. Human blood is 70% water by volume. Regulation of the water volume is directly related to blood pressure regulation. An excessive amount of water in the body raises blood pressure, and if the body excretes more water than it needs to maintain proper functions, blood pressure will drop. Kidney failure has a direct effect on blood pressure. The main functional unit in the kidney is the nephron. Figure 3.21 illustrates a schematic of nephron functions.

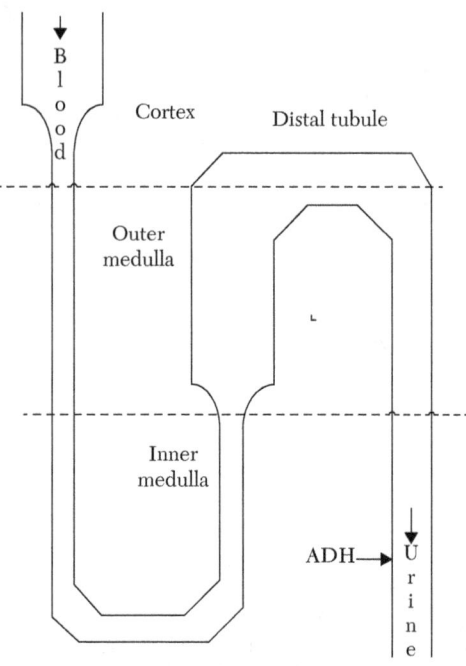

FIGURE 3.21 Nephron function in the human body.

Nephrons are tiny tubules through which blood flows. In nephrons, some components in the blood, such as sodium and potassium, are reabsorbed by the body, and other components, such as urea, are excreted because they are toxic to the body. Any extra water that the body does not need is also excreted as urine. Several hormones control the amount of water excreted. One of those hormones is ADH. The function of ADH is summarized as follows:

- The biological action of ADH is to conserve body water and regulate tonicity of body fluids.

- ADH is released by the hypothalamic cells in the brain.

- Water deprivation (and subsequent low blood pressure) stimulates ADH release. Conversely, excess water (and subsequent high blood pressure) decreases ADH release.

- The major target of ADH is the renal cells, specifically, the collecting ducts of the nephrons.

- ADH causes the kidney to reabsorb (conserve) water. Absence of ADH causes the kidney to excrete water as urine.

- Alcohol and caffeine inhibit ADH release and promote more urine.

Figure 3.22 describes a simplified possible representation of the relationship between the concentration of ADH and blood pressure (BP). Assume that the relationship is linear, and BP takes only positive integer values.

The HDL code is shown in Listing 3.18. It is assumed that the body samples its blood pressure at intervals; each interval is represented in the code by the period of the clock. The major sequential statement in the code is Else-IF. For simplification, the blood pressure and ADH are allowed to take only integer-positive values. In VHDL, this means that BP and ADH are declared as `natural`, allowing the application of the equation `ADH = BP * (-4) + 180.0`. If BP and ADH are declared as `std_logic_vectors`,

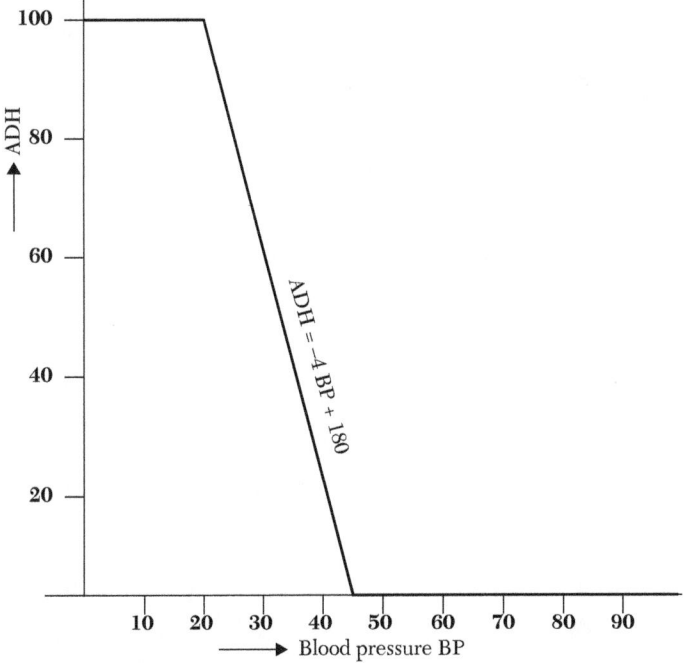

FIGURE 3.22 Concentration of ADH versus blood pressure (units are arbitrary).

VHDL cannot directly multiply or add. In contrast, Verilog allows for direct addition and multiplication if BP and ADH are declared as bit vectors. Figure 3.23 shows the simulation waveform of an ADH-BP relationship.

LISTING 3.18   Antidiuretic Hormone Mechanism: VHDL and Verilog

VHDL Description

```
library IEEE;
use IEEE.STD_LOGIC_1164.ALL;
use IEEE.STD_LOGIC_ARITH.ALL;
use IEEE.STD_LOGIC_UNSIGNED.ALL;
entity ADH_BEH is
     port (clk : in std_logic; BP : in natural;
ADH : out natural);
-- Assume BP takes only positive integer values
end;
architecture ADH_XT of ADH_BEH is
begin
ADHP : process (clk)
variable resADH : natural := 0;
begin
if (clk = '1') then
if Bp <= 20 then resADH := 100;
elsif Bp > 45 then resADH := 0;
else
     resADH := Bp * (-4) + 180;
end if;
end if;
ADH <= resADH;
end process ADHP;
end ADH_XT;
```

Verilog Description

```
module ADH_BEH (clk, BP, ADH);
input clk;
input [8:0] BP;
// Assume BP takes only positive integer values
output [8:0] ADH;
reg [8:0] ADH;
always @ (clk)
begin
if (clk == 1)
```

```
begin
    if (BP <= 20) ADH = 100;
    else if (BP > 45.0) ADH = 0;
    else
    ADH = BP * (-4) + 180.0;
end
end
endmodule
```

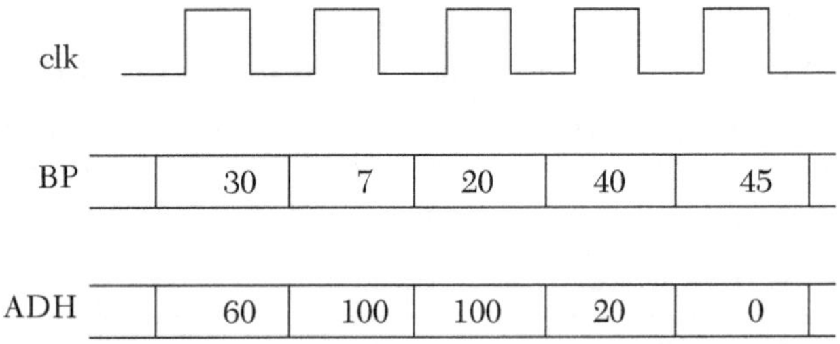

FIGURE 3.23 Simulation waveform of ADH versus blood pressure.

3.5 Common Programming Errors

This section discusses some common programming errors. Additional common errors are discussed in Chapter 2.

3.5.1 Common VHDL Programming Errors

The following is a brief discussion of some common syntax and semantic errors in writing VHDL programs. Table 3.7 considers Listing 3.16 (VHDL) and some possible errors if the code is modified.

TABLE 3.7 Possible Errors in Modified VHDL Listing 3.13

Modified Code	Error
`process (Z)`	Sensitivity list cannot include output ports

`process (N)`
`begin`
`variable y, i : natural;`
`port (N : in integer; z :`
`out natural);`
	Variable declaration should be before begin
	The syntax is correct, but if N is forced to a negative value, the loop will not terminate, causing the program to hang up

`y <= y * i;`	y has been declared as variable; the variable-assignment operator := should be used instead of the signal-assignment operator <=

(contd.)

Modified Code	Error
`Z := y * i;`	Z has been declared as signal; the variable-assignment operator := cannot be used

`while (i < N) loop`
`i := i + 1;`
`y := y * i;`
`end;`
	end; should be written as
	`end loop;`

3.5.2 Common Verilog Programming Errors

Here, some common Verilog syntax and semantic errors are briefly discussed. One of the most common errors for beginners is not adhering to Verilog's case-sensitive nature. Table 3.8 considers Listing 3.16 (Verilog) and discusses some possible errors if the code is modified.

TABLE 3.8 Possible Errors in Modified Verilog Listing 3.16

Modified Code	Error

`module factr (N, z);`
`input [15:0] N;`
`output [15:0] z;`
`integer i;`
`  always @(N)`
	Because z is an output, it has to be declared as reg

`always @ (N)`	To end `always`, write only end.

`Begin`
`z = 1;`
`.........`
`end always;`
	without semicolon

`while (i <= N)`	There is no syntax error, but the result of the program are not correct: try N = 2 and find z

3.6 Summary

In this chapter, the basics of behavioral description have been covered, including the statements process (VHDL) and always (Verilog). Some sequential statements have also been discussed such as IF, wait, case, and Loop. These sequential statements have to appear inside process in VHDL or inside always or initial in Verilog. In VHDL, all signal-assignment statements inside process are executed sequentially. Here, sequentially means calculating the values of the left-hand side of the statements in the order in which they are written. After calculation, the values are assigned taking into consideration any delay times. In Verilog, all statements inside `always` are executed concurrently, based on events. Execution of variable-assignment statements inside `process` in VHDL, in contrast to signal-assignment statements, does not involve any timing delays; execution here is the same as in C language. Table 3.9 shows a list of the VHDL statements covered in this chapter along with their Verilog counterparts (if any).

TABLE 3.9 Summary of VHDL Behavioral Statements and Their Verilog Counterparts

VHDL	Verilog
`process`	`always`
`variable`	`------`
`-------`	`reg`
`if;else;endif`	`if;else;begin end`
`if;elsif;else;endif`	`if;else if;else;begin end`
`case endcase`	`case begin end`
`for loop`	`for`
`while loop`	`while`
`next, exit`	`-----`
`-------`	`repeat, forever`
`MOD`	`%`
`signed`	`signed`
`srl 1`	`>> 1`
`integer`	`integer`
`wait for 10 ns`	`#10`

3.7   Exercises

1. Add asynchronous clear signal to the JK flip-flop discussed in Example 3.8. Write both VHDL and Verilog to describe the flip-flop and simulate the code.

2. Write VHDL and Verilog code for a T flip-flop and simulate.

3. Modify Listing 3.15 to include rotate and arithmetic shift.

4. In Example 3.8, a JK flip-flop was described by using a `case` statement on `JK`. Change the code to describe the flip-flop by using `case` on `Q`. Simulate and verify your description.

5. Use binary-to-integer conversion to describe a four-bit even counter with active low clear and synchronous load (load from external P to Q). Use Verilog, simulate, and verify.

6. Using the Booth algorithm (see Case Study 3.1), modify the code to satisfy all the following requirements:

- The multiplier and the multiplicand are five bits each.

- If the multiplier or the multiplicand is 0, the product should be 0 without going through the multiplication steps.

- If the multiplier or the multiplicand is 1 (decimal), the product should be equal to the multiplicand or the multiplier, respectively, without going through the multiplication steps.

7. In Case Study 3.2, it was assumed that the relationship between ADH and BP is linear: `Bp * (-4) + 180` (VHDL). Change this relationship to be exponential: `ADH = a exp (b * BP)`. The value of ADH is 100 for $BP \leq 20$ and stays at 10 for $BP \geq 45$. Write the VHDL code using the `case` statement to describe this relationship. You can approximate the values of ADH to be integers but be as accurate as possible.

8. Design an arithmetic and logical unit (ALU) that performs addition, subtraction, multiplication, and integer division. The input to the ALU is two signals, A and B, of integer type. The output is signal Z of integer type. The ALU performs the operations according to a signal called op_ code. This op_code is of character type, and Table 3.10 shows the value of the op_code (in character) and the selected operation.

TABLE 3.10 op_code and the selected operation

op_code	Operation
add	Add A to B and store the result in Z
sub	Subtract B from A and store the result in Z
multply	Multiply A x B and store the result in Z
dvdInt	Divide A by B and store the result in Z

CHAPTER 4

STRUCTURAL DESCRIPTION

Chapter Objectives

- Understand the concept of structural description, including the binding of modules
- Identify the basic statements of structural description, such as `component, use, and, or, not, xor, nor, generate, generic,` **and** `parameter`
- Review and understand the fundamentals of digital logic design for digital systems, such as adders, multiplexers, decoders, comparators, encoders, latches, flip-flops, counters, shift registers, and memory cells
- Understand the concept of sequential finite-state machines

4.1 Highlights of Structural Description

Structural description is best implemented when the digital logic of the details of hardware components of the system are known. An example of such a system is a 2x1 multiplexer. The components of the system are known: AND, OR, and NOT gates. Structural description can easily describe these components. On the other hand, it is hard (if not impossible) to describe the digital logic of, say, hormone secretion in the blood; therefore, another description such as behavioral or mixed may be implemented. Structural description is very close to schematic simulation.

In this chapter, structural description is covered. Both gate-level and register-level description are discussed for VHDL and Verilog. Highlights of the structural description can be summerized in the following facts.

Facts

▪ Structural description simulates the system by describing its logical components. The components can be gate level (such as AND gates, OR gates, or NOT gates), or components can be in a higher logical level, such as register-transfer level (RTL) or processor level.

▪ It is more convenient to use structural description than behavioral description for systems that require specific design constraints. Consider, for example, a system performing the operation A + B = C. In behavioral description, the addition can be written as C = A + B with no choice in selecting the type of adders used to perform this addition. In structural description, the type of adder, such as look-ahead adders, can be selected.

▪ All statements in structural description are concurrent. At any simulation time, all statements that have an event are executed concurrently.

▪ A major difference between VHDL and Verilog structural description is the availability of components (especially primitive gates) to the user. Verilog recognizes all the primitive gates such as AND, OR, XOR, NOT, and XNOR gates. Basic VHDL packages do not recognize any gates unless the package is linked to one or more libraries, packages, or modules that have the gate description. Usually, the user develops these links, as will be done in this chapter.

▪ Although structural description is implemented in this chapter to simulate digital systems, this does not mean that only one type of description (structural) can be used in a module. In fact, in most descriptions of complex systems, mixed-type descriptions (e.g., data flow, behavioral, structural, or switch-level) are used in the same module (see Chapter 7).

4.2 Organization of Structural Description

Listing 4.1 shows an example of HDL code that describes a half adder under the name of system using structural description. The entity (VHDL) or module (Verilog) name is system; there are two inputs, a and b, and two outputs, sum and cout. The entity or module declaration is the same as in other description styles previously covered (data flow and behavioral).

In the VHDL description, the structural code (inside the architecture) has two parts: declaration and instantiation. In declaration, all of the different types of components are declared. For example, the statements

```
component xor2
port (I1, I2 : in std_logic; O1 : out std_logic);
end component;
```

declare a generic component by the name of xor2; the component has two inputs (I1, I2) and one output (O1). The name (identifier) xor2 is not a reserved or predefined word in VHDL; it is a user-selected name. To specify the type of the component (e.g., AND, OR, XOR, etc.), additional information should be given to the simulator (see Listing 4.2). If the system has two or more identical components, only one declaration is needed. The instantiation part of the code maps the generic inputs/outputs to the actual inputs/outputs of the system. For example, the statement

```
X1 : xor2 port map (a, b, sum);
```

maps input a to input I1 of xor2, input b to input I2 of xor2, and output sum to output O1 of xor2. This mapping means that the logic relationship between a, b, and sum is the same as between I1, I2, and O1. If xor2 is specified through additional statements to be a XOR gate, for example, then sum = a xor b. A particular order of mapping can be specified as:

```
X1 : xor2 port map (O1 => S, I1 => b , I2 => a);
```

S is mapped to O1, b is mapped to I1, and a is mapped to I2. Note that the mapping of S is written before writing the mapping of the inputs; we could have used any other order of mapping. As previously mentioned, structural-description statements are concurrent and are driven by events. This means that their execution depends on events, not on the order in which the statements are placed in the module. So, placing statement A1 before statement X1 in Listing 4.1 does not change the outcome of the VHDL program.

Verilog has a large number of built-in gates. For example, the statement:

```
xor X1 (sum, a, b);
```

describes a two-input XOR gate. The inputs are a and b, and the output is sum. X1 is an optional identifier for the gate; the identifier can be omitted as:

```
xor (sum, a, b);
```

Verilog has a complete list of built-in primitive gates. The output of the gate sum has to be listed before the inputs a and b. Accordingly, the Verilog code in Listing 4.1 is a complete structural description of a half adder. Figure 4.1 shows a list of gates and their code in Verilog. As in structural VHDL, Verilog statements are concurrent; the order of appearance of statements in the module is irrelevant.

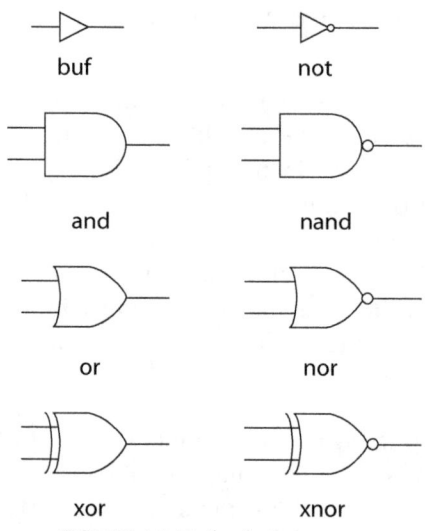

FIGURE 4.1 Verilog built-in gates.

LISTING 4.1 HDL Structural Description

VHDL Description
```
--This code is not complete; binding statements should
--be aaded to recognize components
-- xor2 and and2 as 2-input
-- xor and and gate respectively.
library IEEE;
use IEEE.STD_LOGIC_1164.ALL;
entity system is
port (a, b : in std_logic;
      sum, cout : out std_logic);
  end system;
  architecture struct_exple of system is
  --start declaring all different types of components
  component xor2
  port (I1, I2 : in std_logic;
      O1 : out std_logic);
```

```
end component;
component and2
port (I1, I2 : in std_logic;
    O1 : out std_logic);
end component;
begin
--Start of instantiation statements
X1 : xor2 port map (a, b, sum);
A1 : and2 port map (a, b, cout);
end struct_exple;
```

Verilog Description
```
module system (a, b, sum, cout);
input a, b;
output sum, cout;
xor X1 (sum, a, b);
/* X1 is an optional identifier; it can be omitted.*/
and a1 (cout, a, b);
/* a1 is optional identifier; it can be omitted.*/
endmodule
```

EXAMPLE 4.1   HDL STRUCTURAL DESCRIPTION OF A HALF ADDER

The logic and symbol diagrams of the half adder have been shown before (see Figure 1.1). Listing 4.2 shows the HDL structural code for the half adder. As mentioned before, VHDL does not have built-in gates. To specify xor2 as an EXCLUSIVE-OR gate, bind (link) the component xor2 with an entity bearing the same name. By having the same name, all information in the entity is visible to the component. The entity specifies the relationship between I1, I2, and O1 as EXCLUSIVE-OR; accordingly, the inputs and output of xor2 behave as EXCLUSIVE-OR. The same is done for component and2; it is bound to the entity and2.

LISTING 4.2 HDL Code of Half Adder: VHDL and Verilog

VHDL Description
```
library IEEE;
use IEEE.STD_LOGIC_1164.ALL;
entity xor2 is
port(I1, I2 : in std_logic; O1 : out std_logic);
end xor2;
architecture Xor2_0 of xor2 is
```

```vhdl
begin
    O1 <= I1 xor I2;
end Xor2_0;

library IEEE;
use IEEE.STD_LOGIC_1164.ALL;
entity and2 is
    port (I1, I2 : in std_logic; O1 : out std_logic);
end and2;
architecture and2_0 of and2 is
begin
    O1 <= I1 and I2;
end and2_0;
library IEEE;
use IEEE.STD_LOGIC_1164.ALL;

entity half_add is
    port (a, b : in std_logic; S, C : out std_logic);
end half_add;

architecture HA_str of half_add is
component xor2
port (I1, I2 : in std_logic; O1 : out std_logic);
end component;
component and2
port (I1, I2 : in std_logic; O1 : out std_logic);
end component;
begin
    X1 : xor2 port map (a, b, S);
    A1 : and2 port map (a, b, C);
end HA_str;
```

Verilog Description
```verilog
Module system (a, b, sum, cout);
input a, b;
output sum, cout;
xor X1 (sum, a, b);
/* X1 is an optional identifier; it can be omitted.*/
and a1 (cout, a, b);
/* a1 is optional identifier; it can be omitted.*/
endmodule
```

The VHDL code looks much longer than the Verilog code. This is due to the assumption that the basic VHDL packages do not have built-in libraries or packages for logical gates. The binding method above becomes impractical when the number of gates becomes large. Every time a new description is written, the entities of all gates used must also be written. In the following sections, more efficient ways of binding are discussed.

4.3 Binding

Binding in HDL is common practice. Binding (linking) `segment1` in HDL code to `segment2` makes all information in `segment2` visible to `segment1`. Consider the VHDL code in Listing 4.3.

LISTING 4.3 Binding Between Entity and Architecture in VHDL

```
entity one is
port (I1, I2 : in std_logic; O1 : out std_logic);
end one;
architecture A of one is
signal s : std_logic;
..........
end A;
architecture B of one is
signal x : std_logic;
.......
end B;
```

Architecture A is bound to entity one through the predefined word of. Also, architecture B is bound to entity one through the predefined word of. Accordingly, I1, I2, and O1 can be used in both architecture A and architecture B. Architecture A is not bound to architecture B, so signal s is not recognized in architecture B. Likewise, signal x is not recognized in architecture A.

Now consider Listing 4.4, where an entity is bound to a component.

LISTING 4.4 Binding Between Entity and Component in VHDL

```
entity orgate is
    port (I1, I2 : in std_logic; O1 : out std_logic);
end orgate;

architecture Or_dataflow of orgate is
```

```
begin
    O1 <= I1 or I2;
end Or_dataflow;

entity system is
    port (x, y, z : in std_logic;
    out r : std_logic_vector (3 downto 0);
end system;

architecture system_str of system is
component orgate
port (I1, I2 : in std_logic; O1 : std_logic);
end component;
begin
orgate port map (x, y, r(0));
. . . . . . .
end system_str;
```

The component orgate is bound to the entity orgate because it has the same name. Architecture Or_dataflow is bound to entity orgate by the word of. All information in the entity is now visible to the component. Accordingly, the relationship between I1, I2, and O1 defined in the architecture or_dataflow is visible to the component orgate; hence, the component orgate is an OR gate.

Now consider another way of VHDL binding where a library or a package is bound to a module. Listing 4.5 shows how a library can be bound to a module.

LISTING 4.5 Binding Between Library and Module in VHDL

```
library IEEE;
use IEEE.STD_LOGIC_1164.ALL;
entity system is
port (I1, I2 : in std_logic;
O1 : out std_logic_vector (3 downto 0));

end system;
architecture lib_bound of system is
signal s : std_logic;
. . . . . . . . . . . .
end lib_bound;
```

IEEE is the name of the library, library and use are a predefined words, and IEEE.STD_LOGIC_1164.ALL refers to the part of the library to be linked. Library IEEE provides the definition for the standard_logic type. By entering the name of the library and the statement use, all information in the library is visible to the whole module. If the first two statements are not written in Listing 4.5, the standard_logic type cannot be recognized. Libraries can also be generated by the user. The HDL simulator generates a library named work every time it compiles HDL code. This library can be bound to another module by using the statement use, as follows:

```
use entity work.gates (or_gates);
```

The entity to be bound to the module is gates; gates has an architecture by the name of or_gates, and all information in this architecture is visible to the module wherever the use statement is written. Listing 4.6 shows an example of binding architecture in one module to a component written in another module.

LISTING 4.6 Binding Between a Library and Component in VHDL

```
--First, write the code that will be bound to another
-- module
library IEEE;
use IEEE.STD_LOGIC_1164.ALL;
entity bind2 is
    port (I1, I2 : in std_logic; O1 : out std_logic);
end bind2;

architecture xor2_0 of bind2 is
begin
O1 <= I1 xor I2;
end xor2_0;

architecture and2_0 of bind2 is
begin
    O1 <= I1 and I2;
end and2_0;

architecture and2_4 of bind2 is
begin
    O1 <= I1 and I2 after 4 ns;
end and2_4;
```

```
--After writing the above code; compile it and store it
-- in a known location. Now, open another module
--where the above information is to be used.

library IEEE;
use IEEE.STD_LOGIC_1164.ALL;
entity half_add is
port (a, b : in std_logic; S, C : out std_logic);
end half_add;
architecture HA_str of half_add is
component xor2
    port (I1, I2 : in std_logic; O1 : out std_logic);
end component;
component and2
    port (I1, I2 : in std_logic; O1 : out std_logic);
end component;
for all : xor2 use entity work.bind2 (xor2_0);
for all : and2 use entity work.bind2 (and2_4);
begin
X1 : xor2 port map (a, b, S);
A1 : and2 port map (a, b, C);
end HA_str;
```

The statement `for all : xor2 use entity work.bind2 (xor2_0)` binds the architecture `xor2_0` of the entity `bind2` to the component `xor2`. By this binding, component `xor2` behaves as a two-input XOR gate with zero propagation delay. The statement `for all : and2 use entity work.bind2 (and2_4)` binds the architecture `and2_4` of the entity `bind2` to the component `and2`. By this binding, component `and2` behaves as a two-input AND gate with a 4-ns propagation delay. In Listing 4.6, it is assumed that both entities `bind2` and `half_add` have the same path (stored in the same directory). Otherwise, the path of the library work has to be specified.

Throughout this chapter, the binding shown in Listing 4.6 is adopted. The codes for all the gates expected are written, and the module is compiled and stored. Whenever we want to use any component from the stored module, we bind it to the current module. Listing 4.31 shows the VHDL binding code used in all examples in this chapter. As previously mentioned, Verilog has all primitive gates built in and ready to use. Verilog modules can be bound by just writing the name of the module to be bound. Listing 4.7 shows such binding.

LISTING 4.7 Binding Between Two Modules in Verilog

```
module one (O1, O2, a, b);
input [1:0] a;
input [1:0] b;
output [1:0] O1, O2;
two M0 (O1[0], O2[0], a[0], b[0]);
two M1 (O1[1], O2[1], a[1], b[1]);
endmodule

module two (s1, s2, a1, b1);
input a1;
input b1;
output s1, s2;
xor (s1, a1, b1);
and (s2, a1, b1);
endmodule
```

The statement: `two M0 (O1[0], O2[0], a[0], b[0]);` written in module `one` binds module `two` to module `one`. Accordingly, the relationship between `O1`, `O2`, `a`, and `b` is as follows:

`O1[0]` is the output of a two-input XOR gate with `a[0]` and `b[0]` as the inputs

`O2[1]` is the output of a two-input AND gate with `a[1]` and `b[1]` as the inputs

Other methods of binding are discussed in Chapters 6 and 8. The following examples cover binding and structural descriptions.

EXAMPLE 4.2 STRUCTURAL DESCRIPTION OF A 2x1 MULTIPLEXER WITH ACTIVE LOW ENABLE

The truth table and logic diagram of this multiplexer have been covered in Chapter 2. The logic diagram is redrawn here for convenience (see Figure 4.2).

From Figure 4.2, the components of the multiplexer are: two three-input AND gates, three inverters, and one two-input OR gate. Each gate, including the inverter, is assumed to have a 7-ns propagation delay time.

For VHDL, the binding method shown in Listing 4.6 is used. The code to describe these gates is written, compiled, and then stored. Some other

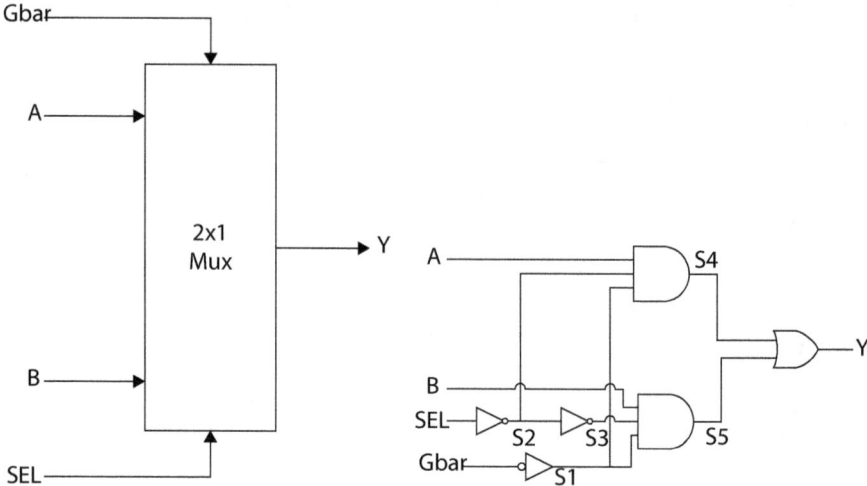

FIGURE 4.2 Multiplexer. a) Logic diagram. b) Logic symbol.

gates are included here that might be used for other examples. Listing 4.8 shows the code for several gates.

LISTING 4.8 VHDL Code for Several Gates

```vhdl
library IEEE;
use IEEE.STD_LOGIC_1164.ALL;

entity bind1 is
    port (I1 : in std_logic; O1 : out std_logic);
end bind1;
architecture inv_0 of bind1 is
begin
O1 <= not I1; --This is an inverter with zero delay
end inv_0;

architecture inv_7 of bind1 is
begin
O1 <= not I1 after 7 ns; --This is an inverter with a
                         -- 7-ns delay
end inv_7;
library IEEE;
use IEEE.STD_LOGIC_1164.ALL;
```

```
entity bind2 is
    port (I1, I2 : in std_logic; O1 : out std_logic);
end bind2;

architecture xor2_0 of bind2 is
begin
O1 <= I1 xor I2; --This is exclusive-or with zero
                 -- delay.
end xor2_0;

architecture and2_0 of bind2 is
begin
O1 <= I1 and I2; --This is a two input and gate with
                 -- zero delay.
end and2_0;

architecture and2_7 of bind2 is
begin
O1 <= I1 and I2 after 7 ns; -- This is a two input and
                            -- gate with 7-ns delay.
end and2_7;

architecture or2_0 of bind2 is
begin
O1 <= I1 or I2; -- This is a two input or gate with
                -- zero delay.
end or2_0;

architecture or2_7 of bind2 is
begin
O1 <= I1 or I2 after 7 ns; -- This is a two input or
                           -- gate with 7-ns delay.
end or2_7;

library IEEE;
use IEEE.STD_LOGIC_1164.ALL;

entity bind3 is
port (I1, I2, I3 : in std_logic; O1 : out std_logic);
end bind3;

architecture and3_0 of bind3 is
begin
```

```
O1 <= I1 and I2 and I3; -- This is a three input and
                        -- gate with zero delay.
end and3_0;

architecture and3_7 of bind3 is
begin
O1 <= I1 and I2 and I3 after 7 ns; --This is a three
                -- input and gate with 7-ns delay.
                --
end and3_7;

architecture or3_0 of bind3 is
begin
O1 <= I1 or I2 or I3; --This is a three input OR gate
                      --with zero delay.
end or3_0;

architecture or3_7 of bind3 is
begin
O1 <= I1 or I2 or I3 after 7 ns; --This is a three
                    --input or gate with 7-ns delay.
end or3_7;
```

After compilation of the above code, it is stored in a known directory (path). Listing 4.9 shows the HDL code for a 2x1 multiplexer with active low enable. The Verilog description is straightforward using the predefined gates.

LISTING 4.9 HDL Description of a 2x1 Multiplexer with Active Low Enable: VHDL and Verilog

```
VHDL Description
library IEEE;
use IEEE.STD_LOGIC_1164.ALL;
entity mux2x1 is
    port (A, B, SEL, Gbar : in std_logic;
        Y : out std_logic);
end mux2x1;

architecture mux_str of mux2x1 is
--Start components Declaration
component and3
port (I1, I2, I3 : in std_logic; O1 : out std_logic);
end component;
```

```
--Only different types of components need be declared.
--Since the multiplexer has two identical AND gates,
--only one is declared.

component or2
port (I1, I2 : in std_logic; O1 : out std_logic);
end component;
component Inv
port (I1 : in std_logic; O1 : out std_logic);
end component;

signal S1, S2, S3, S4, S5 : std_logic;
for all : and3 use entity work.bind3 (and3_7);
for all : Inv use entity work.bind1 (inv_7);
for Or1 : or2 use entity work.bind2 (or2_7);
begin
--Start instantiation
A1 : and3 port map (A,S2, S1, S4);
A2 : and3 port map (B,S3, S1, S5);
IV1 : Inv port map (SEL, S2);
IV2 : Inv port map (Gbar, S1);
IV3 : Inv port map (S2, S3);
or1 : or2 port map (S4, S5, Y);
end mux_str;
```

Verilog Description
```
module mux2x1 (A, B, SEL, Gbar, Y);
input A, B, SEL, Gbar;
output Y;
and #7 (S4, A, S2, S1);
or #7 (Y, S4, S5);
and #7 (S5, B, S3, S1);
not #7 (S2, SEL);
not #7 (S3, S2);
not #7 (S1, Gbar);
endmodule
```

Referring to Listing 4.9, because the multiplexer has two identical AND gates (both three-input AND gates), only one of them is declared in the VHDL description by the statements:

```
component and3
```

```
port (I1, I2, I3 : in std_logic; O1 : out std_logic);
end component;
```

Similarly, only one inverter is declared. If the two AND gates do not have the same delay time (say, A1 has 0 ns and A2 has 7 ns) then instead of `all` in the `use` statement, write:

```
for A1 : and3 use entity work.bind3 (and3_0);
for A2 : and3 use entity work.bind3 (and3_7);
```

For the Verilog description, the statement

```
and #7 (S4, A, S2, S1);
```

declares a three-input (`A`, `s2`, `s1`) AND gate with propagation delay of seven simulation screen units. Note that `s2` or `s1` do not need to be declared as `wire`; Verilog assumes that they are of the same type as `A`. If a four-input AND gate is needed, the code will be:

```
and (o1, in1, in2, in3, in4)
```

where `o1` is the output, and `in1`, `in2`, `in3`, and `in4` are the inputs. The gate in Verilog can have an optional name as:

```
or #7 orgate1 (O1, in1, in2)
```

The statement above describes an OR gate by the name `orgate1`; it has two inputs (`in1`, `in2`) and an output (`o1`). The name is optional and can be omitted. The simulation waveform of the multiplexer is identical to that of Figure 2.10.

EXAMPLE 4.3 STRUCTURAL DESCRIPTION OF A 2x4 DECODER WITH TRI-STATE OUTPUT

A decoder is a combinational circuit. The output is a function of the input only. A 2x4 decoder has two inputs and four outputs. For any input only one output is active; all other outputs are inactive. For an active high output decoder, only one output is high. The output can be deactivated or put in high impedance if the decoder has an enable. For a tri-state output, if the enable is inactive, then all the outputs are in high impedance. The output of an n-bit input decoder is 2^n bits. Table 4.1 shows the truth table of 2x4 decoder.

TABLE 4.1 Truth Table for a 2x4 Decoder with Tri-State Outputs

Inputs			Outputs			
Enable	I1	I2	D3	D2	D1	D0
0	x	x	Z	Z	Z	Z
1	0	0	0	0	0	1
1	0	1	0	0	1	0
1	1	0	0	1	0	0
1	1	1	1	0	0	0

Tri-state buffers are used at the output. If the enable is low, then all outputs are in high impedance (Z). From Table 4.1, we can write the Boolean function of the outputs:

$$D0 = \overline{I0}\ \overline{I1}$$

$$D1 = I0\ \overline{I1}$$

$$D2 = \overline{I0}\ I1$$

$$D3 = I0\ I1$$

Figure 4.3 shows the logic diagram of the decoder.

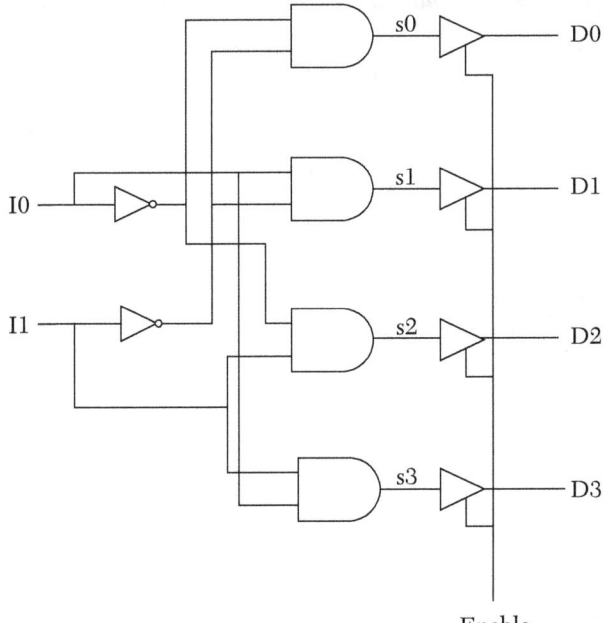

Enable

FIGURE 4.3 Logic diagram of a 2x4 decoder with tri-state output.

To write the VHDL code, we first need to write a description of the tri-state buffer gate. The easiest description type that can be written for the tri-state buffer is behavioral, using the if statement. This description is attached to the entity bind2 (see Listing 4.8). Listing 4.10 shows a behavioral description of a tri-state buffer. The Verilog has built-in buffers (see Figure 4.4).

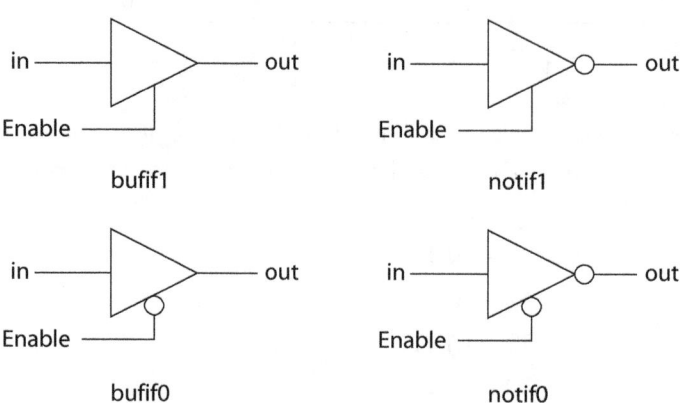

FIGURE 4.4 Verilog built-in buffers.

LISTING 4.10 VHDL Behavioral Description of a Tri-State Buffer

```
entity bind2 is
port (I1, I2 : in std_logic; O1 : out std_logic);
end bind2;
..........
--Add the following architecture to
--the entity bind2 of Listing 4.8
architecture bufif1 of bind2 is
begin
buf : process (I1, I2)
variable tem : std_logic;
begin
if (I2 ='1') then
tem := I1;
else
tem := 'Z';
end if;
O1 <= tem;
end process buf;
end bufif1;
```

Now, write the HDL structural description of the decoder as shown in Listing 4.11. Figure 4.5 shows the simulation waveform of the decoder.

LISTING 4.11 HDL Description of a 2x4 Decoder with Tri-State Output

VHDL Description

```
library IEEE;
use IEEE.STD_LOGIC_1164.ALL;

entity decoder2x4 is
    port (I : in std_logic_vector(1 downto 0);
          Enable : in std_logic;
          D : out std_logic_vector (3 downto 0));
end decoder2x4;

architecture decoder of decoder2x4 is
component bufif1
port (I1, I2 : in std_logic; O1 : out std_logic);
end component;
component inv
port (I1 : in std_logic; O1 : out std_logic);
end component;
component and2
port (I1, I2 : in std_logic; O1 : out std_logic);
end component;
for all : bufif1 use entity work.bind2 (bufif1);
for all : inv use entity work.bind1 (inv_0);
for all : and2 use entity work.bind2 (and2_0);
signal s0, s1, s2, s3 : std_logic;
signal Ibar : std_logic_vector (1 downto 0);
-- The above signals have to be declared before they
-- can be used
begin
    B0 : bufif1 port map (s0, Enable, D(0));
    B1 : bufif1 port map (s1, Enable, D(1));
    B2 : bufif1 port map (s2, Enable, D(2));
    B3 : bufif1 port map (s3, Enable, D(3));
    iv0 : inv port map (I(0), Ibar(0));
    iv1 : inv port map (I(1), Ibar(1));
    a0 : and2 port map (Ibar(0), Ibar(1), s0);
    a1 : and2 port map (I(0), Ibar(1), s1);
    a2 : and2 port map (Ibar(0), I(1), s2);
    a3 : and2 port map (I(0), I(1), s3);
end decoder;
```

Verilog Description
```
module decoder2x4 (I, Enable, D);
input [1:0] I;
input Enable;
output [3:0] D;
wire [1:0] Ibar;
    bufif1 (D[0], s0, Enable);
    bufif1 (D[1], s1, Enable);
    bufif1 (D[2], s2, Enable);
    bufif1 (D[3], s3, Enable);
    not (Ibar[0], I[0]);
    not (Ibar[1], I[1]);
    and (s0, Ibar[0], Ibar[1]);
    and (s1, I[0], Ibar[1]);
    and (s2, Ibar[0], I[1]);
    and (s3, I[0], I[1]);
endmodule
```

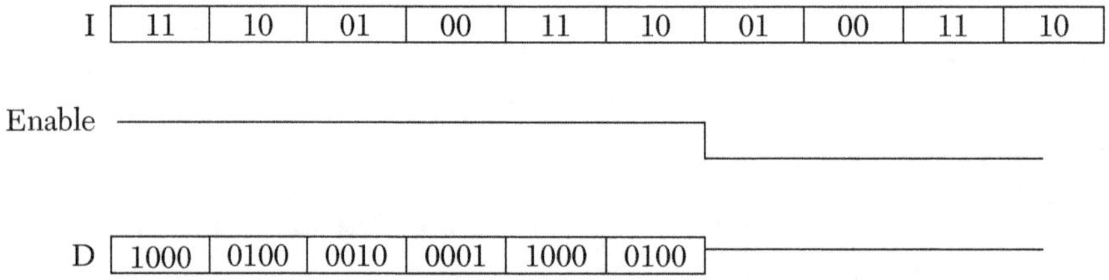

I	11	10	01	00	11	10	01	00	11	10

Enable

D	1000	0100	0010	0001	1000	0100

FIGURE 4.5 Simulation waveform of a 2x1 decoder with tri-state output.

EXAMPLE 4.4   STRUCTURAL DESCRIPTION OF A FULL ADDER

In this example, a full adder (Listing 4.13) is built from two half adders (Listing 4.12). The full adder adds (a + b + cin) to generate sum and carry. A half adder is used to add (a + b) to generate sum1 and carry1. Another half adder is used to add (sum1 + cin) to generate sum and carry2. The carry of the summation (a + b + cin) is the logical OR of carry1 and carry2. Figures 4.6a and 4.6b show the logical symbol and diagram of this full adder, respectively.

For the VHDL code, write the code for half adder. Then, include this code in Listing 4.8. Listing 4.12 shows the code of the half adder as part of Listing 4.8. Now, write the structural description of the full adder as two half adders. Listing 4.13 shows the HDL code for a full adder.

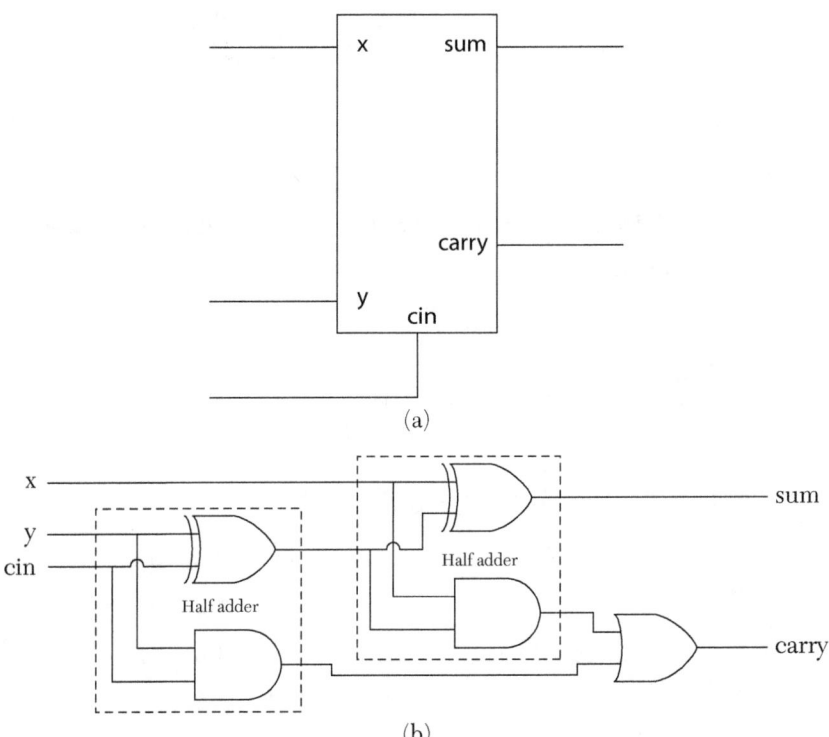

FIGURE 4.6 Full adder as two half adders. a) Logic symbol. b) Logic diagram.

LISTING 4.12 VHDL Description

```
--This code is to be appended to Listing 4.8
library IEEE;
use IEEE.STD_LOGIC_1164.ALL;
entity bind22 is
    Port (I1, I2 : in std_logic;
          O1, O2 : out std_logic);
end bind22;

architecture HA of bind22 is
component xor2
port (I1, I2 : in std_logic; O1 : out std_logic);
end component;
component and2
port (I1, I2 : in std_logic; O1 : out std_logic);
end component;
```

```
for A1 : and2 use entity work.bind2 (and2_0);
for X1 : xor2 use entity work.bind2 (xor2_0);
begin
    X1 : xor2 port map (I1, I2, O1);
    A1 : and2 port map (I1, I2, O2);
end HA;
```

LISTING 4.13 HDL Description of a Full Adder (Figures 4.6a and 4.6b)

VHDL Description

```
library IEEE;
use IEEE.STD_LOGIC_1164.ALL;
entity FULL_ADDER is
    Port (x, y, cin : in std_logic;
          sum, carry : out std_logic);
end FULL_ADDER;
architecture full_add of FULL_ADDER is
component HA
Port (I1, I2 : in std_logic; O1, O2 : out std_logic);
end component;
component or2
Port (I1, I2 : in std_logic; O1 : out std_logic);
end component;

for all : HA use entity work.bind22 (HA);
for all : or2 use entity work.bind2 (or2_0);
signal s0, c0, c1 : std_logic;

begin
    HA1 : HA port map (y, cin, s0, c0);
    HA2 : HA port map (x, s0, sum, c1);
    r1 : or2 port map (c0, c1, carry);
end full_add;
```

Verilog Description

```
module FULL_ADDER (x, y, cin, sum, carry);
input x, y, cin;
output sum, carry;
HA H1 (y, cin, s0, c0);
HA H2 (x, s0, sum, c1);
//The above two statements bind module HA
//to the present module FULL_ADDER
    or (carry, c0, c1);
```

```
endmodule
module HA (a, b, s, c);
input a, b;
output s, c;
xor (s, a, b);
and (c, a, b);
endmodule
```

To use the above VHDL code in future examples, it is added to entity `bind32` in Listing 4.31.

EXAMPLE 4.5   **STRUCTURAL DESCRIPTION OF A THREE-BIT RIPPLE-CARRY ADDER**

In this example, a three-bit ripple-carry adder is described. Then, in Example 4.7, this adder is implemented to build a magnitude comparator. The logic diagram of the adder is as shown in Figure 2.23 of Chapter 2. Listing 4.14 shows the structural description of the three-bit ripple-carry adder.

LISTING 4.14 HDL Description of a Three-Bit Ripple-Carry Adder: VHDL and Verilog

VHDL Description
```
library IEEE;
use IEEE.STD_LOGIC_1164.ALL;

entity three_bit_adder is
    port(x, y : in std_logic_vector (2 downto 0);
         cin : in std_logic;
         sum : out std_logic_vector (2 downto 0);
         cout : out std_logic);
end three_bit_adder;

architecture three_bitadd of three_bit_adder is
component FULL_ADDER
port (I1, I2, I3 : in std_logic;
    O1, O2 : out std_logic);
end component;
for all : FULL_ADDER
        use entity work.bind32 (full_add);
signal carry : std_logic_vector (1 downto 0);
```

```
begin
M0 : FULL_ADDER port map (x(0), y(0), cin, sum(0), carry(0));
M1 : FULL_ADDER port map (x(1), y(1), carry(0), sum(1),
                            carry(1));
M2 : FULL_ADDER port map (x(2), y(2), carry(1), sum(2), cout);
end three_bitadd;
```

Verilog Description
```
module three_bit_adder (x, y, cin, sum, cout);
input [2:0] x, y;
input cin;
output [2:0] sum;
output cout;
wire [1:0] carry;
FULL_ADDER M0 (x[0], y[0], cin, sum[0], carry[0]);
FULL_ADDER M1 (x[1], y[1], carry[0], sum[1], carry[1]);
FULL_ADDER M2 (x[2], y[2], carry[1], sum[2], cout);

/* It is assumed that the module FULL_ADDER
(Listing 4.13) is attached by the simulator to
the module three_bit_adder so, no need to
rewrite the module FULL_ADDER.*/

endmodule
```

Inspection of the code in Listing 4.14 shows that there may be lag time between the steady state of each of the adders and the carryout (cout). This lag time produces transient states before the values of the sum and carryout settle. For example, if the inputs to the adder are 101 and 001, and the previous output of the adder is 1001, some transient states can be 0100 and 1010 before the output settles at 0110. The appearance of these transient states is called *hazards*. These transient states, however, have short duration and may not be noticed.

EXAMPLE 4.6 **STRUCTURAL DESCRIPTION OF A THREE-BIT TWO-STAGE CARRY-SAVE ADDER**

The ripple-carry adder in Example 4.5 has a delay that is proportional to the number of bits added. This is because each full adder has to wait for the generation of the carry out of the preceeding full adder to start adding its input bits to this carry out. If each full adder can add its inputs independently from other full adders, the addition will be proportional to just a delay of a single adder because all full adders would be capable of adding their inputs simultanously. Carry-save adders utilize the concept of independent addition; several of the full adders in the carry-save system, but not all, can add their inputs simultaneously. Figure 4.7 shows the logic diagram of a three-bit four-word carry-save adder. The adder adds four words (a + b + c + d) where each word is three bits. FA1, FA2, and FA3 add a + b + c and generate sum and partial (not final) carryout at the same time. The same is true for FA4, FA5, and FA6; however, these three adders have to wait on the upper-stage adders (FA1, FA2, and FA3) to complete their addition and generate their carryouts (cr0 and cr1). The adders FA7, FA8, and FA9 are connected as ripple-carry adders; each adder of this stage has to wait for carryout from upper-stage and preceeding full adders. These riple-carry adders can be replaced by lookahead adders to decrease the delay associated with them. If each full adder has a delay of d ns, then the first stage takes 1d to finish its task, the second stage takes 1d to finsh its task, and the last stage takes 3d to finsh its task. The total delay to add four three-bit words is (1 + 1 + 3)d = 5d ns, which is faster than using ripple-carry adders.

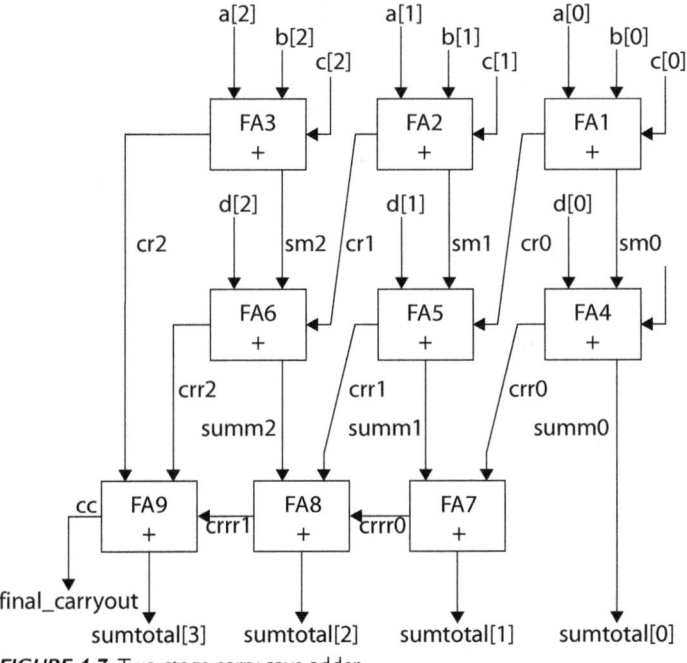

FIGURE 4.7 Two-stage carry-save adder.

Listing 4.15 shows the Verilog code for the adder of Figure 4.7. The Listing contains a main module `carry_saveadder` and another module `full_adder`. The module `full_adder` is bound to the main module by a statement such as:

```
full_adder FA1(a1[0],b1[0],c1[0], sm0,cr0);
```

where a, b, and c in the `full_adder` module is linked (replaced) by a1[0], b1[0], and c1[0]. The result of addition sum and carryout is linked (uploaded) to sm0 and cr0, respectively.

LISTING 4.15 Verilog Description of Carry-Save Adder (Figure 4.7)

```
module carry_saveadder(a1,b1,c1,d1,sum_total,final_carryout);
input[2:0] a1, b1,c1,d1;
output [3:0]sum_total;
output final_carryout;

full_adder FA1(a1[0],b1[0],c1[0], sm0,cr0);
//FA1 is a user-selected label
full_adder FA2(a1[1],b1[1],c1[1], sm1,cr1);
full_adder FA3(a1[2],b1[2],c1[2], sm2,cr2);

full_adder FA4(sm0,d1[0],1'b0, smm0,crr0);
full_adder FA5(sm1,d1[1],cr0, smm1,crr1);
full_adder FA6(sm2,d1[2],cr1, smm2,crr2);
assign sum_total[0] =smm0;

full_adder FA7(crr0,smm1,1'b0, sum_total[1],crrr0);
full_adder FA8(crr1,smm2,crrr0, sum_total[2],crrr1);
full_adder FA9(crr2,crrr1,cr2, sum_total[3],cc);
assign final_carryout = cc;

endmodule
module full_adder (a,b,c,Sum,Carryout);
input a,b,c;
output Sum, Carryout;
not (a_bar,a); // this is an inverter
not (b_bar,b);
not (c_bar,c);
and a1 (s0,a_bar,b_bar, c);/*This is And gate with
                          optional name a1*/
and a2 (s1,a_bar,b, c_bar);
and a3 (s2,a,b_bar, c_bar);
```

```
and a4 (s3,a,b,c);
or o1(Sum, s0,s1,s2,s3);
and a5 ( s5,a,b);
and a6 ( s6,a,c);
and a7 ( s7,b,c);
or o2( Carryout,s5,s6,s7);
endmodule
```

EXAMPLE 4.7   STRUCTURAL DESCRIPTION OF A THREE-BIT MAGNITUDE COMPARATOR USING A THREE-BIT ADDER

Chapter 2 covered a 2x2-bit comparator using truth tables. If the number of bits to be compared is more than two bits, the truth tables become so huge that is too difficult to handle. In this example, a different approach is taken. Consider two numbers X and Y, each of n bits; if X is greater than Y, then:

$$X - Y > 0 \tag{4.2}$$

−Y is the twos complement of $Y = \overline{Y} + 1$; substituting in Equation 4.2, the condition of X > Y is rewritten as:

$$X + \overline{Y} + 1 > 0 \tag{4.3}$$

Or, Equation 4.3 can be rewritten as:

$$X + \overline{Y} > -1 \tag{4.4}$$

For n bits, −1 is a string of n bits; each bit is 1. If n = 5, for example $-1_d = (11111)2$, so Equation 4.4 can be rewritten as:

$$X + \overline{Y} > 1......1111 \tag{4.5}$$

Equation 4.5 states that if X is greater than Y, the sum of X and $\overline{Y}$ should be greater than 1...1111. If n adders are used to add X plus $\overline{Y}$, then for X to be greater than Y, the n-bit sum should be greater than n ones. This can only happen if the n-bit adders have a final carryout of 1. So, if X is added to $\overline{Y}$ using n-bit adders, and the final carryout is 1, then it can be concluded that X > Y. If there is no final carryout, then X ≤ Y. To check for equality, it is noticed that if X = Y then:

$$X + \overline{Y} = 1......1111 \tag{4.6}$$

In this example, n = 3 is being considered. Figure 4.8 shows the logic diagram of the comparator.

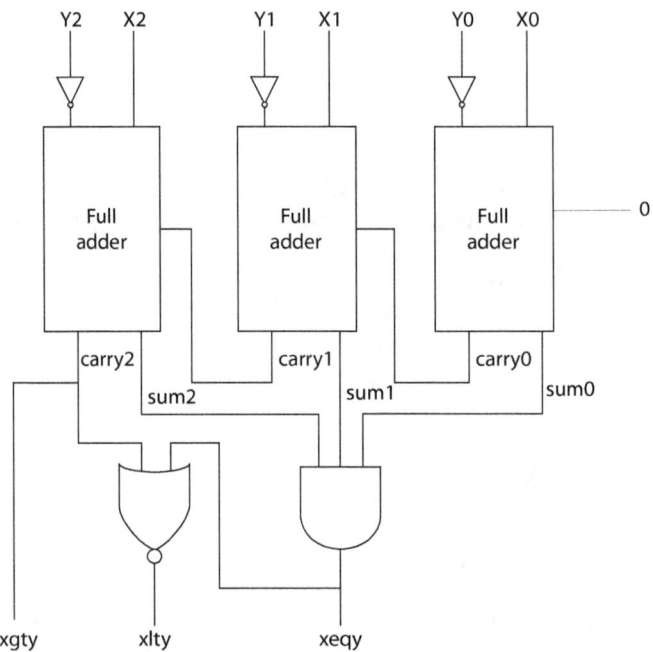

FIGURE 4.8 A full-adder-based comparator.

Listing 4.16 shows the HDL code for the comparator. The HDL code for a full adder has already been written (see Listing 4.13). The full-adder components (macros) are used in Listing 4.16. Because they are identical, only one generic full adder is declared as:

```
component full_adder
port(I1, I2, I3 : in std_logic;
     O1, O2 : out std_logic);
end component;
```

To use these components, link their work library from Listing 4.13 as:

```
for all : full_adder use entity work.bind32 (full_add); --VHDL
```

or, in Verilog, link the module built in Listing 4.13 as:

```
FULL_ADDER M0 (X[0], Yb[0], 1'b0, sum[0], carry[0]); //Verilog
```

LISTING 4.16 HDL Description of a Three-Bit Comparator Using Adders

VHDL Description

```
library IEEE;
use IEEE.STD_LOGIC_1164.ALL;

entity three_bit_cmpare is
port (X, Y : in std_logic_vector (2 downto 0);
    xgty, xlty, xeqy : buffer std_logic);
end three_bit_cmpare;

architecture cmpare of three_bit_cmpare is

--Some simulators will not allow mapping between
--buffer and out. In this
--case, change all out to buffer.

component full_adder
port (I1, I2, I3 : in std_logic;
      O1, O2 : out std_logic);
end component;
component Inv
port (I1 : in std_logic; O1 : out std_logic);
end component;
component nor2
port (I1, I2 : in std_logic; O1 : out std_logic);
end component;
component and3
port (I1, I2, I3 : in std_logic; O1 : out std_logic);
end component;
for all : full_adder use entity work.bind32 (full_add);
for all : Inv use entity work.bind1 (inv_0);
for all : nor2 use entity work.bind2 (nor2_0);
for all : and3 use entity work.bind3 (and3_7);
--To reduce hazards, an AND gate is
--implemented with a 7-ns delay.
signal sum, Yb : std_logic_vector (2 downto 0);
signal carry : std_logic_vector (1 downto 0);
begin
    in1 : inv port map (Y(0), Yb(0));
    in2 : inv port map (Y(1), Yb(1));
    in3 : inv port map (Y(2), Yb(2));
```

```
    F0 : full_adder port map (X(0), Yb(0),
                                '0', sum(0), carry(0));
    F1 : full_adder port map (X(1), Yb(1), carry(0),
                        sum(1), carry(1));
    F2 : full_adder port map (X(2), Yb(2), carry(1),
                        sum(2), xgty);
    a1 : and3 port map (sum(0), sum(1), sum(2), xeqy);
    n1 : nor2 port map (xeqy, xgty, xlty);
end cmpare;
```

Verilog Description
```
module three_bit_cmpare (X, Y, xgty, xlty, xeqy);
input [2:0] X, Y;
output xgty, xlty, xeqy;
wire [1:0] carry;
wire [2:0] sum, Yb;
    not (Yb[0], Y[0]);
    not (Yb[1], Y[1]);
    not (Yb[2], Y[2]);
    FULL_ADDER M0 (X[0], Yb[0], 1'b0, sum[0],
                    carry[0]);
FULL_ADDER M1 (X[1], Yb[1], carry[0], sum[1],
                    carry[1]);
FULL_ADDER M2 (X[2], Yb[2], carry[1], sum[2],
                    xgty);
and #7 (xeqy, sum[0], sum[1], sum[2]);
/* To reduce hazard use an AND gate with a delay of 7 units*/
nor (xlty, xeqy, xgty);
endmodule
```

EXAMPLE 4.8   STRUCTURAL DESCRIPTION OF AN SET-RESET LATCH

A set-reset (SR) latch is a sequential circuit. The output and the next state depends on the input(s) and the current state. It memorizes, as is the case for sequential circuits, one of its states when S = R = 0. Memorization is achieved through feedback between the output Q, its complement $\overline{Q}$, and the inputs. The inputs receive the values of the current output through the feedback lines. The state where S = R = 1 is prohibited because it may lead to unstable output (both Q and $\overline{Q}$ acquire the same logic level). The latch is implemented in digital systems as a switch or memory cell for static random-access memory (SRAM). The excitation table of the latch is shown in Table 4.2.

TABLE 4.2 Excitation Table of an SR-Latch

S	R	Current State	Next State
1	0	x	1
0	1	x	0
0	0	q	q
1	1	x	prohibited

Figures 4.9a and 4.9b show the logic symbol and diagram, respectively, of an SR-latch using NOR gates. Notice the connection (feedback) between the output Q and the input of the NOR gate in Figure 4.9b. Listing 4.17 shows the HDL structural description of an SR-latch based on NOR gates.

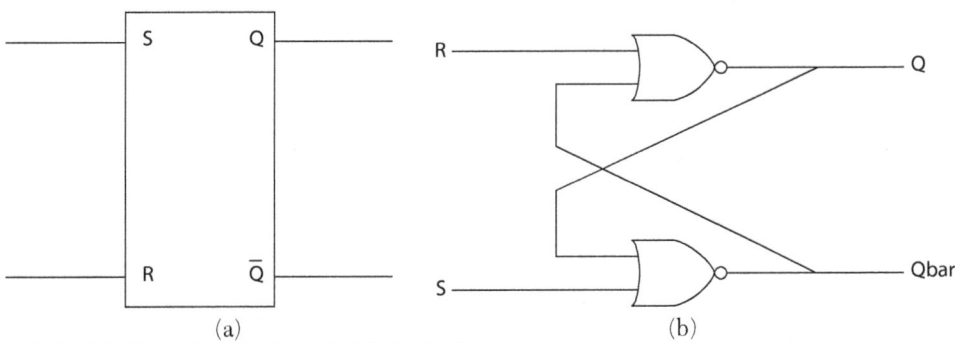

(a)   (b)

FIGURE 4.9 SR-Latch. a) Logic symbol. b) Logic diagram.

LISTING 4.17 HDL Description of an SR-Latch with NOR Gates

VHDL Description
```
library IEEE;
use IEEE.STD_LOGIC_1164.ALL;
entity SR_latch is
    port (R, S : in std_logic;
        Q, Qbar : buffer std_logic);
--Q, Qbar are declared buffer because
--they behave as input and output.

end SR_latch;

architecture SR_strc of SR_latch is
--Some simulators would not allow mapping between
--buffer and out. In this
--case, change all out to buffer.
component nor2
```

```
port (I1, I2 : in std_logic; O1 : out std_logic);
end component;
for all : nor2 use entity work.bind2 (nor2_0);
begin
    n1 : nor2 port map (S, Q, Qbar);
    n2 : nor2 port map (R, Qbar, Q);
end SR_strc;
```

Verilog Description
```
module SR_Latch (R, S, Q, Qbar);
input R, S;
output Q, Qbar;
nor (Qbar, S,Q);
nor (Q, R, Qbar);
endmodule
```

To use the above code in future VHDL examples, it is appended to Listing 4.31. Figure 4.10 shows the simulation waveform of the SR-latch.

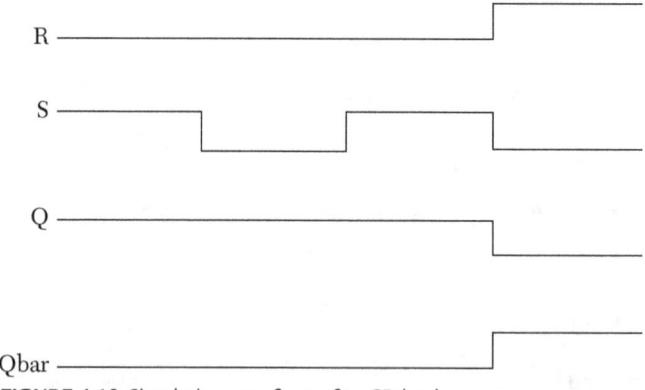

FIGURE 4.10 Simulation waveform of an SR-latch.

EXAMPLE 4.9 **STRUCTURAL DESCRIPTION OF A D-LATCH WITH ACTIVE LOW CLEAR**

A D-latch is a sequential circuit. The output of the latch (Q) follows the input (D) as long as the enable (E) is high. $\overline{Q}$ is the complement of Q. The clear signal is chosen here to be asynchronous active low, which means if the clear signal is low, the output is cleared (Q = 0) momentarily. The latch has been discussed in Chapter 2. The logic symbol and diagram are as shown in Figure 4.11. Listing 4.18 shows the HDL structural description of a D-latch.

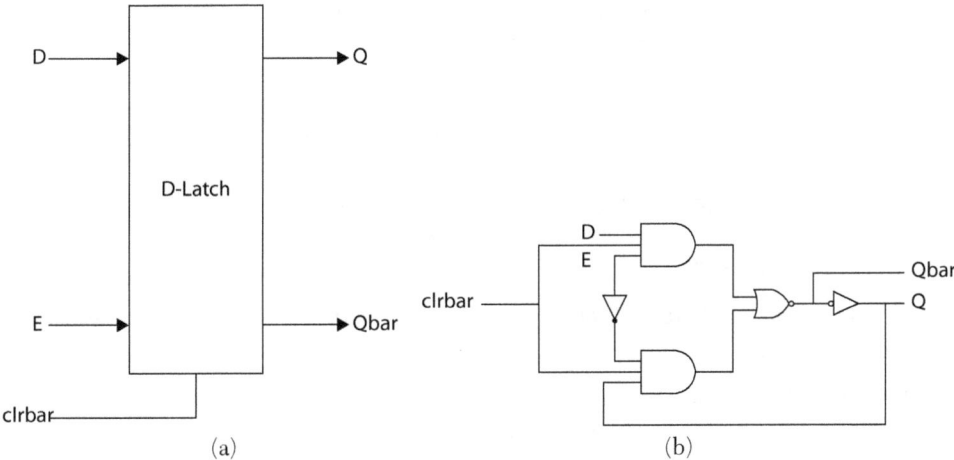

FIGURE 4.11   D-Latch with clear. a) Logic symbol. b) Logic diagram.

LISTING 4.18 HDL Description of a D-Latch with Active Low Clear

VHDL Description
```
library IEEE;
use IEEE.STD_LOGIC_1164.ALL;
entity D_LatchWclr is
    port (D, E,clrbar : in std_logic;
            Q, Qbar : buffer std_logic);
end;
architecture D_latch_str of D_LatchWclr is

--be sure to use buffer rather than output in all
-- components; some simulators will not map output
--to buffer.

component and3
port (I1, I2, I3 : in std_logic;
     O1 : buffer std_logic);
end component;
component nor2
port (I1, I2 : in std_logic; O1 : buffer std_logic);
end component;
component inv
port (I1 : in std_logic; O1 : buffer std_logic);
end component;
for all : and3 use entity work.bind3 (and3_4);
for all : nor2 use entity work.bind2 (nor2_4);
```

```
for all : inv use entity work.bind1 (inv_1);
signal Eb, s1, s2 : std_logic;
begin
    a1 : and3 port map (D, E, clrbar, s1);
    a2 : and3 port map (Eb, D,clrbar, s2);
    in1 : inv port map (E, Eb);
    in2 : inv port map (Qbar, Q);
    n2 : nor2 port map (s1, s2, Qbar);
end D_latch_str;
```

To use the above code in future examples, it is appended to Listing 4.31.

Verilog Description
```
module D_latchWclr(D, E,clrbar, Q, Qbar);
input D, E, clrbar;
output Q, Qbar;
/* assume 4 ns delay for and gate and nor gate,
and 1 ns for inverter */
//The clear is active low; if clrbar = 0, Q=0

and #4 gate1 (s1, D, E, clrbar);

/* the name "gate1" is optional; we could have
   written and #4 (s1, D, E) */
   and #4 gate2 (s2, Eb, Q, clrbar);
   not #1 (Eb, E);
   nor #4 (Qbar, s1, s2);
   not #1 (Q, Qbar);
endmodule
```

The simulation waveform is the same as in Figure 2.19 except for the addition of signal clrbar; if the clrbar signal is low, Q should go low.

EXAMPLE 4.10 **STRUCTURAL DESCRIPTION OF A PULSE-TRIGGERED, MASTER-SLAVE D FLIP-FLOP WITH ACTIVE LOW CLEAR**

The D-latch discussed in Listing 4.18 has a characteristic that may not be desirable in digital circuits such as counters. The D-latch output follows its input as long the enable is high. In counters, for example, the output is desired to change only once during the active phase of the clock. To achieve this, flip-flops are needed. A master-slave D flip-flop is a sequential circuit where the output follows the input only once at the transition of the clock from inactive to active. Figure 4.12 shows the logic symbol of the master-slave D flip-flop. Table 4.3 shows the excitation table of the flip-flop.

TABLE 4.3 Excitation Table for the Master-Slave D Flip-Flop

Input	Current State	Clock	Next State
Clrbar	D Q	clk	Q⁺
0	x x		0
1	0 0	⎍	0
1	0 1	⎍	0
1	1 0	⎍	1
1	1 1	⎍	1

The logic diagram of the master-slave flip-flop is shown in Figure 4.12. The flip-flop consists of two active high enable D-latches; the first latch is called the master, and the second is called the slave. The master latch drives the slave. The clock of the master is the invert of the clock of the slave. Because the clock of one of the latches is the invert of the other, at any time one latch will be active while the other is inactive. At the high level of the clock, the slave is active; its output Q follows its input QM (QM is the output of the master). Because the master is inactive at the high level of the clock, any change in D (the input of the slave) is not transmitted to QM, so QM and Q stay the same during the high level of the clock, unaffected by any change in D. Thus, the flip-flop is sensitive to the clock pulse rather than the level, as in a D-latch.

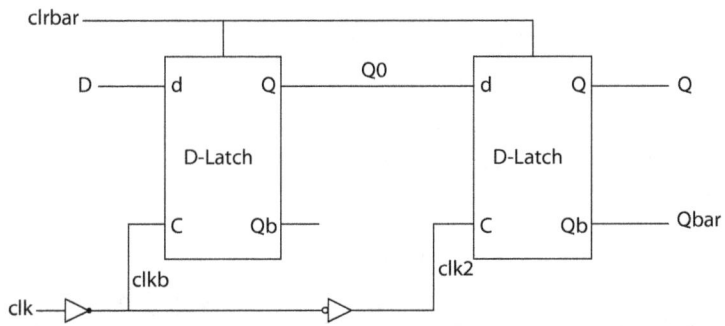

FIGURE 4.12 Logic diagram of a master-slave D flip-flop with active low clear.

Listing 4.19 shows the HDL code of the master-slave D flip-flop. In the VHDL code, there is already code for the D_latchWclrbar (see Listing 4.18); this code is attached to the flip-flop code by the statement

```
for all : D_latchWclrbar use entity work.
            bind32(D_latch_Wclr);
```

which links the architecture D_latch_Wclr to the current module. In Verilog, we link the module D_latchWclr to the module D_FFMasterWclr by the statement

```
D_latchWclr D0 (D, clkb,clrbar, Q0, Qb0);
```

Note that the order of the linked parameters (D, clkb, clrbar, Q0, and Qb0) to match D, E, clrbar, Q, and Qbar of the D_latchWclr module, respectively, for proper mapping.

LISTING 4.19 HDL Description of a Master-Slave D Flip-Flop with Active Low Clear

```
library IEEE;
use IEEE.STD_LOGIC_1164.ALL;

entity D_FFMasterWclr is
    Port (D, clk, clrbar : in std_logic;
          Q, Qbar : buffer std_logic);
end D_FFMasterWclr ;

architecture D_FF_str of D_FFMasterWclr is

component inv
port (I1 : in std_logic; O1 : buffer std_logic);
end component;
component D_latchWclrbar
port (I1, I2, I3 : in std_logic;
      O1, O2 : buffer std_logic);
end component;
for all : D_latchWclrbar use
    entity work. bind32(D_latch_Wclr);
for all : inv use entity work.bind1 (inv_1);
signal clkb, clk2, Q0, Qb0 : std_logic;
begin
    D0 : D_latchWclrbar port map (D, clkb,clrbar, Q0, Qb0);
    D1 : D_latchWclrbar port map (Q0, clk2, clrbar, Q, Qbar);
    in1 : inv port map (clk, clkb);
    in2 : inv port map (clkb, clk2);
end D_FF_str;
```

Verilog Description

```
module D_FFMasterWclr(D, clk,clrbar, Q, Qbar);
input D, clk, clrbar;
output Q, Qbar;
not #1 (clkb, clk);
not #1 (clk2, clkb);
D_latchWclr D0 (D, clkb,clrbar, Q0, Qb0);
D_latchWclr D1 (Q0, clk2,clrbar, Q, Qbar);
endmodule
```

To use the above VHDL code in future examples, it is appended to entity `bind32` in Listing 4.31.

Figure 4.13 shows the simulation waveform of the master-slave D flip-flop. It is clear from the figure that signal D is sampled only at the transition of the clock from low to high. If D changes during the high level (or the low level) of the clock, the output Q remains the same; it does not respond to this change. Compare Figure 4.13 with Figure 2.19 and notice the difference between a latch and a flip-flop. During the high level of the clock (called enable in the latch), Q follows D for the latch. In the flip-flop, Q follows D only at the clock transitions from low to high.

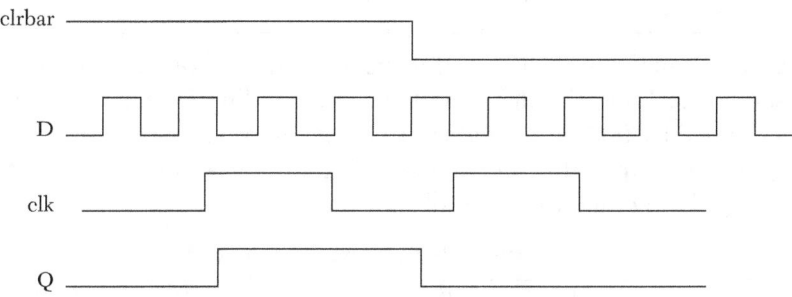

FIGURE 4.13 Simulation waveform of a master-slave D flip-flop.

EXAMPLE 4.11 STRUCTURAL DESCRIPTION OF A PULSE-TRIGGERED MASTER-SLAVE JK FLIP-FLOP WITH ACTIVE LOW CLEAR

A JK flip-flop can be viewed as an extension of the SR-latch. The flip-flop has all the allowed states of the SR. The prohibited state in the SR-latch is replaced by a toggle state where the output of the flip-flop is complemented every time J = K = 1. Table 4.4 shows the excitation table of a pulse-triggered JK flip-flop. Another type of flip-flop is the T flip-flop, where a JK flip-flop with terminal J is connected to terminal K to form terminal T, is shown in Figure 4.14.

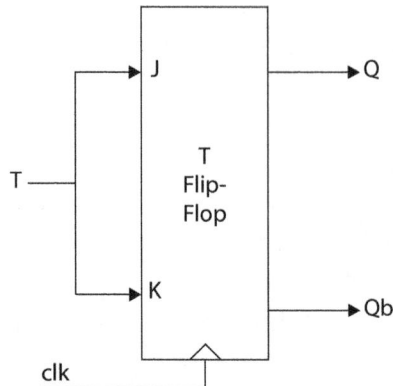

FIGURE 4.14 Logic diagram of a T flip-flop.

TABLE 4.4 Excitation Table for a Pulse-Triggered JK Flip-Flop

J	K	Q	clk	Q$^+$
0	0	Q0	⊓	Q0
0	1	x	⊓	0
1	0	x	⊓	1
1	1	Q0	⊓	$\overline{Q0}$

The Boolean function of a JK flip-flop can be derived from a D flip-flop. Table 4.5 shows the J and K values and the corresponding D values. The D values are obtained by finding the value of D that can produce the transition from Q to Q$^+$ For example, if Q = 0, and Q$^+$ is 1, then D should be 1. In fact, the value of D will be equal to Q$^+$ for all transitions.

TABLE 4.5 Relationship Between JK Flip-Flop and D Flip-Flop

J	K	Q	clk	Q$^+$	D
0	0	0	⊓	0	0
0	0	1	⊓	1	1
0	1	0	⊓	0	0
0	1	1	⊓	0	0
1	0	0	⊓	1	1
1	0	1	⊓	1	1
1	1	0	⊓	1	1
1	1	1	⊓	0	0

To find the Boolean function of D, form K-maps as shown in Figure 4.15.

KQ \\ J	00	01	11	10
00	0	1	0	0
01	1	1	0	1

D

FIGURE 4.15 K-maps of Table 4.5.

From Figure 4.15, the Boolean functions are:

$$D = \overline{K}Q + J\overline{Q} \qquad (4.7)$$

Equation 4.7 is used to build a master-slave JK flip-flop from a master-slave D flip-flop. Figure 4.16 shows a master-slave JK flip-flop generated from a master-slave D flip-flop.

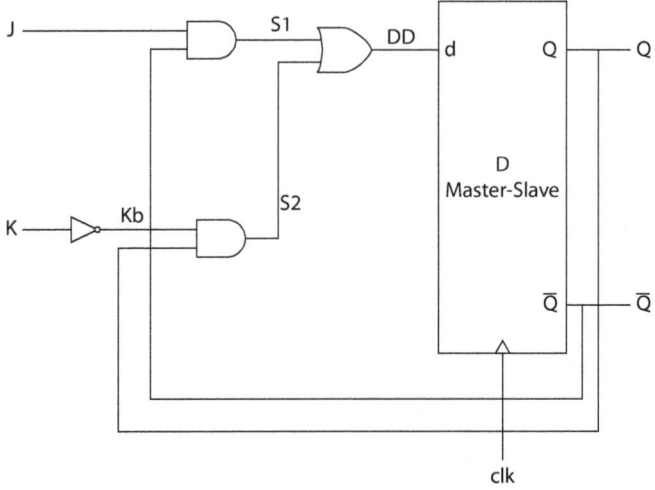

FIGURE 4.16 Pulse-triggered master-slave JK flip-flop.

Listing 4.20 shows the HDL code for the master-slave JK flip-flop illustrated in Figure 4.16.

LISTING 4.20 HDL Description of a Master-Slave JK Flip-Flop: VHDL and Verilog

VHDL Description

```
library IEEE;
use IEEE.STD_LOGIC_1164.ALL;

entity JK_FLFL is
    port (J, K, clk, clrbar : in std_logic;
          Q, Qbar : buffer std_logic);

-- Q and Qbar are declared buffer so they can be input
-- or output

end JK_FLFL;

architecture JK_Master of JK_FLFL is

component and2
port (I1, I2 : in std_logic; O1 : buffer std_logic);
end component;
component or2
port (I1, I2 : in std_logic; O1 : buffer std_logic);
end component;
component inv
port (I1 : in std_logic; O1 : buffer std_logic);
end component;
component D_FFMasterWclr
port (D, clk, clrbar : in std_logic;
      Q, Qbar : buffer std_logic);
end component;
for all : and2 use entity work.bind2 (and2_4);
for all : or2 use entity work.bind2 (or2_4);
for all : inv use entity work.bind1 (inv_1);
for all : D_FFMasterWclr use
          entity work. D_FFMasterWclr (D_FF_str);
signal s1, s2, Kb, DD : std_logic;
begin
    a1 : and2 port map (J, Qbar, s1);
    a2 : and2 port map (Kb, Q, s2);
    in1 : inv port map (K, Kb);
    or1 : or2 port map (s1, s2, DD);
    DFF : D_FFMasterWclr port map (DD, clk,clrbar, Q, Qbar);
```

```
end JK_Master;
```

Verilog Description
```
module JK_FF (J, K, clk,clrbar, Q, Qbar);
input J, K, clk, clrbar;
output Q, Qbar;
wire s1, s2;
    and #4 (s1, J, Qbar);
    and #4 (s2, Kb, Q);
    not #1 (Kb, K);
    or #4 (DD, s1, s2);
D_FFMasterWclr D0 (DD, clk,clrbar, Q, Qbar);
endmodule
module D_FFMasterWclr(D, clk,clrbar, Q, Qbar);

/* no need to rewrite this module here if it has
been already attached to the above module (JK_FF). */

input D, clk, clrbar;
output Q, Qbar;

    not #1 (clkb, clk);
    not #1 (clk2, clkb);
    D_latchWclr D0 (D, clkb,clrbar, Q0, Qb0);
    D_latchWclr D1 (Q0, clk2,clrbar, Q, Qbar);

endmodule

module D_latchWclr(D, E,clrbar, Q, Qbar);
/* no need to rewrite this module here if it has
been already attached to the above module (JK_FF). */

input D, E, clrbar;
output Q, Qbar;
/* assume 4 ns delay for and gate and nor gate,
and 1 ns for inverter */
//The clear is active low; if clrbar = 0, Q=0

and #4 gate1 (s1, D, E, clrbar);

/* the name "gate1" is optional; we could have
    written and #4 (s1, D, E) */
    and #4 gate2 (s2, Eb, Q, clrbar);
```

```
    not #1 (Eb, E);
    nor #4 (Qbar, s1, s2);
    not #1 (Q, Qbar);
endmodule
```

Notice here that the VHDL code in Listing 4.20 is getting shorter (the VHDL code is not shorter but getting shorter) compared to the Verilog code. This is due to the fact that VHDL user-built components are being linked, such as `and2`, `or2`, and `inv`. Their codes do not need to be rewritten because they are linked to the current module.

EXAMPLE 4.12 STRUCTURAL DESCRIPTION OF AN SRAM CELL

A simple memory cell has been designed using an SR-latch; Figure 4.17a shows the symbol diagram of the cell. The cell has tri-state output. If the select line (Sel) is low, the output of the cell is in high impedance. A read/write (R/W) input signal controls the cell's cycle type. If R/W is high, the cell is in read cycle; if it is low, the cell is in write cycle. Table 4.6 shows the excitation table of the cell with inputs (select, R/W, data in, current state) and the corresponding outputs (next state, output). From the current state and next state, S and R of the latch are determined according to Table 4.2. For example, if the current state is 0 and next state 0, then two combinations of SR can generate this transition: S = 0, R = 0, and S = 0, R = 1, so SR = 0x when x is "don't care."

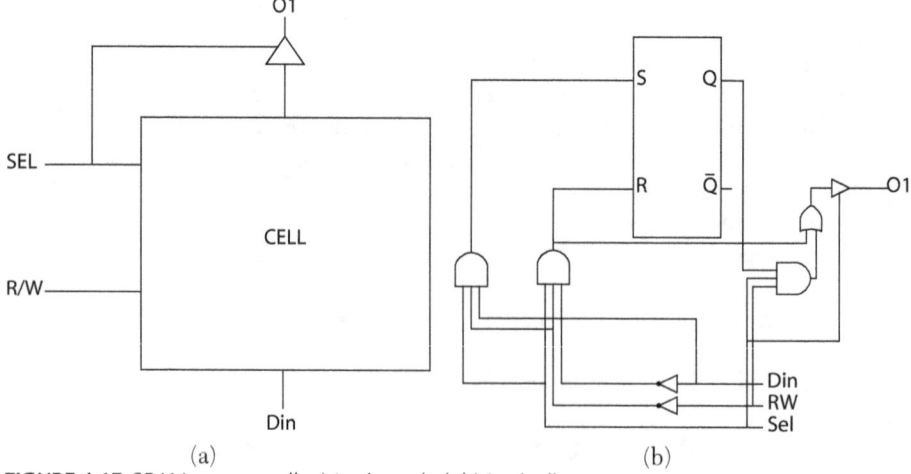

(a)   (b)

FIGURE 4.17 SRAM memory cell. a) Logic symbol. b) Logic diagram.

TABLE 4.6 Excitation Table of an SRAM Memory Cell

Select	R/W	Data In	Current State	Next State	Output	Latch	
Sel	**RW**	**Din**	**Q**	**Q⁺**	**O1**	**S**	**R**
0	x	x	Q	Q	Z	0	0
1	0	0	0	0	0	0	x
1	0	0	1	0	0	0	1
1	0	1	0	1	1	1	0
1	0	1	1	1	1	x	0
1	1	0	0	0	0	0	x
1	1	0	1	1	1	x	0
1	1	1	0	0	0	0	x
1	1	1	1	1	1	x	0

From Table 4.6, K-maps are constructed (see Figure 4.18).

Din Q / Sel R/W	00	01	11	10
00	0	0	0	0
01	0	0	0	0
11	0	x	x	0
10	0	0	x	1

S

Din Q / Sel R/W	00	01	11	10
00	0	0	0	0
01	0	0	0	0
11	x	0	0	x
10	x	1	0	1

R

Din Q / Sel R/W	00	01	11	10
00	Z	Z	Z	Z
01	Z	Z	Z	Z
11	0	1	1	0
10	0	0	1	1

O1

FIGURE 4.18 K-maps for Table 4.6.

From the K-maps:

$$S = Sel \, \overline{RW} \, Din$$

$$R = Sel \overline{RW} \, \overline{Din}$$

$$O1 = Sel \, Din + SelRWQ = R + SelRW \, Q \quad (for \, Sel = 1)$$

$$O1 = Z \, (for \, Sel = 0)$$

The logic diagram of the cell is shown in Figure 4.17b.

The code for the memory cell is shown in Listing 4.21. The VHDL code uses the SR-latch that was designed in Listing 4.17 as a component (macro), using the statement

```
component SR_Latch
port (I1, I2 : in std_logic;
O1, O2 : buffer std_logic);
end component;
```

which declares a generic SR-latch. This latch is linked to the memory-cell code by the statement

```
for all : SR_Latch use entity work.bind22 (SR_Latch);
```

The VHDL statement

```
SR1 : SR_Latch port map (R, S, Q, open);
```

assigns R and S as the inputs of the SR-latch SR1. The noninverted output of the latch is assigned to Q, and the inverted output is left open; open is a VHDL predefined word. For Verilog, link the module of the SR-latch that has been designed in Listing 4.17 to the memory cell code by the statement:

```
SR_Latch RS1 (R, S, Q, Qbar);
```

which links the module SR_Latch to the current module memory.

LISTING 4.21 HDL Description of an SRAM Memory Cell: VHDL and Verilog

VHDL Description
```
library IEEE;
use IEEE.STD_LOGIC_1164.ALL;

entity memory is
```

```
      port (Sel, RW, Din : in std_logic;
            O1: buffer std_logic );

end memory;

architecture memory_str of memory is
--Some simulators will not allow mapping between
--buffer and out. In this
--case, change all out to buffer.

component and3
port (I1, I2, I3 : in std_logic; O1 : out std_logic);
end component;

component inv
port (I1 : in std_logic; O1 : out std_logic);
end component;

component or2
port (I1, I2 : in std_logic; O1 : out std_logic);
end component;

component bufif1
port (I1, I2 : in std_logic; O1 : out std_logic);
end component;

component SR_Latch
port (I1, I2 : in std_logic;
    O1, O2 : buffer std_logic);
end component;
for all : and3 use entity work.bind3 (and3_0);
for all : inv use entity work.bind1 (inv_0);
for all : or2 use entity work.bind2 (or2_0);
for all : bufif1 use entity work.bind2 (bufif1);
for all : SR_Latch use entity work.bind22 (SR_Latch);
signal RWb, Dinb, S, S1, R, O11, Q : std_logic;
begin
    in1 : inv port map (RW, RWb);
    in2 : inv port map (Din, Dinb);
    a1 : and3 port map (Sel, RWb, Din, S);
    a2 : and3 port map (Sel, RWb, Dinb, R);
    SR1 : SR_Latch port map (S, R, Q, open);
--open is a predefined word;
```

```
--it indicates that the port is left open.
    a3 : and3 port map (Sel, RW, Q, S1);
    or1 : or2 port map (S1, S, O11);
    buf1 : bufif1 port map (O11, Sel, O1);
end memory_str;
```

Verilog Description
```
module memory (Sel, RW, Din, O1);
input Sel, RW, Din;
output O1;
    not (RWb, RW);
    not (Dinb, Din);
    and (S, Sel, RWb, Din);
    and (R, Sel, RWb, Dinb);
    SR_Latch RS1 (R, S, Q, Qbar);
    and (S1, Sel, RW, Q);
    or (O11, S1, S);
    bufif1 (O1, O11, Sel);
endmodule
```

EXAMPLE 4.13   STRUCTURAL DESCRIPTION OF A THREE-BIT UNIVERSAL SHIFT REGISTER

Figure 4.19 shows the symbol and logic diagram of a three-bit universal shift register. The register can be loaded externally from a three-bit data P on the positive edge of the clock. The data stored in the register can be right shifted with one-bit DSR replacing the most significant bit of Q every shift. The data stored in the register can also be left shifted with one-bit DSL replacing the least significant bit of Q every shift. The truth table of the register is shown in Table 4.7. Listing 4.22 shows the Verilog code for the shift register. To test the shift function of the register, load external data (P) using the load function and then shift left or right.

TABLE 4.7 Truth Table for the Shift Register

Clrbar	s1	s0	Action
0	x	x	Clear (Q = 0)
1	1	1	Load P into Q (Q = P) at the positive edge of the clock
1	0	1	Shift right, DSR replaces Q2 at the positive edge of the clock
1	0	1	Shift left, DSL replaces Q0 at the positive edge of the clock
1	0	0	Hold (Q retains its current value with the clock)

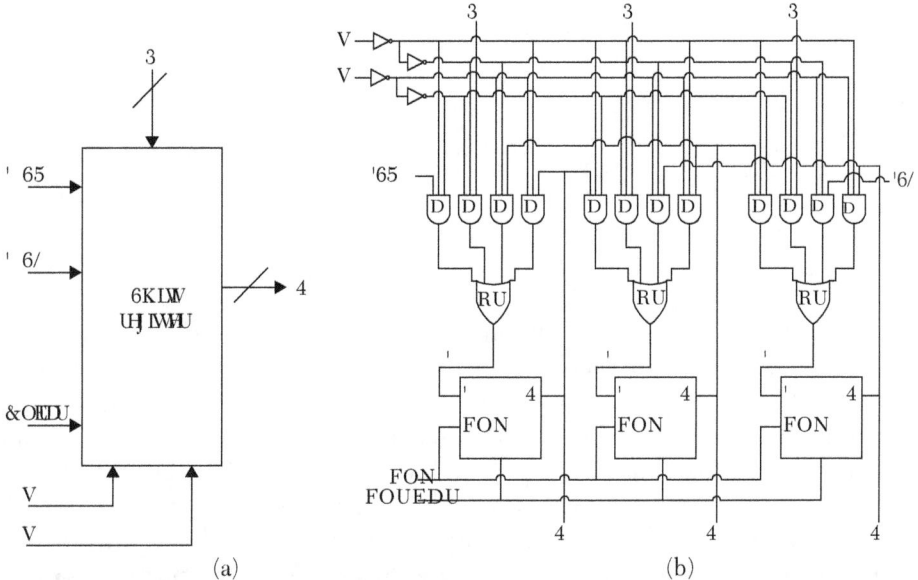

FIGURE 4.19 Universal shift register with clear. a) Symbol diagram. b) Logic diagram.

LISTING 4.22 Verilog Description of a Three-Bit Universal Shift Register

```
module shft_regsterUniv(clk, clrbar,
s0,s1,P,DSR,DSL,Q,Qb);
input clk, clrbar,s0,s1,DSR,DSL;
output [2:0] Q,Qb,P;
not (s0bar, s0);
not (s0t,s0bar);
not (s1bar, s1);
not (s1t, s1bar);

and #4 a0(aa0, DSR, s1bar,s0t);
and #4 a1(aa1, s0t, s1t,P[2]);
and #4 a2(aa2, s0bar, s1t,Q[1]);
and #4 a3(aa3, s0bar, s1bar,Q[2]);
or #4 or2 (D2,aa0,aa1,aa2,aa3);
D_FFMasterWclr DFM0(D2,clk,clrbar,Q[2],Qb[2]);

and #4 a4(aa4, Q[2], s1bar,s0t);
and #4 a5(aa5, s0t, s1t,P[1]);
and #4 a6(aa6, s0bar, s1t,Q[0]);
and #4 a7(aa7, s0bar, s1bar,Q[1]);
or #4 or1 (D1,aa4,aa5,aa6,aa7);
D_FFMasterWclr DFM1(D1,clk, clrbar,Q[1],Qb[1]);
```

```
and #4 a8(aa8, Q[1], s1bar,s0t);
and #4 a9(aa9, s0t, s1t,P[0]);
and #4 a10(aa10, s0bar, s1t,DSL);
and #4 a11(aa11, s0bar, s1bar,Q[0]);
or #4 or0 (D0,aa8,aa9,aa10,aa11);
D_FFMasterWclr DFM2(D0,clk,clrbar,Q[0], Qb[0]);
endmodule
```

4.4   State Machines

Synchronous sequential circuits are called *state machines*. The main components of the state machine are latches and flip-flops; additional combinational components may also be present. Synchronous clock pulses are fed to all flip-flops and latches of the machine. There are two types of synchronous sequential circuits: Mealy and Moore circuits. The output or next state of Mealy circuits depends on the inputs and the present (current) state of the flip-flops/latches. The output or next state of the Moore circuit depends only on the present states. The present state and next state for a particular flip-flop are the same pin (output Q). The current state is the value of Q just before the present clock pulse or edge; the next state is the value of Q after the clock pulse or the edge. To build a state machine, the following steps are performed:

1. Determine the number of states. If the system is n-bit, then the number of flip-flops is n, and the number of states is 2^n. The number of flip-flops here is calculated according to the classical method, where the number of flip-flops is the minimum possible. Another method in which each state is represented by one flip-flop is frequently used when the number of bits is getting too large to handle by the classical method. For example, if the system is three bits, then the classical method requires three flip-flops, while the one flip-flop per state method requires eight flip-flops. In this chapter, the classical method is implemented.

2. Construct a state diagram that shows the transition between states. At each state, consider it as the current state; after the clock is active (edge or pulse), the system moves from current state to next state. Determine the next state according to the input if the system is Mealy or according to the current state only if the system is Moore. Also, determine the output (if any) of the system at this current state.

3. From the state diagram, construct the excitation table that tabulates the inputs and the outputs. The inputs always include the current states, and the outputs always include the next states. The table also includes the inputs of the flip-flops or latches that constitute the state machine. For example, if the flip-flops implemented in a certain machine are JK flip-flops, then the inputs J and K of the flip-flop are determined according to the transition from current to next state. If, for example, the current state is 0 and the next is 0, then J = 0 and K = x (don't care). If the flip-flops are D flip-flops, then the Ds of the flip-flops are the same as the corresponding next states.

4. Find J and K in terms of the inputs and minimize using K-maps or any other appropriate method.

5. If using structural description to simulate the system, draw a logic diagram of the system using appropriate available macros such as latches, adders, and flip-flops.

The following examples are state machines. More examples of state machines and counters will be discussed in Chapters 6 and 7.

EXAMPLE 4.14 **STRUCTURAL DESCRIPTION OF A THREE-BIT SYNCHRONOUS COUNTER WITH ACTIVE LOW CLEAR**

A synchronous counter can be viewed as a simple finite state machine. The logic symbol of the counter is shown in Figure 4.20. The counter is constructed from JK flip-flops.

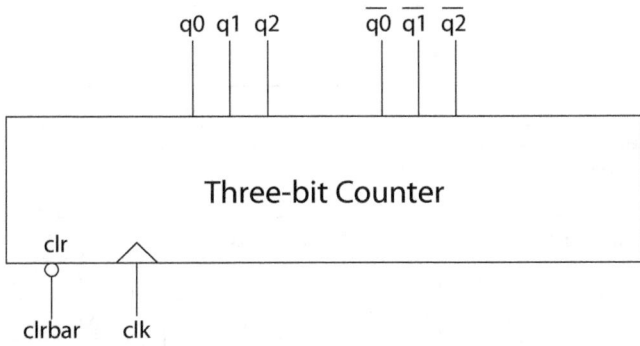

FIGURE 4.20 Logic symbol of a three-bit counter with active low clear.

The state diagram of the counter is shown in Figure 4.21. Because the counter counts from 0 to 7, three flip-flops are needed to cover that count. The transition depends on the current state and the input (`clrbar`). Usually D flip flops are used; however, we will use JK flip flops just to practice with their implementation in the state machine. The next step is to construct the excitation table.

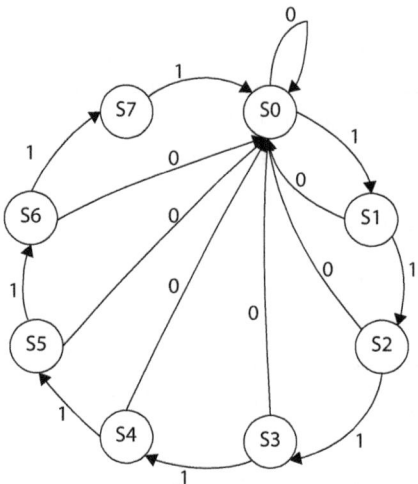

FIGURE 4.21 State diagram of a three-bit counter with active low clear.

Table 4.8a shows the excitation table of a JK flip-flop, and Table 4.8b shows the excitation table of the counter.

TABLE 4.8A Excitation Table for a JK Flip-Flop

		Inputs		Outputs
Current State		**J**	**K**	**Next State**
0		1	x	1
0		0	x	0
1		x	1	0
1		x	0	1

TABLE 4.8B Excitation Table for a Three-Bit Synchronous Counter with Active Low Clear

Inputs				Outputs			Flip-Flops		
Input	Current State			Next State					
clrbar	q2	q1	q0	q2$^+$	q1$^+$	q0$^+$	J2K2	J1K1	J0K0
0	x	x	x	0	0	0	xx	xx	xx
1	0	0	0	0	0	1	0x	0x	1x
1	0	0	1	0	1	0	0x	1x	x1
1	0	1	0	0	1	1	0x	x0	1x
1	0	1	1	1	0	0	1x	x1	x1
1	1	0	0	1	0	1	x0	0x	1x
1	1	0	1	1	1	0	x0	1x	x1

Inputs				Outputs				Flip-Flops		
Input	Current State			Next State				Flip-Flops		
clrbar	q2	q1	q0	$q2^+$	$q1^+$	$q0^+$		J2K2	J1K1	J0K0
1	1	1	0	1	1	1		x0	x0	1x
1	1	1	1	0	0	0		x1	x1	x1

Now, construct the K-maps of the Table 4.8b. The J-K flip-flops with active low clear previously constructed in Example 4.11 are used here. Accordingly, the clear action will be done by just activating the clear function of the JK flip-flops. Figure 4.22 shows the K-maps of Table 4.8b.

q1q0 / q2	– 00	01	11	10
0			1	
1	X	X	X	X

J2 = K2

FIGURE 4.22 K-maps of Table 4.8b.

From Table 4.b and the K-maps:

$$J0 = K0 = 1$$

$$J1 = K1 = q0$$

$$J2 = K2 = q0 \, q1$$

Next, draw the logic diagram of the counter (see Figure 4.23).

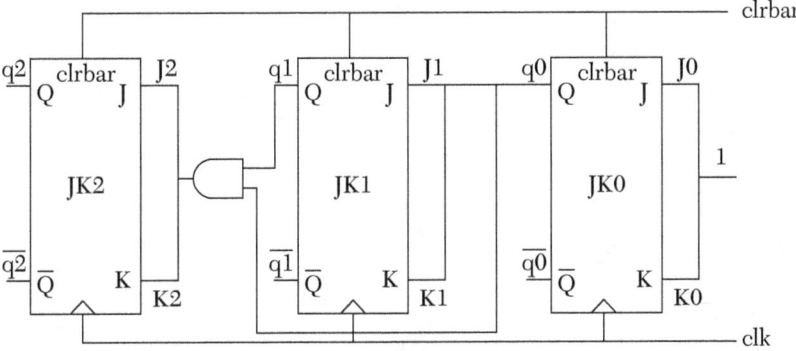

FIGURE 4.23 Logic diagram of a three-bit synchronous counter with active low clear using master-slave JK flip-flops.

Now, write the structural description of the counter. The previously built macros and modules are used, as is the JK flip-flop designed in Listing 4.20. In VHDL, declare it as `component`:

```
component JK_FLFL
port (J, K, clk, clrbar : in std_logic;
    Q, Qbar : buffer std_logic);
end component;
for all : JK_FLFL
    use entity work. JK_FLFL (JK_Master);
```

Be sure to attach all the entities needed, such as entity JK_ FLFL, and be sure to compile all of those entities to generate the work library before using them in the entity `CTStatemachine`.

In Verilog, link the current module to the `JK_FF` module written in Listing 4.20b. As an example of this linking, when J = K = 1:

```
JK_FF FF0(1'b1, 1'b1, clk, clrbar, q[0], qb[0]);
```

Listing 4.23 shows the HDL code of the counter. The basic VHDL package does not include definitions of the components `JK_FLFL` and `and2`. Several CAD vendors can provide packages that contain these definitions; if these packages are included in Listing 4.23, there is no need for component declaration statements for them.

LISTING 4.23 HDL Description of a Three-Bit Synchronous Counter Using Master-Slave JK Flip-Flops: VHDL and Verilog

VHDL Description
```
library IEEE;
use IEEE.STD_LOGIC_1164.ALL;

entity CTStatemachine is
    port( clk, clrbar : in std_logic;
        Q, Qbar: buffer std_logic_vector (2 downto 0));
end CTStateMachine;

architecture ct_3 of CTStateMachine is
component and2
port (I1, I2 : in std_logic; O1 : buffer std_logic);
end component;
component JK_FLFL
port (J, K, clk, clrbar : in std_logic;
    Q, Qbar : buffer std_logic);
```

```
end component;
for all : and2 use entity work.bind2 (and2_4);
for all : JK_FLFL use entity work.
        JK_FLFL (JK_Master );

--Be sure to attach the entity-architectures
-- shown above
signal J2,K2 : std_logic;

begin
JK0 : JK_FLFL port map ('1', '1', clk, clrbar, Q(0), Qbar(0));
JK1 : JK_FLFL port map (q(0), q(0), clk, clrbar, Q(1), Qbar(1));
A1: and2 port map (q(0), q(1), J2);
A2: and2 port map (q(0), q(1), K2);
JK2 : JK_FLFL port map (J2, K2, clk, clrbar, Q(2), Qbar(2));
end ct_3;
```

Verilog Description
```
module CTstatemachine(clk, clrbar, q, qb);
input clk, clrbar;
output [2:0] q, qb;

JK_FF FF0(1'b1, 1'b1, clk, clrbar, q[0], qb[0]);

assign J1 = q[0]; /* a buffer could have been used here
        and in all assign statement in this module*/

assign K1 = q[0];
JK_FF FF1 (J1, K1, clk, clrbar, q[1], qb[1]);

and A1 (J2, q[0], q[1]);
assign K2 = J2;
JK_FF FF2(J2, K2, clk, clrbar, q[2], qb[2]);
endmodule
```

The simulation waveform of the counter is shown in Figure 4.24.

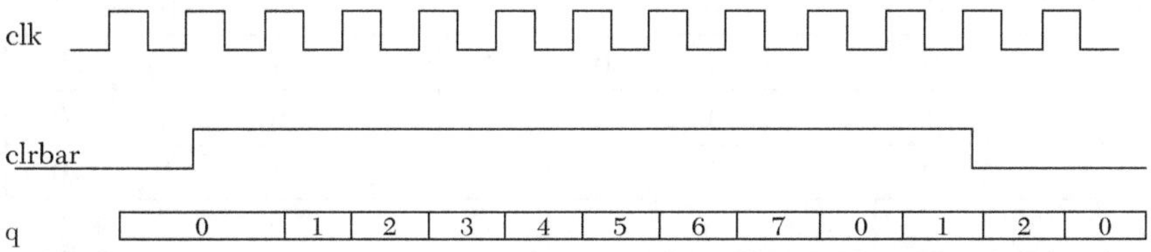

FIGURE 4.24 Simulation waveform of a three-bit synchronous counter with active low clear.

EXAMPLE 4.15 STRUCTURAL DESCRIPTION OF A THREE-BIT SYNCHRONOUS EVEN COUNTER WITH ACTIVE HIGH HOLD

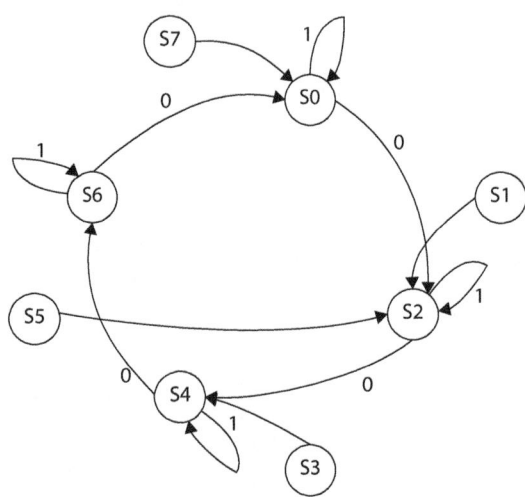

Assume the counter here is counting up. The number of flip-flops is three. First, draw the state diagram of the counter as shown in Figure 4.25. For all even current states, the next state is the next even. For example, if the current state is 010 (2), then the next state is 100 (4). For any odd state (invalid state), the next state can be selected to be any state that ensures the continuity of the count. For example, if the current state is the invalid state 001, the next state can be 000. In the case of invalid states, choose the next state that yields the minimum number of components or minterms; this is done by assigning "don't cares" to the next state of invalid state and selecting 1 or 0 instead of the "don't care" that yields to more minimizations. This will be explained when the excitation table is formed.

FIGURE 4.25 State diagram of an even three-bit counter. The Hold is shown in the diagram as only input.

From the state diagram, generate the excitation table. Table 4.9 shows the excitation table of the counter using D flip-flops. The Ds of the flip-flop are the same as the next state.

TABLE 4.9 Excitation Table for a Three-Bit Even Counter

Inputs					Outputs						
Current state					Next State				Flip-Flops		
H	Q2	Q1	Q0		$Q2^+$	$Q1+$	$Q0^+$		D2	D1	D0
0	0	0	0		0	1	0		0	1	0
0	0	0	1		0	1	0		0	0	0
0	0	1	0		1	0	0		1	0	0
0	0	1	1		1	0	0		0	0	0
0	1	0	0		1	1	0		1	1	0
0	1	0	1		0	1	0		0	0	0

Inputs				Outputs				Flip-Flops		
Current state				Next State						
H	**Q2**	**Q1**	**Q0**	**Q2⁺**	**Q1+**	**Q0⁺**		**D2**	**D1**	**D0**
0	1	1	0	0	0	0		0	0	0
0	1	1	1	0	0	0		0	0	0
1	0	0	0	0	0	0		0	0	0
1	0	0	1	0	0	0		0	0	0
1	0	1	0	0	1	0		0	1	0
1	0	1	1	0	0	0		0	0	0
1	1	0	0	1	0	0		1	0	0
1	1	0	1	0	0	0		0	0	0
1	1	1	0	1	1	0		1	1	0
1	1	1	1	0	0	0		0	0	0

From the excitation table, generate the K-maps. Figure 4.26 shows the K-maps of the counter. Referring to the K-maps, for odd states any next state can be assigned because odd states are not valid. The only restriction is that the next state should yield a valid state. Select the next state that yields elimination of more terms. For example, if the current state is 101, select the next state 100; this yields less minterms.

Q1Q0 / H Q2	00	01	11	10
00	1	1	0	0
01	1	1	0	0
11	0	0	0	1
10	0	0	0	1

D1

Q1Q0 / H Q2	00	01	11	10
00	0	0	1	1
01	1	0	0	0
11	1	0	0	1
10	0	0	0	0

D2

FIGURE 4.26 K-maps of an even three-bit counter.

From the K-maps, find the Boolean functions:

$$D0 = 0$$

$$D1 = \overline{Q1}\,\overline{H} + HQ1\,\overline{Q0}$$

$$D2 = Q2\,\overline{Q1}\,\overline{Q0} + \overline{Q0}\,HQ2 + \overline{H}\,\overline{Q2}\,Q1$$

Using the above Boolean functions, draw the logic diagram of the counter. Figure 4.27 shows the logic symbol and logic diagram of the counter.

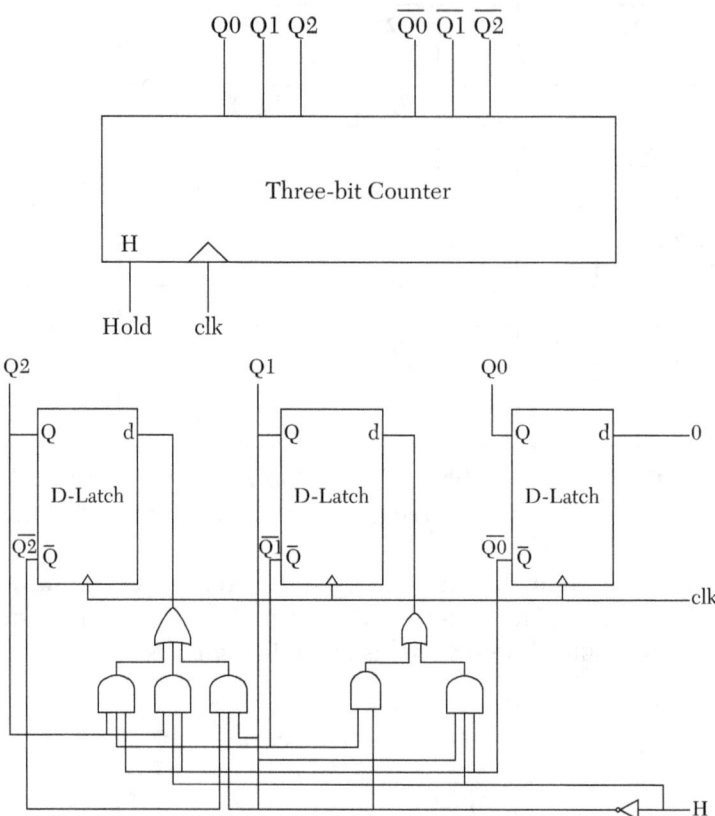

FIGURE 4.27 Three-bit even counter. a) Logic symbol. b) Logic diagram.

Next, write the HDL code for the counter. The macros for the D master-slave flip-flops developed in Listing 4.19 are used. In VHDL code:

```
component D_FFMasterWclr
port (D, clk, clrbar : in std_logic;
        Q, Qbar : buffer std_logic);
end component;
for all : D_FFMasterWclr use
        entity work. D_FFMasterWclr (D_FF_str);
```

In Verilog, write the code that links the D_FFMaster designed in Listing 4.19 to the new module:

```
D_FFMasterWclr DFF0 (1'b0, clk, clrbar, Q[0], Qbar[0]);
```

Listing 4.24 shows the HDL code of the counter.

LISTING 4.24 HDL Description of a Three-Bit Synchronous Even Counter with Hold: VHDL and Verilog

VHDL Description

```
library IEEE;
use IEEE.STD_LOGIC_1164.ALL;

entity CTR_EVEN is
port (H, clk, clrbar : in std_logic;
Q, Qbar : buffer std_logic_vector (2 downto 0));

-- Input clrbar is added to help in testing;
-- set clrbar to low initially when testing
--to clear the output and then set it back to high

end CTR_EVEN;

architecture Counter_even of CTR_EVEN is
--Some simulators will not allow mapping between
--buffer and out. In this
--case, change all out to buffer.

component inv
port (I1 : in std_logic; O1 : buffer std_logic);
end component;

component and2
port (I1, I2 : in std_logic; O1 : buffer std_logic);
end component;

component or2
port (I1, I2 : in std_logic; O1 : buffer std_logic);
end component;

component and3
port (I1, I2, I3 : in std_logic;
O1 : buffer std_logic);
end component;

component or3
port (I1, I2, I3 : in std_logic;
O1 : buffer std_logic);
end component;
```

```
component D_FFMasterWclr
port (D, clk, clrbar : in std_logic;
        Q, Qbar : buffer std_logic);
end component;

for all : D_FFMasterWclr use
        entity work. D_FFMasterWclr (D_FF_str);
for all : inv use entity work.bind1 (inv_0);
for all : and2 use entity work.bind2 (and2_0);
for all : and3 use entity work.bind3 (and3_0);
for all : or2 use entity work.bind2 (or2_0);
for all : or3 use entity work.bind3 (or3_0);
signal Hbar, a1, a2, a3, a4,
        a5, OR11, OR22 : std_logic;
begin
DFF0 : D_FFMasterWclr port map ('0', clk, clrbar, Q(0),Qbar(0));
inv1 : inv port map (H, Hbar);
an1 : and2 port map (Hbar, Qbar(1), a1);
an2 : and3 port map (H, Q(1), Qbar(0), a2);
r1 : or2 port map (a2, a1, OR11);

DFF1 : D_FFMasterWclr port map (OR11, clk, clrbar,
                                Q(1), Qbar(1));
an3 : and3 port map    (Q(2), Qbar(1), Qbar(0), a3);
an4 : and3 port map (Qbar(0), H, Q(2), a4);
an5 : and3 port map (Hbar, Qbar(2), Q(1), a5);
r2 : or3 port map (a3, a4, a5, OR22);

DFF2 : D_FFMasterWclr port map (OR22, clk, clrbar,
                                Q(2), Qbar(2));
end Counter_even;
```

Verilog Description

```
module CTR_EVEN(H, clk, clrbar, Q, Qbar);
// Input clrbar is added to help in testing;
//set clrbar to low initially when testing
//to clear the output and then set it back to high

input H, clk, clrbar;
output [2:0] Q, Qbar;

D_FFMasterWclr DFF0 (1'b0, clk, clrbar, Q[0], Qbar[0]);
not (Hbar, H);
```

```
and (a1, Qbar[1], Hbar);
and (a2, H, Q[1], Qbar[0]);
or (OR1, a1, a2);

D_FFMasterWclr DFF1 (OR1, clk, clrbar, Q[1], Qbar[1]);
and (a3, Q[2], Qbar[1], Qbar[0]);
and (a4, Qbar[0], H, Q[2]);
and (a5, Hbar, Qbar[2], Q[1]);
or (OR2, a3, a4, a5);

D_FFMasterWclr DFF2 (OR2, clk, clrbar, Q[2], Qbar[2]);
endmodule
```

The simulation waveform of the counter is shown in Figure 4.28. As shown in the figure, the Hold is active high. If it is high and the clock pulse is present, the counter holds its output Q to the present value. Some transient states may appear in the simulation due to hazards.

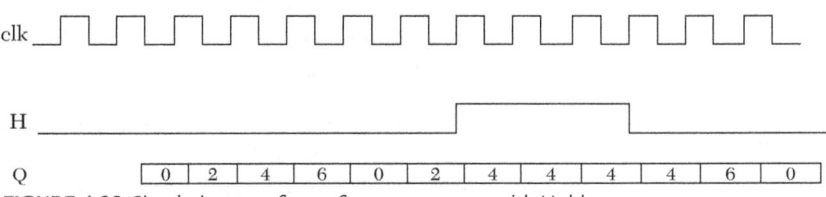

FIGURE 4.28 Simulation waveform of an even counter with Hold.

EXAMPLE 4.16 STRUCTURAL DESCRIPTION OF A THREE-BIT SYNCHRONOUS UP/DOWN COUNTER WITH ACTIVE HIGH CLEAR

The logic symbol of the three-bit synchronous up/down counter is shown in Figure 4.29. The number of flip-flops is three. TC is a terminal count; it is active when the counter completes its count. In this example, TC is high when the count is up to seven or down to zero. The clear here is active high; if it is high, the output of the counter is set to zero. Again just to practice with JK flip-flops we will use them here rather than using D flip-flops.

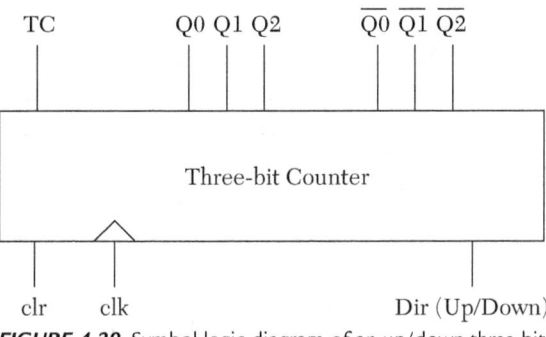

FIGURE 4.29 Symbol logic diagram of an up/down three-bit counter.

The state diagram of the counter is shown in Figure 4.30. The input signal, Dir, determines whether the counter counts up or down. If Dir = 0, the counter counts down, if Dir = 1, the counter counts up. From the state diagram, generate the excitation table of the counter (see Table 4.10).

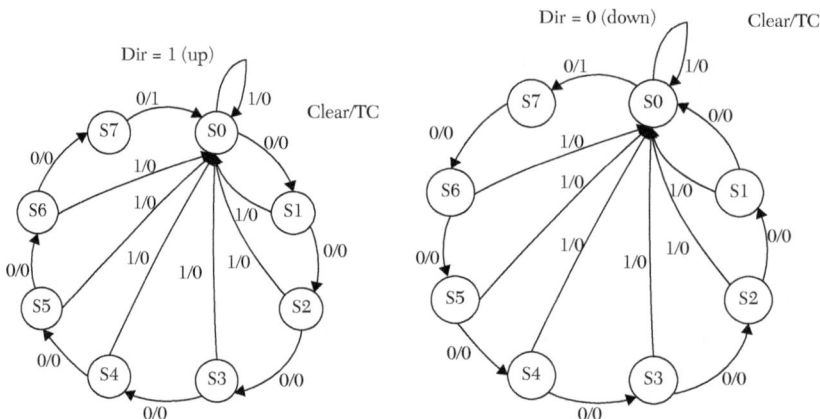

FIGURE 4.30 State diagram of three-bit synchronous up/down counter.

TABLE 4.10 Excitation Table for a Three-Bit Up/Down Counter with a Terminal Count Using Master-Slave JK Flip-Flops

Inputs					Outputs						
Input	Current State				Next State		Output		Flip-Flop		
Clr	Dir	Q2	Q1	Q0	Q2$^+$	Q1$^+$	Q0$^+$	TC	J2K2	J1K1	J0K0
1	x	x	x	x	0	0	0	0	xx	xx	xx
0	0	0	0	0	1	1	1	1	1x	1x	1x
0	0	0	0	1	0	0	0	0	0x	0x	x1
0	0	0	1	0	0	0	1	0	0x	x1	1x
0	0	0	1	1	0	1	0	0	0x	1x	x1
0	0	1	0	0	0	1	1	0	x1	1x	1x
0	0	1	0	1	1	0	0	0	x0	0x	x1
0	0	1	1	0	1	0	1	0	x0	x1	1x
0	0	1	1	1	1	1	0	0	x0	x0	x1
0	1	0	0	0	0	0	1	0	0x	0x	1x
0	1	0	0	1	0	1	0	0	0x	1x	x1
0	1	0	1	0	0	1	1	0	0x	x0	1x
0	1	0	1	1	1	0	0	0	1x	x1	x1
0	1	1	0	0	1	0	1	0	x0	0x	1x
0	1	1	0	1	1	1	0	0	x0	1x	x1
0	1	1	1	0	1	1	1	0	x0	x0	1x
0	1	1	1	1	0	0	0	1	x1	x1	x1

Next, use K-maps to find the Boolean function of the outputs. The clear function will be provided by activating the clear of the JK flip-flop that was covered in Example 4.11. Accordingly, the clear input (clr) in Table 4.10 is not included in the Boolean function of the outputs. Figure 4.31 shows the K-maps from which the following Boolean functions are obtained.

Dir Q2 \ Q1Q0	00	01	11	10
00	1	1	1	1
01	1	1	1	1
11	1	1	1	1
10	1	1	1	1

J0

Dir Q2 \ Q1Q0	00	01	11	10
00	1	1	1	1
01	1	1	1	1
11	1	1	1	1
10	1	1	1	1

K0

Dir Q2 \ Q1Q0	00	01	11	10
00	1	0	1	1
01	1	0	1	1
11	0	1	1	1
10	0	1	1	1

J1

Dir Q2 \ Q1Q0	00	01	11	10
00	1	0	0	1
01	1	0	0	1
11	0	1	1	0
10	0	1	1	0

K1

Dir Q2 \ Q1Q0	00	01	11	10
00	1	0	0	0
01	1	0	0	0
11	0	0	1	0
10	0	0	1	0

J2

Dir Q2 \ Q1Q0	00	01	11	10
00	1	0	0	0
01	1	0	0	0
11	0	0	1	0
10	0	0	1	0

K2

FIGURE 4.31 K-maps of a three-bit synchronous up/down counter.

$$J0 = K0 = 1$$
$$J1 = \overline{Dir}\ \overline{Q0} + Q1 + DirQ0, \quad K1 = \overline{Dir}\ \overline{Q0} + DirQ$$
$$J2 = Q0Q1Dir + \overline{Dir}\ \overline{Q0}\ \overline{Q1}, \quad K2 = J2$$
$$TC = \overline{Dir}\ \overline{Q0}\ \overline{Q1}\ \overline{Q2} + QoQ1A2Dir$$

From the above Boolean functions, draw the logic diagram of the counter, as shown in Figure 4.32.

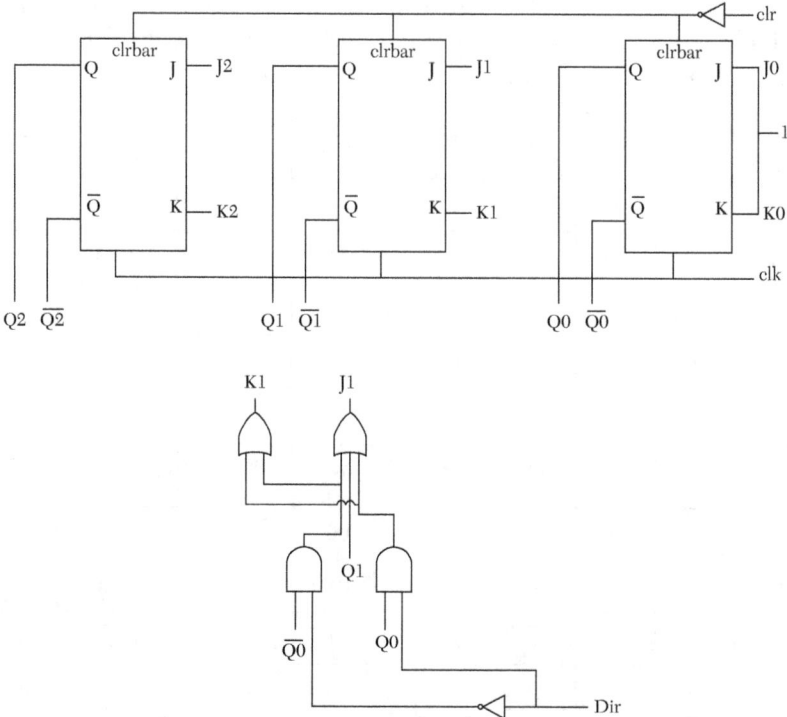

FIGURE 4.32 Logic diagram of a three-bit synchronous up/down counter (for only J0, J1, K0, and K1).

Listing 4.25 shows the HDL code for the counter. To reduce the hazards use gates with a propagation delay. Four nanoseconds are assigned for all primitive gates except for the inverter which is assigned 1 ns.

LISTING 4.25 HDL Description of a 3-Bit Synchronous Up/Down Counter with Clear and Terminal Count—VHDL and Verilog

VHDL Description

```
library IEEE;
use IEEE.STD_LOGIC_1164.ALL;

entity up_down is
    port (clr, Dir, clk : in std_logic;
          TC : buffer std_logic;
       Q, Qbar : buffer std_logic_vector (2 downto 0));
end up_down;
```

```
architecture Ctr_updown of up_down is
--Some simulators will not allow mapping between
--buffer and out. In this
--case, change all out to buffer.

component inv
port (I1 : in std_logic; O1 : buffer std_logic);
end component;

component and2
port (I1, I2 : in std_logic; O1 : buffer std_logic);
end component;
component or2
port (I1, I2 : in std_logic; O1 : buffer std_logic);
end component;
component or3
port (I1, I2,I3 : in std_logic; O1 : buffer std_logic);
end component;
component and3
port (I1, I2, I3 : in std_logic;
        O1 : buffer std_logic);
end component;

component and4
port (I1, I2, I3, I4 : in std_logic;
      O1 : buffer std_logic);
end component;

component JK_FLFL
port (J, K, clk, clrbar : in std_logic;
      Q, Qbar : buffer std_logic);
end component;

for all : JK_FLFL use entity work.
        JK_FLFL (JK_Master );
for all : inv use entity work.bind1 (inv_1);
for all : and2 use entity work.bind2 (and2_4);
for all : and3 use entity work.bind3 (and3_4);
for all : and4 use entity work.bind4 (and4_4);
for all : or2 use entity work.bind2 (or2_4);
for all : or3 use entity work.bind3 (or3_4);
--Be sure that all the reference entities
```

```
--above such as JK_FLFL
--are attached in the project.
signal clrbar, Dirbar, J1, K1, J2, K2 : std_logic;
signal s : std_logic_vector (5 downto 0);
begin
    in1 : inv port map (clr, clrbar);
    in2 : inv port map (Dir, Dirbar);
    an1 : and2 port map (Dirbar, Qbar(0), s(0));
    an2 : and2 port map (Dir, Q(0), s(1));
    an3 : and3 port map (Dirbar, Qbar(1), Qbar(0), s(2));
    an4 : and3 port map (Dir, Q(1), Q(0), s(3));
    an5 : and4 port map (Dir, Q(1), Q(0), Q(2), s(4));
    an6 : and4 port map (Dirbar, Qbar(1),
                            Qbar(0), Qbar(2), s(5));

    r0 : or3 port map (s(0), s(1), Q(1), J1);
    r1 : or2 port map (s(0), s(1), K1);
    r2 : or2 port map (s(2), s(3), J2);
    K2 <= J2;
    r3 : or2 port map (s(4), s(5), TC);

    JKFF0 : JK_FLFL port map
                ('1', '1', clk, clrbar, Q(0), Qbar(0));
    JKFF1 : JK_FLFL port map
                (J1, K1, clk, clrbar, Q(1), Qbar(1));
    JKFF2 : JK_FLFL port map
                (J2, K2, clk, clrbar, Q(2), Qbar(2));
end Ctr_updown;
```

Verilog Description

```
module up_down(clr, Dir, clk, Q, Qbar, TC);

input clr, Dir, clk;
output [2:0] Q, Qbar;
output TC;
not #1 (clrbar, clr);
not #1 (Dirbar, Dir);
and #4 a1(s0, Dirbar, Qbar[0]);
and #4 a2(s1, Dir, Q[0]);
and #4 a3(s2, Dirbar, Qbar[0], Qbar[1]);
and #4 a4(s3, Q[0], Q[1], Dir);
and #4 a5(s4, Dirbar, Qbar[0], Qbar[1], Qbar[2]);
and #4 a6(s5, Q[0], Q[1], Q[2],Dir);
```

```
or #4 r1(J1, s0, Q[1], s1);
or #4 r2(K1, s0, s1);
or #4 r3(J2, s2, s3);
assign K2 = J2;// a buffer can be
//used to generate the above statement
or #4 r4(TC, s4, s5);

JK_FF JKFF0 (1'b1, 1'b1, clk, clrbar, Q[0], Qbar[0]);
JK_FF JKFF1 (J1, K1, clk, clrbar, Q[1], Qbar[1]);
JK_FF JKFF2 (J2, K2, clk,clrbar, Q[2], Qbar[2]);
/*Be sure that all the reference entities above
such as JK_FLFL are attached in the project.*/
endmodule
```

The simulation waveform of the counter is shown in Figure 4.33. When the count is three, the Dir (up/down) is changed from down to up count. Due to the synchronous nature of the Dir signal, the counter continues counting down to two, then starts counting up to three, four, five, and so forth.

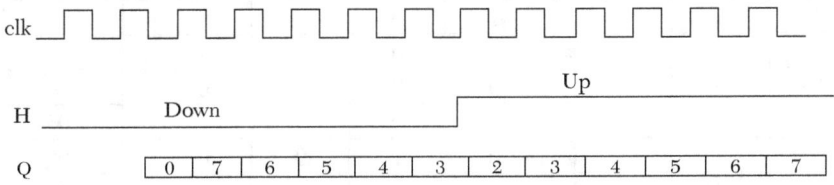

FIGURE 4.33 The simulation waveform of an up/down counter.

EXAMPLE 4.17   STRUCTURAL DESCRIPTION OF A THREE-BIT SYNCHRONOUS DECADE COUNTER

A cecade up counter counts from zero to nine, and the number of flip-flops to cover all counts is four. The state diagram of the counter is shown in Figure 4.34a. There are invalid states from 10 to 14. If any one of these invalid states is a current state, the next state can be any state that restores continuity of the count. As before, the next state selected should be the ones

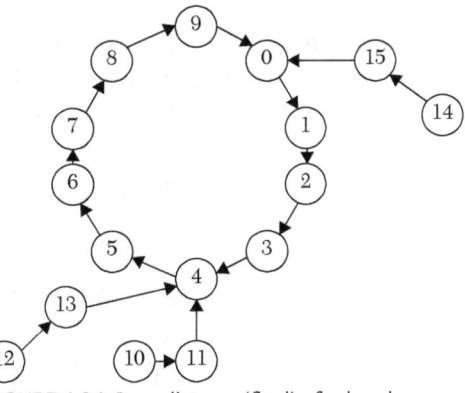

FIGURE 4.34 State diagram (final) of a decade counter.

that yields more minimization. This is determined when the K-maps are generated. Figure 4.34 shows the final state diagram after taking into consideration the K-maps.

Next, construct the excitation table. Table 4.11 shows the excitation table of the decade counter.

TABLE 4.11 Excitation Table for a Decade Counter with a Terminal Count Using D Master-Slave Flip-Flops

Inputs				Outputs				Output
Current State				Next State				
Q3	Q2	Q1	Q0	Q3⁺	Q2⁺	Q1⁺	Q0⁺	TC
0	0	0	0	0	0	0	1	0
0	0	0	1	0	0	1	0	0
0	0	1	0	0	0	1	1	0
0	0	1	1	0	1	0	0	0
0	1	0	0	0	1	0	1	0
0	1	0	1	0	1	1	0	0
0	1	1	0	0	1	1	1	0
0	1	1	1	1	0	0	0	0
1	0	0	0	1	0	0	1	0
1	0	0	1	0	0	0	0	1
1	0	1	0	1	0	1	1	0
1	0	1	1	0	1	0	0	0
1	1	0	0	1	1	0	1	0
1	1	0	1	0	1	0	0	0
1	1	1	0	1	1	1	1	0
1	1	1	1	0	0	0	0	0

K-maps of the outputs are shown in Figure 4.35.

All Ds of the D flip-flops are equal to the corresponding next state. For example, when the current state is 0101 (5), the next state is 0110 (6), and $D0 = 0$, $D1 = 1$, $D2 = 1$, and $D3 = 0$. Applying K-maps (Figure 4.35) to Table 4.11 gives:

$$D0 = \overline{Q0}$$

$$D1 = \overline{Q3}\ \overline{Q1}\ Q0 + Q1\overline{Q0}$$

$$D2 = Q2\overline{Q1} + Q2\overline{Q0} + Q1Q0\overline{Q2}$$

$$D3 = Q3\overline{Q0} + Q0Q1Q2\overline{Q3}$$

$$TC = Q0\overline{Q1}\ \overline{Q2}\ Q3$$

Q3Q2 \ Q1Q0	00	01	11	10
00	1	0	0	1
01	1	0	0	1
11	1	0	0	1
10	1	0	0	1

D0

Q3Q2 \ Q1Q0	00	01	11	10
00	0	1	0	1
01	0	1	0	1
11	0	0	0	1
10	0	0	0	1

D1

Q3Q2 \ Q1Q0	00	01	11	10
00	0	0	1	0
01	1	1	0	1
11	1	1	0	1
10	0	0	1	0

D2

Q3Q2 \ Q1Q0	00	01	11	10
00	0	0	0	0
01	0	0	1	0
11	1	0	0	1
10	1	0	0	1

D3

FIGURE 4.35 K-maps for a decade counter. All "don't cares" have been assigned 0 or 1 to yield to minimum components.

From the Boolean functions, draw the logic diagram of the counter. Figure 4.36 shows the logic diagram of the counter. Listing 4.26 shows the HDL code for the counter.

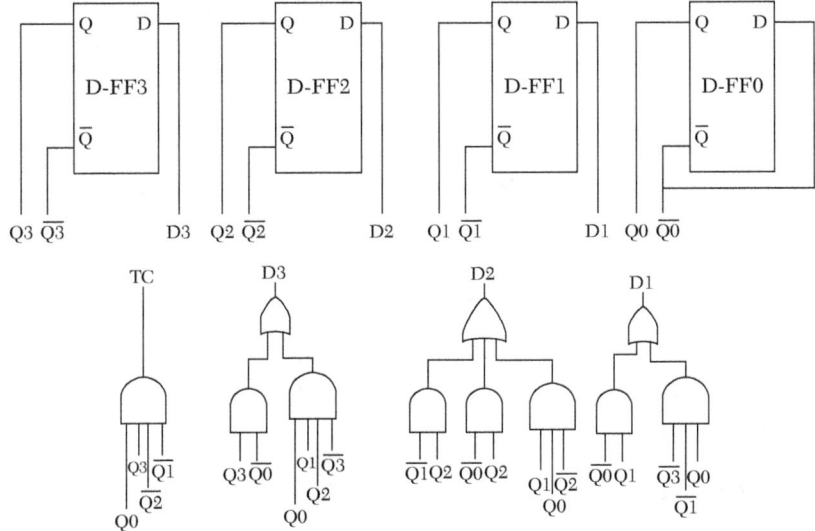

FIGURE 4.36 Logic diagram of a decade counter.

LISTING 4.26 HDL Description of a Three-Bit Synchronous Decade Counter with Terminal Count: VHDL and Verilog

VHDL Description

```
library IEEE;
use IEEE.STD_LOGIC_1164.ALL;
entity decade_ctr is
    port (clk, clr : in std_logic;
    Q, Qbar : buffer std_logic_vector (3 downto 0);
    TC : buffer std_logic);
end decade_ctr;

architecture decade_str of decade_ctr is
--Some simulators will not allow mapping between
--buffer and out. In this
--case, change all out to buffer.
component inv
port (I1 : in std_logic; O1 : buffer std_logic);
end component;
component buf
port (I1 : in std_logic; O1 : buffer std_logic);
end component;
component and2
port (I1, I2 : in std_logic; O1 : buffer std_logic);
end component;
component and3
port (I1, I2, I3 : in std_logic;
    O1 : buffer std_logic);
end component;
component and4
port (I1, I2, I3, I4 : in std_logic;
    O1 : buffer std_logic);
end component;
component or2
port (I1, I2 : in std_logic; O1 : buffer std_logic);
end component;
component or3
port (I1, I2, I3 : in std_logic;
    O1 : buffer std_logic);
end component;

component D_FFMasterWclr
port (D, clk, clrbar : in std_logic;
    Q, Qbar : buffer std_logic);
```

```
end component;

for all : D_FFMasterWclr use entity
          work. D_FFMasterWclr (D_FF_str);
for all : inv use entity work.bind1 (inv_1);
for all : buf use entity work.bind1 (buf_1);
for all : and2 use entity work.bind2 (and2_4);
for all : and3 use entity work.bind3 (and3_4);
for all : and4 use entity work.bind4 (and4_4);
for all : or2 use entity work.bind2 (or2_4);
for all : or3 use entity work.bind3 (or3_4);
signal s : std_logic_vector (6 downto 0);
signal D : std_logic_vector (3 downto 0);
signal clrbar : std_logic;
begin
i1 : inv port map( clr, clrbar);
b1 : buf port map (Qbar(0), D(0));
DFF0 : D_FFMasterWclr port map (D(0), clk, clrbar,
                                Q(0), Qbar(0));

--Assume AND gates and OR gates have 4 ns propagation
--delay and invert has 1 ns.
a1 : and3 port map (Qbar(3), Qbar(1), Q(0), s(0));
a2 : and2 port map (Q(1), Qbar(0), s(1));
r1 : or2 port map (s(0), s(1), D(1));
DFF1 : D_FFMasterWclr port map (D(1), clk, clrbar,
                                Q(1), Qbar(1));

a3 : and2 port map (Q(2), Qbar(1), s(2));
a4 : and2 port map (Q(2), Qbar(0), s(3));
a5 : and3 port map (Q(1), Q(0), Qbar(2), s(4));
r2 : or3 port map (s(2), s(3), s(4), D(2));
DFF2 : D_FFMasterWclr port map (D(2), clk, clrbar,
                                Q(2), Qbar(2));

a6 : and2 port map (Q(3), Qbar(0), s(5));
a7 : and4 port map (Q(0), Q(1), Q(2), Qbar(3), s(6));
r3 : or2 port map (s(5), s(6), D(3));
DFF3 : D_FFMasterWclr port map (D(3), clk, clrbar,
                                Q(3), Qbar(3));
a8 : and4 port map (Q(0), Qbar(1), Qbar(2), Q(3), TC);

end decade_str;
```

Verilog Description
```
module decade_ctr(clk, clrbar,Q, Qbar, TC );
input clk,clrbar;
//use clrbar input to clear the counter when simulting
output [3:0] Q, Qbar;
output TC;
wire [3:0] D;
wire [6:0] s;
buf #1 (D[0], Qbar[0]);

D_FFMasterWclr DFF0 (D[0], clk, clrbar, Q[0], Qbar[0]);
/*Assume and gates and or gates have 4 ns propagation
delay and invert has 1 ns.*/

and #4 (s[0], Qbar[3], Qbar[1], Q[0]);
and #4 (s[1], Q[1], Qbar[0]);

or #4 (D[1], s[0], s[1]);
D_FFMasterWclr FF1 (D[1], clk, clrbar, Q[1], Qbar[1]);

and #4 (s[2],Q[2], Qbar[1]);
and #4 (s[3],Q[2], Qbar[0]);
and #4 (s[4],Q[1], Q[0], Qbar[2]);
or #4 (D[2], s[2], s[3], s[4]);
D_FFMasterWclr FF2 (D[2], clk,clrbar, Q[2], Qbar[2]);
and #4 (s[5], Q[3], Qbar[0]);
and #4 (s[6], Q[0], Q[1], Q[2], Qbar[3]);
or #4 (D[3], s[5], s[6]);
D_FFMasterWclr FF3 (D[3], clk,clrbar, Q[3], Qbar[3]);
and #4 (TC, Q[0], Qbar[1], Qbar[2], Q[3]);

endmodule
```

Figure 4.37 shows the simulation waveform of the decade counter.

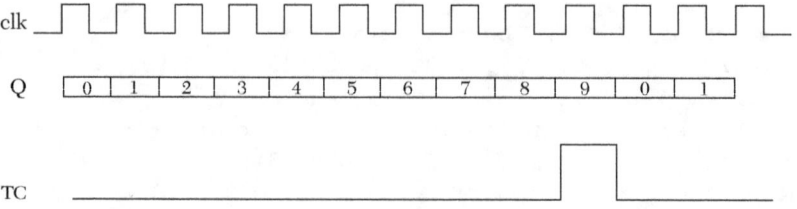

FIGURE 4.37 Simulation waveform of the decade counter.

4.5 `generate` **(HDL),** `generic` **(VHDL), and** `parameter` **(Verilog)**

The predefined word `generate` is mainly used for repetition of concurrent statements. Its counterpart in behavioral description is the `For-Loop`, and it can be used to replicate structural or gate-level description statements. `generate` has several formats, one of which is covered here. See Chapter 7 for more formats.

In VHDL, the format for the `generate` statement is:

```
L1 : for i in 0 to N generate
v1 : inv port map (Y(i), Yb(i));
--other concurrent statements can be entered here
end generate;
```

The above statement describes N + 1 inverters (assuming `inv` was declared as an inverter component with input `Y` and output `Yb`). The input to inverter is `Y(i)`, and the output is `Yb(i)`. `L1` is a required label for the `generate` statement.

An equivalent `generate` statement in Verilog is:

```
generate
genvar i;
for (i = 0; i <= N; i = i + 1)
begin : u
not (Yb[i], Y[i]);
end
endgenerate
```

The statement `genvar i` declares the index `i` of the `generate` statement; `genvar` is a predefined word. `u` is a label for the predefined word `begin`; and `begin` must have a label.

The words generic (in VHDL) and parameter (in Verilog) are used to define global constants. The generic statement can be placed within entity, component, or instantiation statements. The following generic VHDL statement inside the entity declares N as a global constant of value 3:

```
entity compr_genr is
generic (N : integer := 3);
port (X, Y : in std_logic_vector (N downto 0);
  xgty, xlty, xeqy : buffer std_logic);
```

The following Verilog statement declares N as a global constant with a value of 3:

```
parameter N = 3;
input [N:0] X, Y;
```

The following examples cover `generate`, `generic`, and `parameter`.

EXAMPLE 4.18 **STRUCTURAL DESCRIPTION OF (N+1)-BIT MAGNITUDE COMPARATOR USING THE GENERATE STATEMENT**

In Listing 4.16, a three-bit comparator has been described. In this section, an (N+1)-bit comparator using the `generate` statement is introduced. Listing 4.28 shows the HDL code for the (N+1)-bit comparator. Referring to Listing 4.28, the following statements generate N+1 inverters, N+1 full adders, and N+1 two-input *and* gates:

```
G1 : for i in 0 to N generate
v1 : inv port map (Y(i), Yb(i));
FA : full_adder port map (X(i), Yb(i),
carry(i), sum(i), carry(i+1));
a1 : and2 port map (eq(i), sum(i), eq(i+1));
end generate G1;
```

The following Verilog statements also generate N+1 inverters, *N*+1 full adders, and N+1 two-input *and* gates:

```
generate
genvar i;
for (i = 0; i <= N; i = i + 1)
begin : u
not (Yb[i], Y[i]);
FULL_ADDER FA (X[i], Yb[i], carry [i], sum [i], carry[i+1]);
and (eq[i+1], sum[i], eq[i]);
end
```

LISTING 4.28 HDL Description of N-Bit Magnitude Comparator Using the generate Statement: VHDL and Verilog

VHDL Description
```
library IEEE;
use IEEE.STD_LOGIC_1164.ALL;

entity compr_genr is
generic (N : integer := 3);
    port (X, Y : in std_logic_vector (N downto 0);
```

```
    xgty, xlty, xeqy : buffer std_logic);
end compr_genr;


architecture cmpare_str of compr_genr is
--Some simulators will not allow mapping between
--buffer and out. In this
--case, change all out to buffer.

component full_adder
port (I1, I2, I3 : in std_logic;
      O1, O2 : buffer std_logic);
end component;
component inv
port (I1 : in std_logic; O1 : buffer std_logic);
end component;
component nor2
port (I1, I2 : in std_logic; O1 : buffer std_logic);
end component;
component and2
port (I1, I2 : in std_logic; O1 : buffer std_logic);
end component;
signal sum, Yb : std_logic_vector (N downto 0);
signal carry, eq : std_logic_vector (N + 1 downto 0);

for all : full_adder use entity work.bind32 (full_add);
for all : inv use entity work.bind1 (inv_1);
for all : nor2 use entity work.bind2 (nor2_7);
for all : and2 use entity work.bind2 (and2_7);
begin
     carry(0) <= '0';
    eq(0) <= '1';

    G1 : for i in 0 to N generate
    v1 : inv port map (Y(i), Yb(i));
    FA : full_adder port map (X(i), Yb(i), carry(i),
    sum(i), carry(i+1));
    a1 : and2 port map (eq(i), sum(i), eq(i+1));
end generate G1;
xgty <= carry(N+1);
xeqy <= eq(N+1);
n1 : nor2 port map (xeqy, xgty, xlty);

end cmpare_str;
```

Verilog Description

```
module Compr_genr(X, Y, xgty, xlty, xeqy);
parameter N = 3;
input [N:0] X, Y;
output xgty, xlty, xeqy;
wire [N:0] sum, Yb;
wire [N+1 : 0] carry, eq;
assign carry[0] = 1'b0;
assign eq[0] = 1'b1;

generate

genvar i;
for (i = 0; i <= N; i = i + 1)
    begin : u
    not (Yb[i], Y[i]);
/* The above statement is equivalent to assign Yb = ~Y if outside
the generate loop */

    FULL_ADDER FA(X[i], Yb[i], carry [i], sum [i], carry[i+1]);
        /*be sure that the module FULL_ADDER
         is entered (attached) in the project*/
    and (eq[i+1], sum[i], eq[i]);
    end
endgenerate
assign xgty = carry[N+1];
assign xeqy = eq[N+1];
nor (xlty, xeqy, xgty);

endmodule
```

**EXAMPLE 4.19   STRUCTURAL DESCRIPTION OF AN N-BIT ASYNCHRONOUS
DOWN COUNTER USING THE GENERATE STATEMENT**

Asynchronous counters differ from synchronous counters in the way
the clock is connected to each flip-flop. In synchronous counters, all flip-
flops are driven by the same clock. In asynchronous counters, each flip-flop
may be driven by a different clock. Figure 4.38 shows an n-bit asynchronous counter using JK flip-flops. The clock of the first flip-flop is the main
clock. The clock of the second flip-flop is the output of the first JK flip-flop.
This pattern is repeated where the clock of the ith flip-flop is driven by the
output of $(i-1)$th flip-flop.

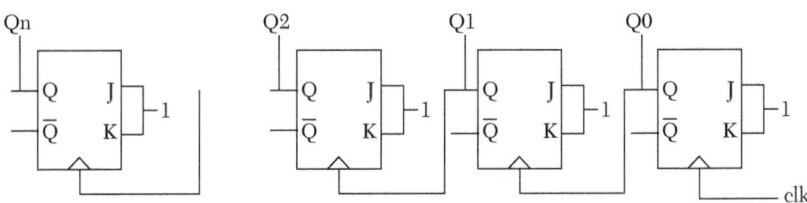

FIGURE 4.38 Logic diagram of n-bit asynchronous down counter.

Asynchronous counters suffer more from hazards than synchronous counters. This is due to the way the clock of each flip-flop is connected. Each flip-flop has to wait until the output of the preceding flip-flop settles. During the period before the flip-flop settles, there will be transient states. Listing 4.29 shows the HDL code for an n-bit asynchronous counter. To use `generate` effectively, the n flip-flops should be described by a general statement that will be replicated. All flip-flops, except the first, have a repeated pattern: the clock of the ith flip-flop is the output of the $(i–1)$th. To bring the first flip-flop into this pattern, concatenate the clock and the Qs of all flip-flops in one vector, S, that represents all the clocks:

```
s <= (Q & clk); --VHDL
assign s = {Q, clk}; //Verilog
```

LISTING 4.29 HDL Description of an N-Bit Asynchronous Down Counter Using `generate`: VHDL and Verilog

VHDL Description
```
library IEEE;
use IEEE.STD_LOGIC_1164.ALL;
entity asynch_ctr is
Generic (N : integer := 3);

-- This is a 3-bit counter. If a different number of
-- bits is needed, simply change the
-- value of N here only.
    port (clk, clrbar : in std_logic;
    Q, Qbar : buffer std_logic_vector (N-1 downto 0));

end asynch_ctr;

architecture CT_strgnt of asynch_ctr is
--Some simulators will not allow mapping between
--buffer and out. In this
--case, change all out to buffer.
```

```
component JK_FLFL
port (J, K, clk, clrbar : in std_logic;
      Q, Qbar : buffer std_logic);
end component;

for all : JK_FLFL use entity work.
                JK_FLFL (JK_Master );

-- For bind32, see Listing 4.17a

signal h, l : std_logic;
signal s : std_logic_vector (N downto 0);
begin
h <= '1';
l <= '0';
s <= (Q & clk);

-- s is the concatenation of Q and clk. We need
-- this concatenation to
-- describe the clock of each JK flip-flop.
Gnlop : for i in (N-1) downto 0 generate

G1 : JK_FLFL port map (h, h, s(i), clrbar,
                        Q(i), Qbar(i));
end generate GnLop;
end CT_strgnt;
```

Verilog Description
```
module asynch_ctr(clk,clrbar, Q, Qbar);

parameter N = 3;
/* This is a 3-bit counter. If a different number of
bits is needed, simply change the value
of N here only.*/

input clk, clrbar;
output [N-1:0] Q, Qbar;
wire [N:0] s;
assign s = {Q, clk};
/* s is the concatenation of Q and clk.
   This concatenation is needed to describe the clock
   of each JK flip-flop. */
```

```
generate
genvar i;
for (i = 0; i < N; i = i + 1)

begin : u

JK_FF JKFF0 (1'b1, 1'b1, s[i],clrbar, Q[i],
                Qbar[i]);
// JK_FF is as shown in Listing 4.17b
end
endgenerate
```

```
endmodule
```

Figure 4.39 shows the simulation waveform of the counter with N = 3. The waveform may contain several transient states.

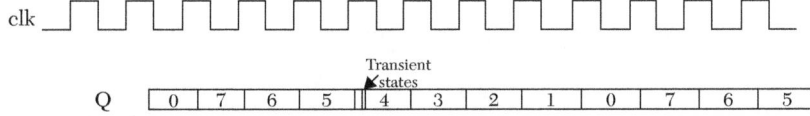

FIGURE 4.39 Simulation waveform of n-bit asynchronous down counter (n = 3).

EXAMPLE 4.20   STRUCTURAL DESCRIPTION OF AN N-BIT MEMORY WORD USING GENERATE

In Listing 4.21, a single memory cell is described. The cell here is expanded to n bits using the `generate` statement. Listing 4.30 shows the HDL code for the n-bit memory word. Referring to Listing 4.30, the VHDL statements

```
G1 : for i in 0 to N generate
M : memory_cell port map (sel, R_W, Data_in(i),
Data_out(i));
end generate;
```

and the Verilog statements

```
generate
genvar i;
for (i = 0; i <= N; i = i + 1)
begin : u
memory M1 (sel, R_W, Data_in [i], Data_out[i]);
end
endgenerate
```

replicate the memory cell designed in Listing 4.29 n times.

LISTING 4.30   HDL Description of N-Bit Memory Word Using `generate`:
VHDL and Verilog

VHDL Description
```
library IEEE;
use IEEE.STD_LOGIC_1164.ALL;
entity Memory_word is
Generic (N : integer := 7);
    port (Data_in : in std_logic_vector (N downto 0);
          sel, R_W : in std_logic; Data_out : out
          std_logic_vector (N downto 0));
end Memory_word;

architecture Word_generate of Memory_word is
component memory_cell
Port (Sel, RW, Din : in std_logic;
      O1 : buffer std_logic );
end component;

for all : memory_cell use entity
          work.memory (memory_str);
begin
G1 : for i in 0 to N generate
M : memory_cell port map (sel, R_W, Data_in(i),
                          Data_out(i));
end generate;
end Word_generate;
```

Verilog Description
```
module Memory_Word (Data_in, sel, R_W, Data_out);

parameter N = 7;
input [N:0] Data_in;
input sel, R_W;
output [N:0] Data_out;

generate
genvar i;
for (i = 0; i <= N; i = i + 1)
begin : u
```

```
memory M1 (sel, R_W, Data_in [i], Data_out[i]);

end

endgenerate
endmodule
```

LISTING 4.31 VHDL Code for Components Used for Binding in Chapter 4

```
--Some simulators will not allow mapping between
--buffer and out. In this
--case, change all out to buffer as it is done here.
library IEEE;
use IEEE.STD_LOGIC_1164.ALL;
entity bind1 is
port (O1 : buffer std_logic; I1 : in std_logic);
end bind1;
architecture inv_0 of bind1 is
begin
O1 <= not I1;
end inv_0;
architecture inv_1 of bind1 is
begin
O1 <= not I1 after 1 ns;
end inv_1;

architecture inv_7 of bind1 is
begin
O1 <= not I1 after 7 ns;
end inv_7;
library IEEE;
use IEEE.STD_LOGIC_1164.ALL;

entity bind2 is
port (O1 : buffer std_logic; I1, I2 : in std_logic);
end bind2;

architecture xor2_0 of bind2 is
begin
O1 <= I1 xor I2;
end xor2_0;
architecture and2_0 of bind2 is
begin
```

```
O1 <= I1 and I2;
end and2_0;
architecture and2_4 of bind2 is
begin
O1 <= I1 and I2 after 4 ns;
end and2_4;

architecture and2_7 of bind2 is
begin
O1 <= I1 and I2 after 7 ns;
end and2_7;

architecture or2_0 of bind2 is
begin
O1 <= I1 or I2;
end or2_0;

architecture or2_7 of bind2 is
begin
O1 <= I1 or I2 after 7 ns;
end or2_7;

architecture nor2_0 of bind2 is
begin
O1 <= I1 nor I2;
end nor2_0;

architecture nor2_7 of bind2 is
begin
O1 <= I1 nor I2 after 7 ns;
end nor2_7;

architecture nor2_4 of bind2 is
begin
O1 <= I1 nor I2 after 4 ns;
end nor2_4;

architecture bufif1 of bind2 is
begin
buf : process (I1, I2)
variable tem : std_logic;
begin
```

```
if (I2 ='1')then
tem := I1;
else
tem := 'Z';
end if;
O1 <= tem;
end process buf;
end bufif1;

library IEEE;
use IEEE.STD_LOGIC_1164.ALL;
entity bind3 is
port (O1 : buffer std_logic;
I1, I2, I3 : in std_logic);
end bind3;

architecture and3_0 of bind3 is
begin
O1 <= I1 and I2 and I3;
end and3_0;

architecture and3_4 of bind3 is
begin
O1 <= I1 and I2 and I3 after 4 ns;
end and3_4;

architecture and3_7 of bind3 is
begin
O1 <= I1 and I2 and I3 after 7 ns;
end and3_7;

architecture or3_0 of bind3 is
begin
O1 <= I1 or I2 or I3;
end or3_0;

architecture or3_7 of bind3 is
begin
O1 <= I1 or I2 or I3 after 7 ns;
end or3_7;

library IEEE;
```

```
use IEEE.STD_LOGIC_1164.ALL;
entity bind22 is
Port (O1, O2 : buffer std_logic;
I1, I2 : in std_logic);
end bind22;

architecture HA of bind22 is
component xor2
port (I1, I2 : in std_logic; O1 : buffer std_logic);
end component;
component and2
port (I1, I2 : in std_logic; O1 : buffer std_logic);
end component;
for A1 : and2 use entity work.bind2 (and2_0);
for X1 : xor2 use entity work.bind2 (xor2_0);
    begin
    X1 : xor2 port map (I1, I2, O1);
    A1 : and2 port map (I1, I2, O2);
    end HA;

architecture SR_Latch of bind22 is
component nor2
port (I1, I2 : in std_logic; O1 : buffer std_logic);
end component;
for all : nor2 use entity work.bind2 (nor2_0);

begin
n1 : nor2 port map (I1, O1, O2);
n2 : nor2 port map (I2, O2, O1);
end SR_Latch;

architecture D_latch of bind22 is
component and2
port (I1, I2 : in std_logic; O1 : buffer std_logic);
end component;
component nor2
port (I1, I2 : in std_logic; O1 : buffer std_logic);
end component;
component inv
port (I1 : in std_logic; O1 : buffer std_logic);
end component;
for all : and2 use entity work.bind2 (and2_4);
```

```
for all : nor2 use entity work.bind2 (nor2_4);
for all : inv use entity work.bind1 (inv_1);
signal I2b, s1, s2 : std_logic;
begin
a1 : and2 port map (I1, I2, s1);
a2 : and2 port map (I2b, O1, s2);
in1 : inv port map (I2, I2b);
in2 : inv port map (O2, O1);
n2 : nor2 port map (s1, s2, O2);
end D_latch;

library IEEE;
use IEEE.STD_LOGIC_1164.ALL;
entity bind32 is
port (I1, I2, I3 : in std_logic;
O1, O2 : buffer std_logic);

end bind32;

architecture full_add of bind32 is
component HA
port (I1, I2 : in std_logic;
      O1, O2 : buffer std_logic);
end component;
component or2
port (I1, I2 : in std_logic; O1 : buffer std_logic);
end component;
for all : HA use entity work.bind22 (HA);
for all : or2 use entity work.bind2 (or2_0);
signal s0, c0, c1 : std_logic;

begin
HA1 : HA port map (I2, I3, s0, c0);
HA2 : HA port map (I1, s0, O1, c1);
r1 : or2 port map (c0, c1, O2);
end full_add;

architecture D_latch_Wclr of bind32 is
component and3
port (I1, I2, I3 : in std_logic;
      O1 : buffer std_logic);
end component;
```

```
component nor2
port (I1, I2 : in std_logic; O1 : buffer std_logic);
end component;
component inv
port (I1 : in std_logic; O1 : buffer std_logic);
end component;
for all : and3 use entity work.bind3 (and3_4);
for all : nor2 use entity work.bind2 (nor2_4);
for all : inv use entity work.bind1 (inv_1);
signal I2b, s1, s2 : std_logic;
begin
a1 : and3 port map (I1, I2, I3, s1);
a2 : and3 port map (I2b, O1,I3, s2);
in1 : inv port map (I2, I2b);
in2 : inv port map (O2, O1);
n2 : nor2 port map (s1, s2, O2);
end D_latch_Wclr;


library IEEE;
use IEEE.STD_LOGIC_1164.ALL;
entity D_LatchWclr is
port (D, E,clrbar : in std_logic;
      Q, Qbar : buffer std_logic);
end;


architecture D_latch_str of D_LatchWclr is
component and3
port (I1, I2, I3 : in std_logic;
      O1 : buffer std_logic);
end component;
component nor2
port (I1, I2 : in std_logic; O1 : buffer std_logic);
end component;
component inv
port (I1 : in std_logic; O1 : buffer std_logic);
end component;
for all : and3 use entity work.bind3 (and3_4);
for all : nor2 use entity work.bind2 (nor2_4);
for all : inv use entity work.bind1 (inv_1);
signal Eb, s1, s2 : std_logic;
begin
a1 : and3 port map (D, E, clrbar, s1);
```

```
a2 : and3 port map (Eb, D,clrbar, s2);
in1 : inv port map (E, Eb);
in2 : inv port map (Qbar, Q);
n2 : nor2 port map (s1, s2, Qbar);
end D_latch_str;

library IEEE;
use IEEE.STD_LOGIC_1164.ALL;

entity D_FFMasterWclr is
Port (D, clk, clrbar : in std_logic;
     Q, Qbar : buffer std_logic);
end D_FFMasterWclr ;

architecture D_FF_str of D_FFMasterWclr is
component inv
port (I1 : in std_logic; O1 : buffer std_logic);
end component;
component D_latchWclrbar
port (I1, I2, I3 : in std_logic;
     O1, O2 : buffer std_logic);
end component;
for all : D_latchWclrbar use entity
       work. bind32(D_latch_Wclr);
for all : inv use entity work.bind1 (inv_1);
signal clkb, clk2, Q0, Qb0 : std_logic;
begin
D0 : D_latchWclrbar port map (D, clkb,clrbar, Q0, Qb0);
D1 : D_latchWclrbar port map (Q0, clk2, clrbar, Q,
Qbar);
in1 : inv port map (clk, clkb);
in2 : inv port map (clkb, clk2);
end D_FF_str;
```

4.6 Summary

In this chapter, the fundamentals of structural description have been covered. Gate-level description was discussed and implemented to build more complex structures (macros). Verilog has built-in gates such as and, or, nand, nor, and buf. Basic VHDL does not have built-in gates, but these gates can be built by using the predefined word component and binding it to written behavioral descriptions. Both VHDL and Verilog have the predefined command generate for replicating structural macros. Table 4.11 shows a list of the VHDL statements covered in this chapter, along with their Verilog counterparts (if any).

TABLE 4.11 Summary of VHDL Statements and Their Verilog Counterparts

VHDL	Verilog
generate	generate
port map	Built in
and2, or2, xor2, nor2,	and, or, xor, nor,
xnor2, inv	xnor, not
(The above VHDL gates are user-built)	
use library	Built in

4.7 Exercises

1. Design a four-bit parity generator. The output is 0 for even parity and 1 for odd parity. Write both the VHDL and Verilog codes.

2. Design a counter that counts 0, 1, 3, 6, 7, 0, 1... using the state-machine approach. Show all details of your answer. Write both the VHDL and Verilog codes.

3. Referring to Listing 4.26 (Verilog), change the count from down to up and rewrite the code.

4. Translate the VHDL code shown in Listing 4.32 to Verilog. What is the logic function of the system?

   LISTING 4.32   Code for Exercise 4.4

```
library IEEE;
use IEEE.STD_LOGIC_1164.ALL;
entity system is
Port (a, b, c : in std_logic;
      d, e : buffer std_logic );
end system;

architecture prob_6 of system is
component xor2
port (I1, I2 : in std_logic; O1 : buffer std_logic);
end component;
```

```
component and2
port (I1, I2 : in std_logic; O1 : buffer std_logic);
end component;

component or3
port (I1, I2, I3 : in std_logic;
O1 : buffer std_logic);

end component;

component inv
port (I1 : in std_logic; O1 : buffer std_logic);
end component;

for all : xor2 use entity work.bind2 (xor2_0);
for all : and2 use entity work.bind2 (and2_0);
for all : inv use entity work.bind1 (inv_0);
for all : or3 use entity work.bind3 (or3_0);
signal s1, s2, s3, s4, abar, bbar, cbar : std_logic;
begin
x1 : xor2 port map (a, b, s1);
x2 : xor2 port map (s1, c, d);
c1 : inv port map (a, abar);
c2 : inv port map (b, bbar);
c3 : inv port map (a, cbar);
a1 : and2 port map (abar, b, s2);
a2 : and2 port map (abar, c, s3);
a3 : and2 port map (b, c, s4);
r1 : or3 port map (s2, s3, s4, e);
end prob_6;
```

5. Construct a two-digit decade counter that counts from 0 to 99. Use the module of the decade counter in Listing 4.26. Write both the VHDL and Verilog codes. (Hint: use the terminal count, TC, to cascade the decade counters.)

6. Write VHDL description for the universal shift register discussed in Example 4.13.

7. Repeat Example 4.14 using D flip-flops.

8. Repeat Example 4.16 using D flip-flops.

9. Use `generate` and `parameter` to write a Verilog code for an n-bit subtractor.

CHAPTER 5

SWITCH-LEVEL DESCRIPTION

Chapter Objectives

- Understand the concept of describing and simulating digital systems using transistors
- Identify the basic statements of switch-level description in Verilog, such as `nmos`, `pmos`, `cmos`, `supply1`, `supply0`, `tranif0`, `tran`, and `tranif0`
- Develop a counterpart VHDL switch-level package that matches the switch-level functions of the Verilog description
- Review and understand the fundamentals of transistors and how they can be implemented as switches
- Review Boolean functions for combinational circuits

5.1 Highlights of Switch-Level Description

Highlights of the switch-level description can be summarized in the following facts.

Facts

- Switch-level description implements switches (transistors) to describe relatively small-scale digital systems.

- Switch-level description is usually implemented in very-large-scale integrated (VLSI) circuit layouts.

- Switch-level description is the lowest HDL logic level that can be used to simulate digital systems.

- Only small-scale systems can be simulated using pure switch-level description. If the system is not small, a huge number of switches are needed, which may render the simulation impractical.

- Switch-level description is routinely used along with other types of modeling to describe digital systems.

- The switches used in this chapter are assumed to be perfect; they are either open (high impedance) or closed (zero impedance).

- In contrast to Verilog, basic VHDL does not have built-in statements such as nmos, pmos, and cmos. To use these statements in VHDL, user-built packages must be developed or supplied by the vendor.

Before discussing the HDL code for transistor-level description, let's review some facts

5.2   Useful Definitions

- **MOS:** Metal oxide semiconductor.

- **N-type semiconductor:** The free carriers are negatively charged electrons.

- **P-type semiconductor:** The free carriers are positively charged holes.

- **Valence electrons:** Electrons in the outer shell of an atom that can interact with the valence electrons of another atom.

5.3   Single NMOS and PMOS Switches

Figure 5.1a shows a single **N-Channel MOS** (NMOS) switch, and Figure 5.1b shows a single **P-Channel MOS** (PMOS) switch. The switch has three signals: drain, gate, and source. If the gate is at logic 1, then the NMOS is closed (ON), and the PMOS is open (OFF). If the gate is at logic 0, then the NMOS is open (OFF), and the PMOS is closed (ON).

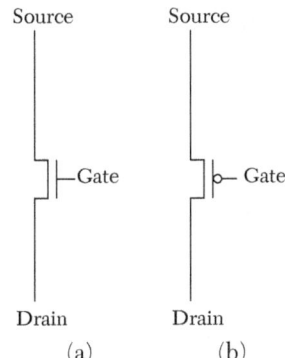

FIGURE 5.1 MOS switch. a) NMOS. b) PMOS.

5.3.1 Verilog Description of NMOS and PMOS Switches

Verilog has built-in code for NMOS and PMOS switches. In Verilog, there are four logical levels: 1, 0, X ("don't care"), and Z (high impedance). Table 5.1a shows the relationship between the drain, source, and gate of a NMOS switch, and Table 5.1b shows the same for a PMOS switch.

TABLE 5.1A Relationship Between Source, Drain, and Gate in NMOS Switches

		Gate			
		0	1	X	Z
	0	Z	0	L	L
Drain	1	Z	1	H	H
	X	Z	X	X	X
	Z	Z	Z	Z	Z

TABLE 5.1B Relationship Between Source, Drain, and Gate in PMOS Switches

		Gate			
		0	1	X	Z
	0	0	Z	L	L
Drain	1	1	Z	H	H
	X	X	Z	X	X
	Z	Z	Z	Z	Z

For an NMOS switch, the Verilog code is:

```
nmos n1 (drain, source, gate) //The switch name "n1" is optional.
```

The code can be written as:

```
nmos n1 (O1, I1, I2);
```

For the PMOS switch, the Verilog code is:

```
pmos p1 (drain, source, gate) //The switch name "p1" is optional.
```

or the code can be written as:

```
pmos p1 (O1, I1, I2);
```

5.3.2   VHDL Description of NMOS and PMOS Switches

Basic VHDL does not have built-in descriptions for NMOS or NMOS switches. Switches are built using behavioral description. Listing 5.1 shows the code, which does not include any consideration of delay times.

LISTING 5.1 VHDL Behavioral Code for NMOS and PMOS Switches

```
library IEEE;
use IEEE.STD_LOGIC_1164.ALL;

entity mos is
    Port (O1 : out std_logic; I1, I2 : in std_logic);
end mos;

architecture nmos_behavioral of mos is

-- All switches presented here do not include any
-- time parameters, such as rise time and fall time.
-- They only mimic the logical functions of their
-- Verilog counterparts.

begin
switch : process (I1, I2)
variable temp : std_logic;
begin
case I2 is
when '0'=> temp := 'Z';
when '1' => temp := I1;
when others => case I1 is
    when '0' => temp := 'L';
    when '1' => temp := 'H';
    when others => temp := I1;
```

```
    end case;
end case;
O1 <= temp;
end process switch;
end nmos_behavioral;
architecture pmos_behavioral of mos is

begin
switch : process (I1, I2)
variable temp : std_logic;
begin

case I2 is
when '1'=> temp := 'Z';
when '0' => temp := I1;
when others => case I1 is
    when '0' => temp := 'L';
    when '1' => temp := 'H';
    when others => temp := I1;
    end case;
end case;
O1 <= temp;
end process switch;
end pmos_behavioral;
```

To write the NMOS and PMOS codes as components, bind the entity of Listing 5.1 to a component statement. Listing 5.2 shows such binding.

LISTING 5.2 VHDL Code for NMOS and PMOS Switches as Components

```
architecture nmos of mos is
component nmos
port (O1 : out std_logic; I1, I2 : in std_logic);
end component;

component pmos
port (O1 : out std_logic; I1, I2 : in std_logic);
end component;

for all : pmos use entity work.mos (pmos_behavioral);
for all : nmos use entity work.mos (nmos_behavioral);
```

5.3.3   Serial and Parallel Combinations of Switches

Consider two NMOS switches connected in serial as shown in Figure 5.2a. Assume the gates g1 and g2 can only take logic 0 or logic 1. If g1 or g2 is at 0, then the path between y and d is open (OFF). If g1 and g2 are at 1, then the path between y and d is closed (ON), and y = d.

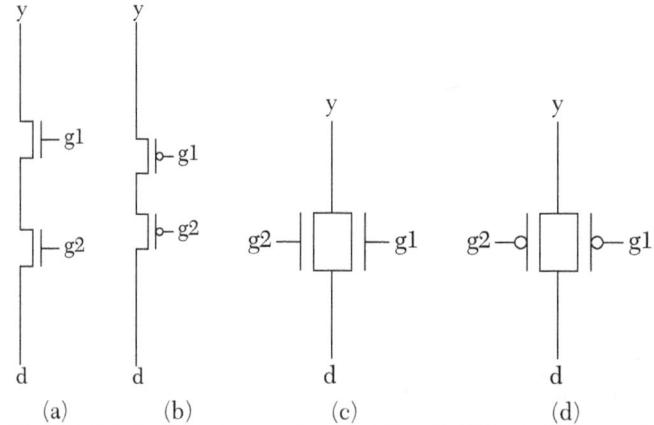

| (a) | (b) | (c) | (d) |

FIGURE 5.2 Combination of switches. a) Two NMOS switches in serial. b) Two PMOS switches in serial. c) Two NMOS switches in parallel. d) Two PMOS switches in parallel.

Table 5.2a summarizes the relationship between y, d, g1, and g2.

TABLE 5.2A Two NMOS Switches Connected in Serial (Figure 5.2a)

g1	g2	y
1	1	d
0	1	Z
1	0	Z
0	0	Z

Now, consider two PMOS switches connected in serial (Figure 5.2b). The path between y and d is closed (ON) only when both g1 and g2 are at 0; at this instant, y = d. The path is open (OFF) if g1 or g2 is at 1. Table 5.2b summarizes the relationship between y and d.

TABLE 5.2B Two PMOS Switches Connected in Serial (Figure 5.2b)

g1	g2	y
1	1	Z
0	1	Z
1	0	Z
0	0	d

When two NMOS switches are connected in parallel (Figure 5.2c), the path between y and d is open only when both g1 and g2 are 0. Otherwise, it is closed, and y = d, as shown in Table 5.2c

TABLE 5.2C Two NMOS Switches Connected in Parallel (Figure 5.2c)

g1	g2	y
1	1	d
0	1	d
1	0	d
0	0	Z

For two PMOS switches connected in parallel (Figure 5.2d), the path between y and d is open only when both g1 and g2 are at 1. Otherwise, it is closed, and y = d, as shown in Table 5.2d.

TABLE 5.2D Two PMOS Switches Connected in Parallel (Figure 5.2d)

g1	g2	y
1	1	Z
0	1	d
1	0	d
0	0	d

5.4 Switch-Level Description of Primitive Gates

This section describes the design of primitive gates from switches (transistors). The approach here is a simple one, but it may yield a greater number of transistors. Tables 5.2a–d are used to build the gate from a combination of switches. After constructing the switches, the code is written using Listings 5.1 and 5.2.

EXAMPLE 5.1 SWITCH-LEVEL DESCRIPTION OF AN INVERTER

The truth table of the inverter is shown in Table 5.3.

TABLE 5.3 Truth Table for an Inverter

Input	Output
a	y
0	1
1	0

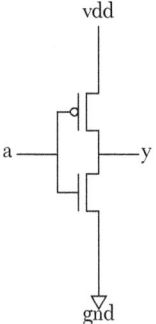

FIGURE 5.3 An inverter.

For any switch circuit, level 1 is represented by the power supply voltage (vdd), and level 0 is represented by the ground (gnd). To design the inverter, two complementary switches are needed: one to pull y down to 0 (gnd) and the other to pull y up to 1 (vdd). Figure 5.3 shows this connection.

The HDL code is shown in Listing 5.3, and the statement

```
pmos port map (y, vdd, a);
```

represents a PMOS switch with source y, drain vdd, and gate a. In the Verilog statements

```
supply1 vdd;
supply0 gnd;
```

`supply1` and `supply0` are predefined words that represent high voltage and ground, respectively. In VHDL, these two voltage levels are created by using `constant` declaration statements:

```
constant vdd : std_logic := '1';
constant gnd : std_logic := '0';
```

LISTING 5.3 HDL Code for an Inverter: VHDL and Verilog

VHDL Description
```
library IEEE;
use IEEE.STD_LOGIC_1164.ALL;

entity Inverter is
    port (y : out std_logic; a : in std_logic);
end Inverter;

architecture Invert_switch of Inverter is
component nmos
port (O1 : out std_logic; I1, I2 : in std_logic);
end component;

component pmos
port (O1 : out std_logic; I1, I2 : in std_logic);
end component;

for all : pmos use entity work.mos (pmos_behavioral);
for all : nmos use entity work.mos (nmos_behavioral);
constant vdd : std_logic := '1';
```

```
constant gnd : std_logic := '0';
begin
p1 : pmos port map (y, vdd, a);
n1 : nmos port map (y, gnd, a);
end Invert_switch;
```

Verilog Description
```
module invert (y, a);
input a;
output y;
supply1 vdd; /*supply1 is a predefined word for the
              high voltage.*/
supply0 gnd; /*supply0 is a predefined word for the
              ground.*/
pmos p1 (y, vdd, a); /*the name "p1" is optional; it
                      can be omitted.*/
nmos n1 (y, gnd, a); /*the name "n1" is optional; it can
                      be omitted. */
endmodule
```

EXAMPLE 5.2   SWITCH-LEVEL DESCRIPTION OF A TWO-INPUT AND GATE

In this example, a two-input AND gate is described. The truth table of the two-input AND gate is shown in Table 5.4.

TABLE 5.4 Truth Table for a Two-Input AND Gate

Input		Output
a	**b**	**y**
0	0	0
1	0	0
0	1	0
1	1	1

From Table 5.4, two switch combinations are needed: one to pull y up to vdd only when both gates of the combination are at level 1 (Table 5.2a satisfies this requirement), and another combination to pull y to ground whenever one of the gates is at level 0 (Table 5.2b satisfies this requirement). The final design is composed of two serial NMOS switches and two parallel PMOS switches. Figure 5.4 shows the switch-level

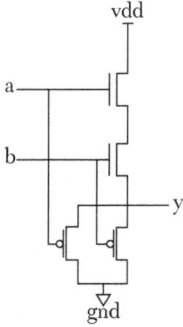

FIGURE 5.4 Switch-level logic diagram of an AND gate with week output.

diagram of the AND gate. The design here will yield to week output, see Example 5.6 for details.

From Figure 5.4, write the HDL code. Listing 5.4 shows the HDL code of a two-input AND gate.

LISTING 5.4 HDL Code for a Two-Input AND Gate: VHDL and Verilog

VHDL Description
```
library IEEE;
use IEEE.STD_LOGIC_1164.ALL;

entity and2gate is
    port (y : out std_logic; a, b : in std_logic);
end and2gate;

architecture and_switch of and2gate is
component nmos
port (O1 : out std_logic; I1, I2 : in std_logic);
end component;

component pmos
port (O1 : out std_logic; I1, I2 : in std_logic);
end component;

for all : pmos use entity work.mos (pmos_behavioral);
for all : nmos use entity work.mos (nmos_behavioral);
constant vdd : std_logic := '1';
constant gnd : std_logic := '0';
signal s1 : std_logic;
begin

n1 : nmos port map (s1, vdd, a);
n2 : nmos port map (y, s1, b);
p1 : pmos port map (y, gnd, a);
p2 : pmos port map (y, gnd, b);
end and_switch;
```

Verilog Description
```
module and2gate (y, a, b);
input a, b;
output y;
supply1 vdd;
```

```
supply0 gnd;

nmos (s1, vdd, a);
nmos (y, s1, b);
pmos (y, gnd, a);
pmos (y, gnd, b);
endmodule
```

As shown in Figure 5.4, the PMOS switches pull y down to ground level, and the NMOS switches pull y up to vdd level. This arrangement results in degraded output and should be avoided. When cascaded, degraded outputs can deteriorate the final outputs and render them unrecognizable. To generate strong outputs, the NMOS switches should pull the output down to ground, and the PMOS switches should pull the output up to vdd. To design a switch-level AND gate with strong output, a different approach should be followed (see Section 5.5).

EXAMPLE 5.3 SWITCH-LEVEL DESCRIPTION OF A TWO-INPUT OR GATE

In this example, a two-input OR gate is designed. The truth table of a two-input OR gate is shown in Table 5.5.

TABLE 5.5 Truth Table for a Two-Input OR Gate

Input		Output
a	b	y
0	0	0
1	0	1
0	1	1
1	1	1

From the table, notice that to design switch-level circuits for the OR gate, two complementary combinations are needed (see Table 5.2). The first combination pulls y down to ground level only when both gates are at level 0 (Table 5.2b satisfies this requirement). The second combination pulls y up to vdd when either g1 or g2 is at level 1 (Table 5.2c satisfies this requirement). The switch-level OR gate consists of two complementary combinations: two serial PMOS switches and two parallel NMOS switches. Figure 5.5 shows the switch-level diagram of a two-input OR gate.

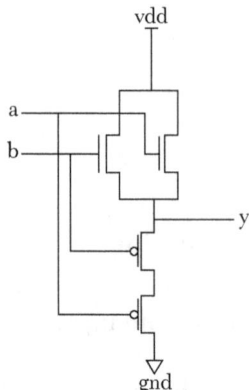

FIGURE 5.5 Switch-level logic diagram of an OR gate.

From Figure 5.5, the HDL code is written using the macros `pmos` and `nmos`. Listing 5.5 shows the HDL code of the switch-level two-input OR gate.

LISTING 5.5 HDL Code of a Two-Input OR Gate: VHDL and Verilog

VHDL Description
```
library IEEE;
use IEEE.STD_LOGIC_1164.ALL;

entity or2gate is
    port (y : out std_logic; a, b : in std_logic);
end or2gate;

architecture or_switch of or2gate is
component nmos
port (O1 : out std_logic; I1, I2 : in std_logic);
end component;

component pmos
port (O1 : out std_logic; I1, I2 : in std_logic);
end component;

for all : pmos use entity work.mos (pmos_behavioral);
for all : nmos use entity work.mos (nmos_behavioral);
constant vdd : std_logic := '1';
constant gnd : std_logic := '0';
signal s1 : std_logic;
begin
```

```
n1 : nmos port map (y, vdd, a);
n2 : nmos port map (y, vdd, b);
p1 : pmos port map (y, s1, a);
p2 : pmos port map (s1, gnd, b);
end or_switch;
```

Verilog Description
```
module OR2gate (a, b, y);

input a, b;
output y;

supply1 vdd;
supply0 gnd;

nmos (y, vdd, a);
nmos (y, vdd, b);
pmos (y, s1, a);
pmos (s1, gnd, b);
endmodule
```

As shown in Figure 5.5, the PMOS switches pull y down to ground level, and the NMOS switches pull y up to vdd level. This arrangement results in degraded outputs and should be avoided. Degraded outputs, when cascaded, can deteriorate the final outputs and render them unrecognizable. To generate strong outputs, the NMOS switches should pull the output down to ground, and the PMOS switches should pull the output up to vdd. If we want to design a switch-level OR gate with strong output, we should follow a different approach (see Section 5.5).

EXAMPLE 5.4   SWITCH-LEVEL DESCRIPTION OF A TWO-INPUT NAND GATE

In this example, a switch-level NAND two-input gate is designed. The truth table of the two-input NAND gate is shown in Table 5.6.

TABLE 5.6 Truth Table for a Two-Input NAND Gate

Input		Output
a	**b**	**y**
0	0	1
1	0	1
0	1	1
1	1	0

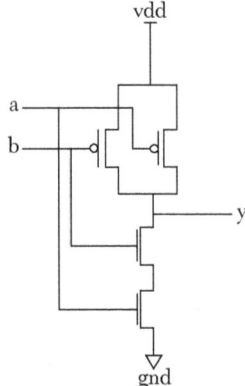

vdd

a

b

y

gnd

FIGURE 5.6 Switch-level logic diagram of an NAND gate.

Referring to Table 5.6, the gate can be designed using two complementary switch combinations (see Table 5.2). The first combination pulls y up to vdd when either of the two switch gates is at level 0 (Table 5.2d satisfies this requirement). The second combination pulls y down to ground level only when both gates are at level 1 (Table 5.2a satisfies this requirement). The final design consists of two complementary combinations: two parallel PMOS switches and two serial NMOS switches. Figure 5.6 shows the switch-level logic diagram of an NAND gate.

From Figure 5.6, write the HDL code. Listing 5.6 shows the HDL code of the NAND gate using the two macros pmos and nmos.

LISTING 5.6 HDL Code for a Two-Input NAND Gate: VHDL and Verilog

VHDL Description

```
library IEEE;
use IEEE.STD_LOGIC_1164.ALL;

entity nand2gate is
    port (y : out std_logic; a, b : in std_logic);
end nand2gate;

architecture nand_switch of nand2gate is
component nmos
port (O1 : out std_logic; I1, I2 : in std_logic);
end component;

component pmos
port (O1 : out std_logic; I1, I2 : in std_logic);
end component;

for all : pmos use entity work.mos (pmos_behavioral);
for all : nmos use entity work.mos (nmos_behavioral);
constant vdd : std_logic := '1';
constant gnd : std_logic := '0';
signal s1 : std_logic;
begin
n1 : nmos port map (s1, gnd, b);
n2 : nmos port map (y, s1, a);
```

```
p1 : pmos port map (y, vdd, a);
p2 : pmos port map (y, vdd, b);
end nand_switch;
```

Verilog Description
```
module NAND2gate (a, b, y);
input a, b;
output y;
supply1 vdd;
supply0 gnd;
nmos (s1, gnd, b);
nmos (y, s1, a);
pmos (y, vdd, a);
pmos (y, vdd, b);
endmodule
```

EXAMPLE 5.5 SWITCH-LEVEL DESCRIPTION OF A TWO-INPUT NOR GATE

Here, a switch-level two-input NOR gate is designed. The truth table of the two-input NOR gate is shown in Table 5.7.

TABLE 5.7 Truth Table for a Two-Input NOR Gate

Input		Output
a	b	y
0	0	1
1	0	0
0	1	0
1	1	0

Referring to Table 5.7, we can design the gate using two complementary switch combinations (see Table 5.2). The first combination pulls y up to vdd when the gate levels of both switches are at 0 (Table 5.2b satisfies this requirement). The second combination pulls y down to ground level when either switch gate is at level 1 (Table 5.2c satisfies this requirement). The final design consists of two complementary combinations: two serial PMOS switches and two parallel NMOS switches. Figure 5.7 shows the switch-level logic diagram of the NOR gate.

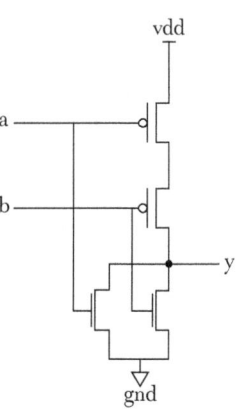

FIGURE 5.7 Switch-level logic diagram of an NOR gate.

From Figure 5.7, write the HDL code. Listing 5.7 shows the HDL code of the NOR gate using the two macros pmos and nmos.

LISTING 5.7 HDL Code for a Two-Input NOR Gate: VHDL and Verilog

VHDL Description

```
library IEEE;
use IEEE.STD_LOGIC_1164.ALL;

entity nor2gate is
    port (y : out std_logic; a, b : in std_logic );
end nor2gate;

architecture nor_switch of nor2gate is
component nmos
port (O1 : out std_logic; I1, I2 : in std_logic);
end component;

component pmos
port (O1 : out std_logic; I1, I2 : in std_logic);
end component;

for all : pmos use entity work.mos (pmos_behavioral);

for all : nmos use entity work.mos (nmos_behavioral);
constant vdd : std_logic := '1';
constant gnd : std_logic := '0';
signal s1 : std_logic;
begin

n1 : nmos port map (y, gnd, a);
n2 : nmos port map (y, gnd, b);
p1 : pmos port map (s1, vdd, a);
p2 : pmos port map (y, s1, b);
end nor_switch;
```

Verilog Description

```
module nor2gate (a, b, y);
input a, b;
output y;
supply1 vdd;
supply0 gnd;
nmos (y, gnd, a);
```

```
nmos (y, gnd, b);
pmos (s1, vdd, a);
pmos (y, s1, b);
endmodule
```

5.5   Switch-Level Description of Simple Combinational Logics

In this section, simple combinational circuits will be designed using single PMOS and NMOS switches. The same logic is implemented as in Section 5.4, where Table 5.2 was used to come up with switch-level logics. Unless otherwise mentioned, all switch-level circuits here are designed to produce *strong outputs* (i.e., the output is either the ground or the vdd).

EXAMPLE 5.6   SWITCH-LEVEL DESCRIPTION OF A TWO-INPUT AND GATE WITH STRONG OUTPUT

As mentioned in Section 5.4, to produce strong output, the NMOS switches should pull the output down to ground, and the PMOS should pull the output up to vdd. The design of NAND, invert, and NOR systems discussed in Section 5.4 satisfy this requirement. One approach is to convert the AND gate to a NAND and inverter. Figure 5.8 shows a switch-level logic diagram of an AND gate constructed from a NAND gate and an inverter.

Listing 5.8 shows the HDL code for the AND gate. The code is longer than that of Listing 5.4, but it should produce strong outputs.

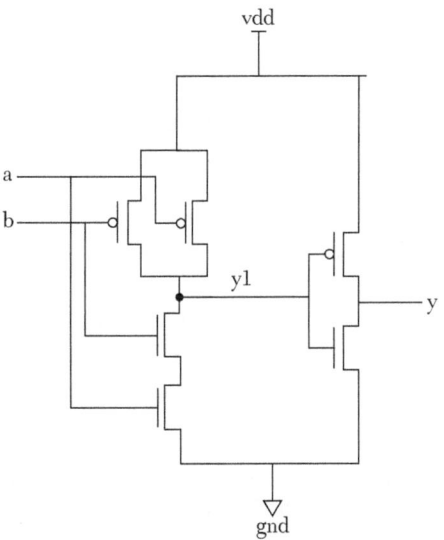

FIGURE 5.8 Switch-level logic diagram of an AND gate constructed from an NAND gate and an inverter.

LISTING 5.8 HDL Code for a Two-Input AND Gate with Strong Output: VHDL and Verilog

VHDL Description
```
library IEEE;
use IEEE.STD_LOGIC_1164.ALL;
```

```
entity and2Sgate is
    port (y : out std_logic; a, b : in std_logic);
end and2Sgate;
architecture and_strong of and2Sgate is

component nmos
port (O1 : out std_logic; I1, I2 : in std_logic);
end component;

component pmos
port (O1 : out std_logic; I1, I2 : in std_logic);
end component;
for all : pmos use entity work.mos (pmos_behavioral);
for all : nmos use entity work.mos (nmos_behavioral);
constant vdd : std_logic := '1';
constant gnd : std_logic := '0';
signal s1, y1 : std_logic;
begin

-- NAND
n1 : nmos port map (s1, gnd, b);
n2 : nmos port map (y1, s1, a);
p1 : pmos port map (y1, vdd, a);
p2 : pmos port map (y1, vdd, b);

-- Invert
n3 : nmos port map (y, gnd, y1);
p3 : pmos port map (y, vdd, y1);

end and_strong;
```

Verilog Description
```
module and2Sgate (a, b, y);

input a, b;
output y;
supply1 vdd;
supply0 gnd;
//NAND
nmos (s1, gnd, a);
nmos (y1, s1, b);
```

```
pmos (y1, vdd, a);
pmos (y1, vdd, b);

//inverter
nmos (y, gnd, y1);
pmos (y, vdd, y1);
endmodule
```

EXAMPLE 5.7   SWITCH-LEVEL DESCRIPTION OF A TWO-INPUT OR GATE WITH STRONG OUTPUT

As was done in Listing 5.8, to produce a strong output, the OR gate is changed to a NOR and inverter. The switch-level logic of both NOR and inverter use NMOS switches to pull the output down to ground level, and NMOS switches to pull the output up to vdd. This generates strong outputs that are not degraded. Figure 5.9 shows the switch-level logic diagram of an OR gate constructed from the NOR gate and inverter.

Listing 5.9 shows the HDL code for an OR gate constructed from the NOR gate and inverter. The code is longer than that of Listing 5.5, but it should produce strong outputs.

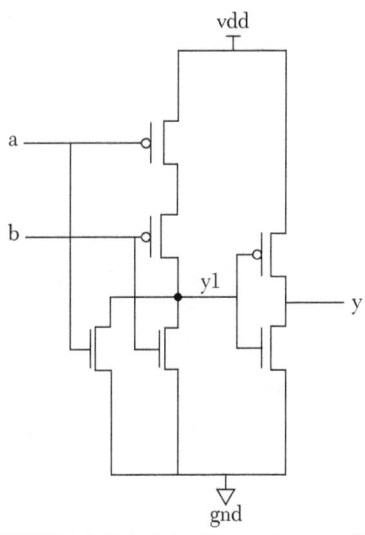

FIGURE 5.9 Switch-level logic diagram of an OR gate constructed from the NOR gate and inverter.

LISTING 5.9 HDL Code of a Two-Input OR Gate with Strong Output: VHDL and Verilog

VHDL Description
```
library IEEE;
use IEEE.STD_LOGIC_1164.ALL;
entity OR2Sgate is
    port (y : out std_logic; a, b : in std_logic);
end OR2Sgate;

architecture orgate_strong of OR2Sgate is
```

```
component nmos
    port (O1 : out std_logic; I1, I2 : in std_logic);
end component;

component pmos
port (O1 : out std_logic; I1, I2 : in std_logic);
end component;

for all : pmos use entity work.mos (pmos_behavioral);
for all : nmos use entity work.mos (nmos_behavioral);
constant vdd : std_logic := '1';
constant gnd : std_logic := '0';
signal s1, y1 : std_logic;
begin

--NOR
n1 : nmos port map (y1, gnd, a);
n2 : nmos port map (y1, gnd, b);
p1 : pmos port map (s1, vdd, a);
p2 : pmos port map (y1, s1, b);

--Invert
n3 : nmos port map (y, gnd, y1);
p3 : pmos port map (y, vdd, y1);

end orgate_strong;
```

Verilog Description
```
module OR2Sgate (a, b, y);

input a, b;
output y;
supply1 vdd;
supply0 gnd;

//NOR
nmos (y1, gnd, a);
nmos (y1, gnd, b);
pmos (s1, vdd, a);
pmos (y1, s1, b);
```

```
//inverter
nmos (y, gnd, y1);
pmos (y, vdd, y1);
endmodule
```

EXAMPLE 5.8 SWITCH-LEVEL DESCRIPTION OF A THREE-INPUT NAND GATE

Here, a three-input NAND gate is described. Table 5.8 shows the truth table of the three-input NAND gate.

TABLE 5.8 Truth Table for a Three-Input NAND Gate

Input			Output
a	**b**	**c**	**y**
0	0	0	1
1	0	0	1
0	1	0	1
1	1	0	1
1	0	1	1
1	0	1	1
1	1	0	1
1	1	1	0

As shown in Table 5.8, the output is 0 only when a, b, and c are 1. Table 5.2a, when extended to three switches, indicates the use of three NMOS switches connected in serial as the pull-down combination. For the pull-up combination, Table 5.2d, extended to three switches, needs three PMOS switches connected in parallel. Figure 5.10 shows the switch-level logic diagram for a three-input NAND gate.

Listing 5.10 shows the HDL code for the three-input NAND gate using pmos and nmos switches.

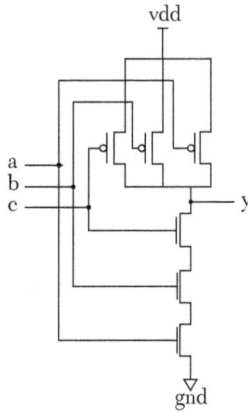

FIGURE 5.10 Switch-level logic diagram of a three-input NAND gate.

LISTING 5.10 HDL Code for a Three-Input NAND Gate: VHDL and Verilog

VHDL Description

```
library IEEE;
use IEEE.STD_LOGIC_1164.ALL;

entity nand3gate is
    port (y : out std_logic; a, b, c : in std_logic);
end nand3gate;

architecture nand3_switch of nand3gate is
component nmos
port (O1 : out std_logic; I1, I2 : in std_logic);
end component;

component pmos
port (O1 : out std_logic; I1, I2 : in std_logic);
end component;

for all : pmos use entity work.mos (pmos_behavioral);
for all : nmos use entity work.mos (nmos_behavioral);
constant vdd : std_logic := '1';
constant gnd : std_logic := '0';
signal s1, s2 : std_logic;
begin

n1 : nmos port map (s1, gnd, a);
n2 : nmos port map (s2, s1, b);
n3 : nmos port map (y, s2, c);
p1 : pmos port map (y, vdd, a);
p2 : pmos port map (y, vdd, b);
p3 : pmos port map (y, vdd, c);
end nand3_switch;
```

Verilog Description

```
module nand3gate (a, b, c, y);
input a, b, c;
output y;
supply1 vdd;
supply0 gnd;

nmos (s1, gnd, a);
nmos (s2, s1, b);
```

```
nmos (y, s2, c);
pmos (y, vdd, a);
pmos (y, vdd, b);
pmos (y, vdd, c);
endmodule
```

EXAMPLE 5.9   SWITCH-LEVEL DESCRIPTION OF A THREE-INPUT NOR GATE

In this example, a three-input NOR gate is described using switch-level description. Table 5.9 shows the truth table of the NOR gate.

TABLE 5.9 Truth Table for a Three-Input NOR Gate

Input			Output
a	b	c	y
0	0	0	1
1	0	0	0
0	1	0	0
1	1	0	0
0	0	1	0
1	0	1	0
0	1	0	0
1	1	1	0

Referring to Table 5.9, the NOR gate has an output of 1 only when a, b, and c are zeros. This is the opposite logic of NAND, so the serial combination of NMOS switches for the pull-down combination for the NAND is converted to a parallel combination of NMOS switches for the NOR gate. Similarly, the parallel combination of PMOS switches for the pull-up in the NAND gate is converted to a serial combination of PMOS switches in the NOR gate. Figure 5.11 shows the switch-level logic diagram of a three-input NOR gate.

Listing 5.11 shows the HDL code for the three-input NOR gate using `pmos` and `nmos` switches.

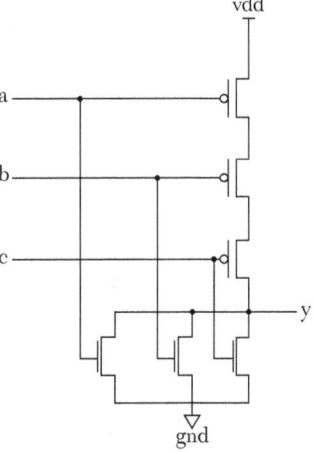

FIGURE 5.11 Switch-level logic diagram of a NOR gate.

LISTING 5.11 HDL Code for a Three-Input NOR Gate: VHDL and Verilog

VHDL Description

```vhdl
library IEEE;
use IEEE.STD_LOGIC_1164.ALL;
entity nor3gate is
    port (y : out std_logic; a, b, c : in std_logic);
end nor3gate;

architecture nor3_switch of nor3gate is

component nmos
port (O1 : out std_logic; I1, I2 : in std_logic);
end component;

component pmos
port (O1 : out std_logic; I1, I2 : in std_logic);
end component;

for all : pmos use entity work.mos (pmos_behavioral);
for all : nmos use entity work.mos (nmos_behavioral);
constant vdd : std_logic := '1';
constant gnd : std_logic := '0';
signal s1, s2 : std_logic;

begin
n1 : nmos port map (y, gnd, a);
n2 : nmos port map (y, gnd, b);
n3 : nmos port map (y, gnd, c);

p1: pmos port map (s1, vdd, a);
p2: pmos port map (s2, s1, b);
p3: pmos port map (y, s2, c);
end nor3_switch;
```

Verilog Description

```verilog
module nor3gate (a, b, c, y);
input a, b, c;
output y;
supply1 vdd;
supply0 gnd;
nmos (y, gnd, a);
```

```
nmos (y, gnd, b);
nmos (y, gnd, c);
pmos (s1, vdd, a);
pmos (s2, s1, b);
pmos (y, s2, c);
endmodule
```

EXAMPLE 5.10   SWITCH-LEVEL DESCRIPTION OF SIMPLE COMBINATIONAL LOGIC

This example discusses the switch-level description of the logic presented by the Boolean function $y = \overline{abc + de}$. A straightforward approach would be to treat the logic as a three-input NAND gate: two two-input NAND gates and an inverter (see Figure 5.12). The number of switches (transistors) for this combination is $6 + (2 \times 4) + 2 = 16$.

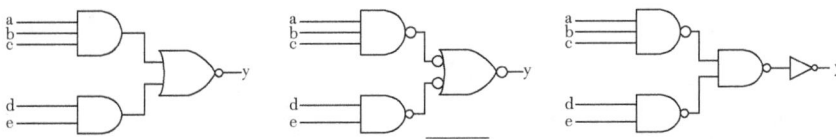

FIGURE 5.12 Gate-level logic diagram for $y = \overline{abc+de}$.

The number of transistors can be reduced by investigating the Boolean function. Note that y is pulled to zero only if abc = 1 or if de = 1. This means that the pull-down combination for abc is three NMOS switches driven by a, b, and c. The three switches are connected in serial. For de, two serial NMOS switches are driven by d and e; the two switches are connected in parallel with the three NMOS switches. The pull-up combination is three PMOS switches driven by a, b, and c connected in parallel; the combination is connected in serial with another two PMOS switches driven by d and e, accounting for the ORing of abc with de. The total number of transistors is 10, in contrast to 16 for the straightforward approach. Figure 5.13 shows the switch-level logic diagram.

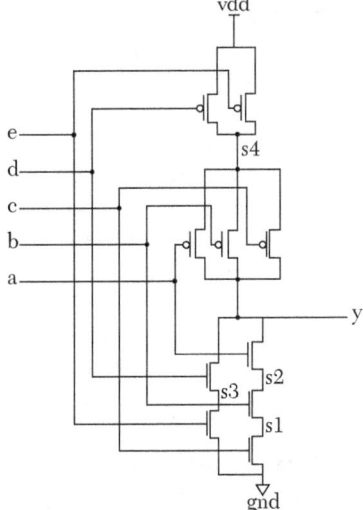

FIGURE 5.13 Switch-level logic diagram for $y = \overline{abc+de}$.

Listing 5.12 shows the HDL code for the logic using pmos and nmos switches.

LISTING 5.12 HDL Code for the Logic $y = \overline{abc+de}$: VHDL and Verilog

VHDL Description

```
library IEEE;
use IEEE.STD_LOGIC_1164.ALL;

    entity simple_logic is
port (y : out std_logic; a, b, c, d : in std_logic);
end simple_logic;

architecture ABC of simple_logic is
component nmos
port (O1 : out std_logic; I1, I2 : in std_logic);
end component;

component pmos
port (O1 : out std_logic; I1, I2 : in std_logic);
end component;

for all : pmos use entity work.mos (pmos_behavioral);
for all : nmos use entity work.mos (nmos_behavioral);
constant vdd : std_logic := '1';
constant gnd : std_logic := '0';
signal s1, s2, s3 : std_logic;

begin
n1 : nmos port map (s1, gnd, c);
n2 : nmos port map (s2, s1, b);
n3 : nmos port map (y, s2, a);
n4 : nmos port map (y, gnd, d);
p1 : pmos port map (y, s3, a);
p2 : pmos port map (y, s3, b);
p3 : pmos port map (y, s3, c);
p4 : pmos port map (s3, vdd, d);
end ABC;
```

Verilog Description

```
module simple_logic (a, b, c, d, e, y);
input a, b, c, d, e;
output y;
```

```
supply1 vdd;
supply0 gnd;
nmos (s1, gnd, c);
nmos (s2, s1, b);
nmos (y, s2, a);
nmos (s3, gnd, e);
nmos (y, s3, d);

pmos (y, s4, a);
pmos (y, s4, b);
pmos (y, s4, c);

pmos (s4, vdd, d);
pmos (s4, vdd, e);
endmodule
```

EXAMPLE 5.11   SWITCH-LEVEL DESCRIPTION OF A XNOR GATE

To satisfy the requirement that NMOS switches pull down to ground (pass 0) and PMOS switches pull up (pass 1) to vdd, the XNOR gate is treated as the inverse of an XOR gate, so the Boolean function of the XNOR gate can be written as:

$$y = \overline{a}\overline{b} + ab = (\overline{b\overline{a} + a\overline{b}}) \qquad (5.1)$$

According to the relationship in Table 5.1, the pull-down combination is active when $b\overline{a}$ or $a\overline{b}$ is equal to 1. For $b\overline{a}$, this is accomplished with two NMOS transistors (switches) connected in serial and driven by b and $\overline{a}$ (see Table 5.2a). The same is true for a $\overline{b}$; two transistors connected in serial are needed. For the OR, the two serial transistors are connected in parallel (see Table 5.2c). For the pull-up combination, the serial and parallel combinations in the pull-down are converted to parallel and serial, respectively. Figure 5.14 shows the transistor switch-level logic diagram of the XNOR gate.

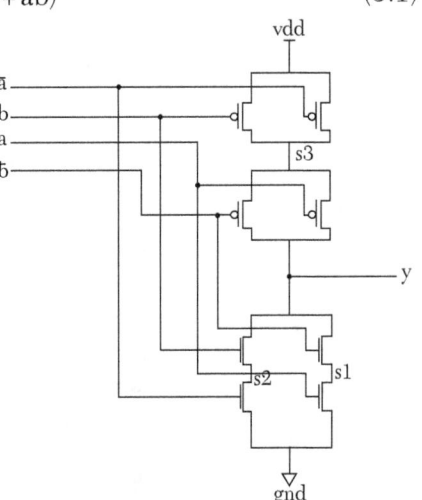

FIGURE 5.14 Switch-level logic diagram for a XNOR gate. Assume both input signal and its complement are available.

Listing 5.13 shows the HDL code for the XNOR gate. Two inverters are used to generate the inverse of a and b. If the true and complement logic of a and b are available, there is no need for the inverters.

LISTING 5.13 HDL Code for a XNOR Gate: VHDL and Verilog

VHDL Description

```
library IEEE;
use IEEE.STD_LOGIC_1164.ALL;
entity XOR_XNOR is
    port (y : out std_logic; a, b : in std_logic);
end XOR_XNOR;

architecture XNORgate of XOR_XNOR is
component nmos
port (O1 : out std_logic; I1, I2 : in std_logic);
end component;

component pmos
port (O1 : out std_logic; I1, I2 : in std_logic);
end component;

for all : pmos use entity work.mos (pmos_behavioral)
for all : nmos use entity work.mos (nmos_behavioral);
constant vdd : std_logic := '1';
constant gnd : std_logic := '0';
signal abar, bbar, s1, s2, s3 : std_logic;
begin

-- Invert a and b. If the complement of a and b
--are available, then there is no need for the
--following two pair (nmos and pmos) switches.
--

p1 : pmos port map (abar, vdd, a);
n1 : nmos port map (abar, gnd, a);
p2 : pmos port map (bbar, vdd, b);
n2 : nmos port map (bbar, gnd, b);

--Write the pull-down combination
n3 : nmos port map (s1, gnd, a);
n4 : nmos port map (y, s1, bbar);
n5 : nmos port map (s2, gnd, abar);
n6 : nmos port map (y, s2, b);
```

```
--Write the pull-up combination
p3 : pmos port map (y, s3, a);
p4 : pmos port map (y, s3, bbar);
p5 : pmos port map (s3, vdd, abar);
p6 : pmos port map (s3, vdd, b);

end XNORgate;
```

Verilog Description
```
module XOR_XNOR (a, b, y);

input a, b;
output y;

supply1 vdd;
supply0 gnd;

/* Invert a and b. If the complement of a and b
   are available, then there is no need for the
   following two pair (nmos and pmos) switches */

pmos (abar, vdd, a);
nmos (abar, gnd, a);
pmos (bbar, vdd, b);
nmos (bbar, gnd, b);

// Write the pull-down combination
nmos (s1, gnd, a);
nmos (y, s1, bbar);
nmos (s2, gnd, abar);
nmos (y, s2, b);

// Write the pull-up combination
pmos (y, s3, a);
pmos (y, s3, bbar);
pmos (s3, vdd, abar);
pmos (s3, vdd, b);
endmodule
```

XOR/XNOR gates are very important because they are the basic components in full adders. They have also been implemented in comparison

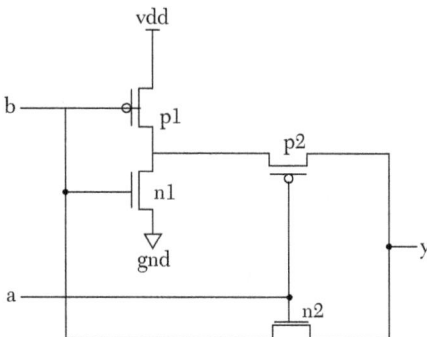

circuits. Several publications can provide more information on reducing the number of transistors used in XOR or XNOR gates (Wang 1995, Weste 2003). Figure 5.15 shows an XNOR gate with four transistors (Wang 1995). Figure 5.15 is based on the fact that the output of the XNOR gate is equal to $\overline{b}$ if a = 0 or is equal to b if a = 1. Listing 5.14 shows the code for such an XNOR gate.

FIGURE 5.15 Switch-level logic diagram for a XNOR gate.

LISTING 5.14 HDL Code for a XNOR Gate: VHDL and Verilog

VHDL Description
```
library IEEE;
use IEEE.STD_LOGIC_1164.ALL;
entity XNOR_degrade is
    Port (y : out std_logic; a, b : in std_logic);
end XNOR_degrade;

architecture XNORgate of XNOR_degrade is
component nmos
port (O1 : out std_logic; I1, I2 : in std_logic);
end component;

component pmos
port (O1 : out std_logic; I1, I2 : in std_logic);
end component;

for all : pmos use entity work.mos (pmos_behavioral
for all : nmos use entity work.mos (nmos_behavioral);
constant vdd : std_logic := '1';
constant gnd : std_logic := '0';
signal s0 : std_logic;

begin
p1 : pmos port map (s0, vdd, b);
p2 : pmos port map (y, s0, a);
n1 : nmos port map (s0, gnd, b);
n2 : nmos port map (y, b, a);
end XNORgate;
```

Verilog Description
```
module gate (a, b, y);

input a, b;
output y;

supply1 vdd;
supply0 gnd;

pmos p1 (s0, vdd, b);
pmos p2 (y, s0, a);

nmos n1 (s0, gnd, b);
nmos n2 (y, b, a);
endmodule
```

**EXAMPLE 5.12   SWITCH-LEVEL DESCRIPTION OF A 2x1 MULTIPLEXER
WITH ACTIVE HIGH ENABLE**

The Boolean function of such a multiplexer is as shown in Equation 5.2:

$$y = E(a \, Sel + b \, \overline{Sel})$$ (5.2)

E is the enable, a and b are the inputs, Sel is the select, and y is the output.
If E = 0, the multiplexer is disabled, and the output y is 0. If E = 1 and
Sel = 1, the output y is a; if E = 1 and Sel = 0, the output y = b.

As in Listing 5.14, the output is inversed to satisfy the requirement that
the NMOS switches pull down the output to ground level while the PMOS
switches pull up y to vdd level. Accordingly, the truth table of the multiplexer is as shown in Table 5.10.

TABLE 5.10 Truth Table for the Complement-Output Multiplexer

Input				Output
a	**b**	**Sel**	**E**	**$\overline{y}$**
x	x	x	0	1
0	0	0	1	1
1	0	0	1	1
0	1	0	1	0
1	1	0	1	0
0	0	1	1	1
1	0	1	1	0

(Contd.)

Input					Output
a	b	Sel	E		$\overline{y}$
0	1	1	1		1
1	1	1	1		0

From Table 5.10, note that $\overline{y}$ is 0 when Sel = 1 and a = 1 or when Sel = 0 and b = 1, so the pull-down combination is two NMOS switches in serial driven by a and Sel connected in parallel, with two serial switches driven by b and $\overline{Sel}$. When E = 0, $\overline{y}$ = 1; this is a PMOS (pull-up) switch. The switch level of the multiplexer is shown in Figure 5.16.

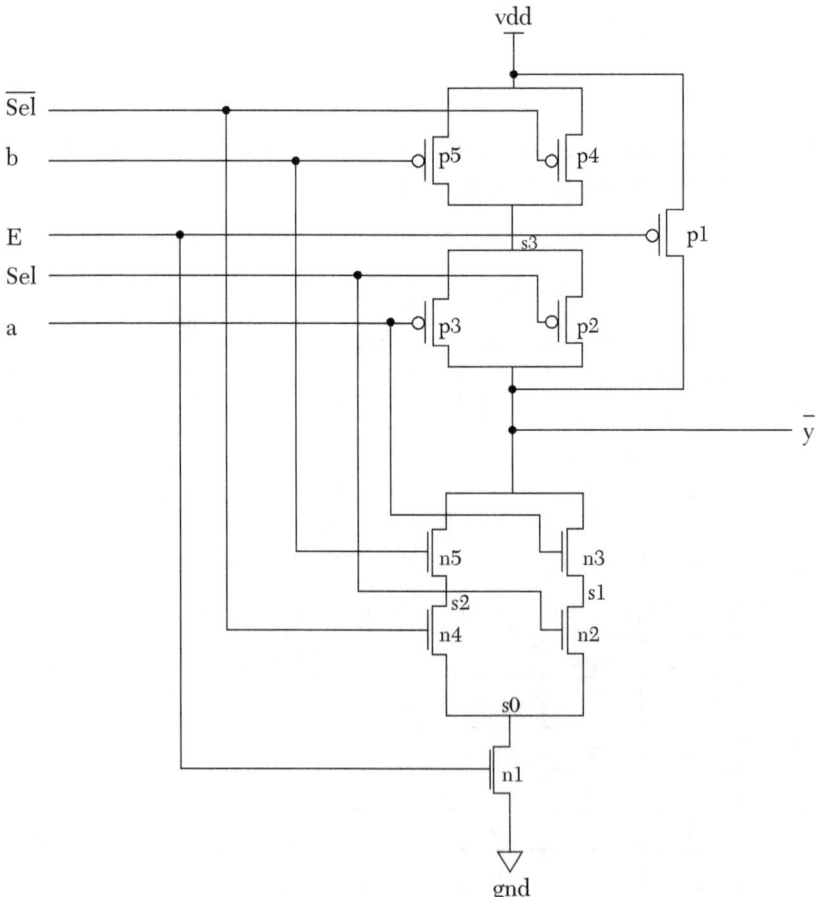

FIGURE 5.16 Switch-level logic diagram for a 2x1 multiplexer with active high enable and complement output. Assume $\overline{Sel}$ signal is available.

Listing 5.15 shows the HDL code for the multiplexer.

LISTING 5.15 HDL Code for a 2x1 Multiplexer with Active High Enable and Complement Output: VHDL and Verilog

VHDL Description

```
library IEEE;
use IEEE.STD_LOGIC_1164.ALL;
entity mux2x1 is
port (a, b, Sel, E : in std_logic;
      ybar : out std_logic);
end mux2x1;
architecture mux2x1switch of mux2x1 is
component nmos
port (O1 : out std_logic; I1, I2 : in std_logic);
end component;

component pmos
port (O1 : out std_logic; I1, I2 : in std_logic);
end component;

for all : pmos use entity work.mos (pmos_behavioral);
for all : nmos use entity work.mos (nmos_behavioral);
constant vdd : std_logic := '1';
constant gnd : std_logic := '0';
signal Selbar, s0, s1, s2, s3 : std_logic;

begin
--Invert signal Sel. If the complement of Sel is
--available then, there is no need for
--the following pair of transistors.

v1 : pmos port map (Selbar, vdd, Sel);
--All instantiation statements should be
--labeled
v2 : nmos port map (Selbar, gnd, Sel);
--Write the pull-down combination
n1 : nmos port map (s0, gnd, E);
n2 : nmos port map (s1, s0, Sel);
n3 : nmos port map (ybar, s1, a);
n4 : nmos port map (s2, s0, Selbar);
n5 : nmos port map (ybar, s2, b);
--Write the pull-up combination
p1 : pmos port map (ybar, vdd, E);
p2 : pmos port map (ybar, s3, Sel);
```

```
p3 : pmos port map (ybar, s3, a);
p4 : pmos port map (s3, vdd, Selbar);
p5 : pmos port map (s3, vdd, b);

end mux2x1switch;
```

Verilog Description
```
module mux2x1 (a, b, Sel, E, ybar);
input a, b, Sel, E;
output ybar;
supply1 vdd;
supply0 gnd;

/* Invert signal Sel. If the complement of Sel
   is available then, there is no need for
   the following pair of transistors */

pmos (Selbar, vdd, Sel);
nmos (Selbar, gnd, Sel);

//Write the pull-down combination
nmos n1 (s0, gnd, E);
nmos n2 (s1, s0, Sel);
nmos n3 (ybar, s1, a);
nmos n4 (s2, s0, Selbar);
nmos n5 (ybar, s2, b);

//Write the pull-up combination
pmos p1 (ybar, vdd, E);
pmos p2 (ybar, s3, Sel);
pmos p3 (ybar, s3, a);
pmos p4 (s3, vdd, Selbar);
pmos p5 (s3, vdd, b);

endmodule
```

5.6   Switch-Level Description of Simple Sequential Circuits

In Section 5.5, the switch-level description of combinational circuits was discussed. This chapter will cover description of some simple sequential circuits.

EXAMPLE 5.13   SWITCH-LEVEL DESCRIPTION OF AN SR-LATCH

The SR-latch was discussed in Chapter 4, and the gate-level logic diagram of the latch was shown in Figure 4.9. It is redrawn here for convenience in Figure 5.17.

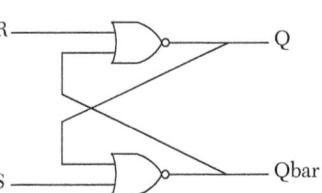

FIGURE 5.17 SR-latch.

As shown in Figure 5.17, the latch consists of two NOR gates. The switch-level logic diagram is designed directly from the gate-level diagram. A switch-level NOR gate was previously built in Listing 5.11. PMOS and NMOS switches are used to build the switch-level logic diagram of the latch (see Figure 5.18).

Listing 5.16 shows the HDL code of the latch. Because there are several transistors (switches), it is preferable to label the Verilog code for each transistor. For example, the Verilog code for switch n1 in Figure 5.18 is `nmos n1 (Qbar, gnd, S)`. The label n1 in Verilog is optional, but because there are several transistors, each transistor has been labeled. In VHDL, the labeling of each instantiation statement is required. For example, the instantiation statement of transistor n1 is `n1: nmos port map (Qbar, gnd, S)`. Label n1 is required. Figure 5.19 shows the simulation waveform of the multiplexer.

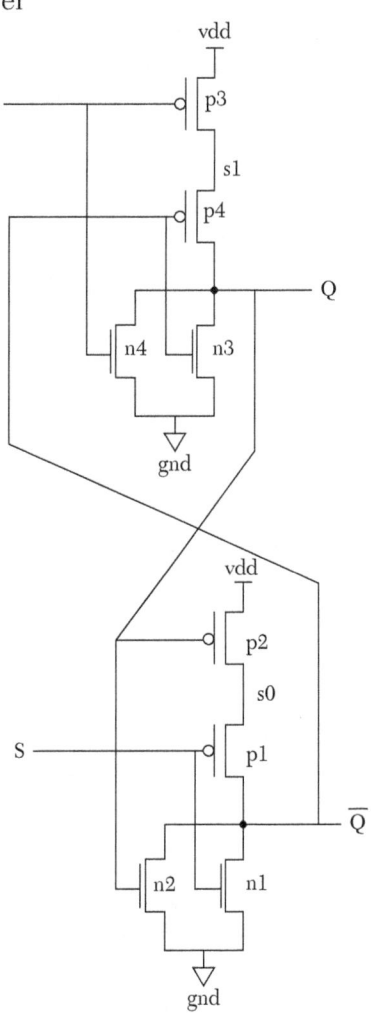

FIGURE 5.18 Switch-level logic diagram of an SR-latch.

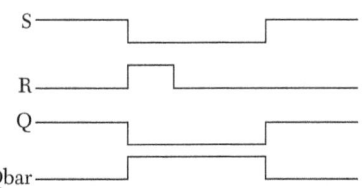

FIGURE 5.19 Simulation waveform of an SR-Latch.

LISTING 5.16 HDL Code for an SR-Latch: VHDL and Verilog

VHDL Description

```
library IEEE;
use IEEE.STD_LOGIC_1164.ALL;
entity SR_Latch is
port (S, R : in std_logic; Q, Qbar : inout std_logic);
end SR_Latch;
architecture SR of SR_Latch is

component nmos
port (O1 : inout std_logic; I1, I2 : in std_logic);
--port O1 is selected here to be inout to match
--its use in the latch circuit
end component;

component pmos
port (O1 : inout std_logic; I1, I2 : in std_logic);
--port O1 is selected here to be inout
--to match its use in the latch circuit
end component;

for all : pmos use entity work.mos (pmos_behavioral);
for all : nmos use entity work.mos (nmos_behavioral);
--In this example only, the mode of Output port O1
--in the entity "mos" should
--be declared as inout instead of out.
constant vdd : std_logic := '1';
constant gnd : std_logic := '0';
signal s0, s1, s2 : std_logic;

begin
n1 : nmos port map (Qbar, gnd, S);
n2 : nmos port map (Qbar, gnd, Q);
p1 : pmos port map (s0, vdd, Q);
p2 : pmos port map (Qbar, s0, S);

n3 : nmos port map (Q, gnd, Qbar);
n4 : nmos port map (Q, gnd, R);
p3 : pmos port map (s1, vdd, R);
p4 : pmos port map (Q, s1, Qbar);
end SR;
```

Verilog Description

```
module SR_latch (S, R, Q, Qbar);
input S, R;
output Q, Qbar;
supply1 vdd;
supply0 gnd;

nmos n1 (Qbar, gnd, S);
nmos n2 (Qbar, gnd, Q);
pmos p1 (s0, vdd, Q);
pmos p2 (Qbar, s0, S);

nmos n3 (Q, gnd, Qbar);
nmos n4 (Q, gnd, R);
pmos p3 (s1, vdd, R);
pmos p4 (Q, s1, Qbar);
endmodule
```

5.6.1 CMOS Switches

In the previous sections, single switches (NMOS or PMOS) were discussed. It has been seen that for a strong signal, the NMOS switch should pass 0, and the PMOS should pass 1. Another family of MOS switches is CMOS. As shown in Figure 5.20, the CMOS switch consists of two switches connected in parallel; one of the switches is NMOS, and the other is PMOS. The gate controls of the two switches, gn and gp, are usually the true and complement of the same signal. In this chapter, gn and gp are always the true and complement of the same signal, respectively. If gn is high (gp is low), the switch becomes conductive (output = input). If gn is low (gp is high), the switch becomes open. The main characteristic of the switch is that it can pass both strong 1 and strong 0.

Verilog has a built-in function to describe a CMOS switch. The following Verilog statement describes a CMOS switch with `input` and `output` and gates `gn` for the NMOS switch and `gp` for the PMOS switch (see Figure 20):

```
cmos (output, input, gn, gp)
```

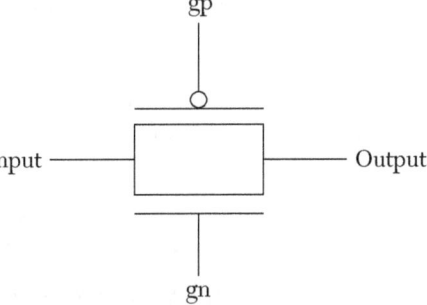

FIGURE 5.20 CMOS switch.

The basic VHDL package does not have a built-in code for CMOS switches, but it can be built as a macro consisting of the NMOS switch and PMOS switch described in Section 5.3.2. Listing 5.17 shows VHDL code for the CMOS switch shown in Figure 5.20.

LISTING 5.17 VHDL Code for the CMOS Switch in Figure 5.19.

```
library IEEE;
use IEEE.STD_LOGIC_1164.ALL;
entity CMOS is
port (output : out std_logic;
      input, gn, gp : in std_logic);
end CMOS;

architecture macro of CMOS is

--All switches presented here do not include any
--time parameters, such as
--rise time and fall times. They only mimic the
--logical functions of their
--Verilog counterparts.

component nmos
port (O1 : out std_logic; I1, I2 : in std_logic);
end component;

component pmos
port (O1 : out std_logic; I1, I2 : in std_logic);
end component;

for all : pmos use entity work.mos (pmos_behavioral);
for all : nmos use entity work.mos (nmos_behavioral);

begin
n1 : nmos port map (output, input, gn);
p1 : pmos port map (output, input, gp);
end macro;
```

EXAMPLE 5.14   SWITCH-LEVEL DESCRIPTION OF A D-LATCH

The D-latch was covered in Chapter 2. Here, no clear signal is considered; adding clear signal is covered in Exercise 5.9. Table 5.11 shows the excitation table the D-latch.

TABLE 5.11 Excitation Table for a D-Latch

Inputs			Next State
E	**D**	**Q**	**Q⁺**
0	x	0	0
0	x	1	1
1	0	x	0
1	1	x	1

The output follows D when the enable is high. When the enable is low, the output retains its previous value. The Boolean function of the output Q is:

$$Q = \overline{E}\, Q + ED \qquad (5.3)$$

Two approaches are taken to find the switch-level logic diagram of the latch. The first approach used is the Boolean function and NMOS and PMOS switches. The second approach uses CMOS switches.

5.6.1.1 Switch-Level Logic Diagram of a D-Latch Using PMOS and NMOS Switches

Equation 5.3 can be rewritten as:

$$\overline{Q} = \overline{(Q\overline{E} + ED)} \quad \text{and Q is the inverse of } \overline{Q}$$

Tables 5.2a–d are used to find the switch-level logic diagram. Figure 5.21 shows the switch-level logic diagram of the D-latch. Listing 5.18 shows the Verilog code for the latch.

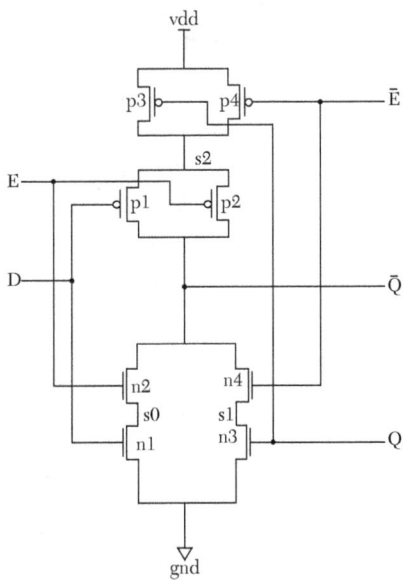

FIGURE 5.21 Switch-level logic diagram of a D-latch using PMOS and NMOS switches. Inverters between Q̄ and Q and between Ē and E are not shown.

LISTING 5.18 Verilog Code for a D-Latch Using NMOS and PMOS Switches

```
module D_latch (D, E, Q, Qbar);
input D, E;
output Q, Qbar;
supply1 vdd;
supply0 gnd;

pmos (Ebar, vdd, E);
nmos (Ebar, gnd, E);
nmos n1 (s0, gnd, D);
nmos n2 (Qbar, s0, E);
nmos n3 (s1, gnd, Q);
nmos n4 (Qbar, s1, Ebar);

pmos p1 (Qbar, s2, D);
pmos p2 (Qbar, s2, E);
pmos p3 (s2, vdd, Q);
pmos p4 (s2, vdd, Ebar);

endmodule
```

5.6.1.2 Switch-Level Logic Diagram of a D-Latch Using CMOS Switches

Figure 5.22 shows the switch-level logic diagram of the D-latch. When enable (E) is high, CMOS switch c1 is closed, CMOS switch c2 is opened, and Q follows D. When E is low, CMOS switch c1 is opened, CMOS switch c2 is closed, and Q retains its previous value.

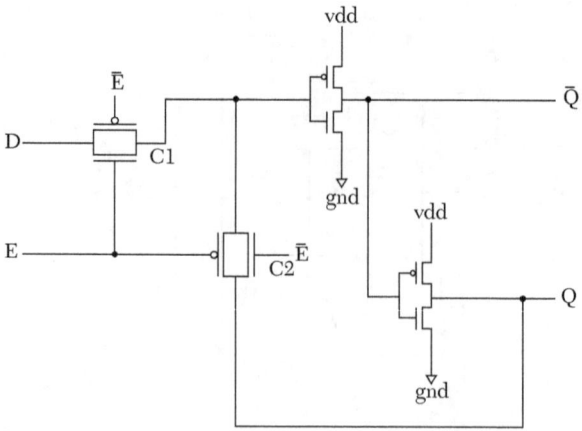

FIGURE 5.22 Switch-level logic diagram of a D-latch using CMOS switches.

Listing 5.19 shows the HDL code for the D-latch. Due to the nature of signal Q, where it is an input and output with more than one source (one CMOS switch and an inverter), Q is declared as `inout`. Because there are three inverters, the inverter module discussed in Listing 5.3 is bound to the current module `D-Latch`, rather than writing three individual inverters. In VHDL, use the statement:

```
for all : invert use entity work.
          inverter (Invert_switch);
```

to bind `inverter` to the current module `D_Latch`. In Verilog, we use the statement:

```
invert inv1 (Ebar, E);
```

which binds the module `invert` to the current module `D_Latch`.

LISTING 5.19 HDL Code for a D-Latch Using CMOS Switches: VHDL and Verilog

```
VHDL Description
library IEEE;
use IEEE.STD_LOGIC_1164.ALL;

entity D_Latch is
port (D, E : in std_logic; Q, Qbar : inout std_logic);
-- Referring to Figure 5.22, signal Q is
--input and output and has multiple
--sources (the inverter and the CMOS switches,
--so Q has to be declared as
--inout. All other ports are also adjusted in
--the following components to be inout.
end D_Latch;

architecture DlatchCmos of D_Latch is
component CMOS
port (output : out std_logic;
      input, gn, gp : in std_logic);
end component;
component invert
port (y : out std_logic; a : in std_logic );
end component;
for all : CMOS use entity work.CMOS (macro);
```

```
for all : invert use entity work.
         inverter( Invert_switch);
signal Ebar, s1 : std_logic;
begin
c1 : cmos port map (Q, D, E, Ebar);
c2 : cmos port map (Q, s1, Ebar, E);
inv1 : invert port map (Ebar, E);
inv2 : invert port map (Qbar, Q);
inv3 : invert port map (s1, Qbar);

end DlatchCmos;
```

Verilog Description
```
module D_latch (D, E, Q, Qbar);
input D, E;
output Q, Qbar;
cmos (Q, D, E, Ebar);
cmos (Q, s1, Ebar, E);
invert inv1 (Ebar, E);
invert inv2 (Qbar, Q);
invert inv3 (s1, Qbar);
endmodule

module invert (y, a);
input a;
output y;
supply1 vdd;
supply0 gnd;
pmos p1 (y, vdd, a);
nmos n1 (y, gnd, a);
endmodule
```

5.7 Bidirectional Switches

Bidirectional switches conduct in both ways, from drain to source and from source to drain. Their main use is as bidirectional buffers (busses). Three types of bidirectional switches are available in Verilog: tran, tranif0, and tranif1. Switch tran has no control; it conducts all the time. Switch tranif1 conducts if control is 1. Otherwise, the nondriving signal (output) is put on high impedance. Switch tranif0 conducts if control is 0. Otherwise, the nonconducting signal (output) is put on high impedance. The Verilog code for the three switches is as follows:

```
tran (dataio1, dataio2);
trannif0 (dataio1, dataio2, control);
tranif1 (dataio1, dataio2, control);
```

VHDL does not have built-in switches, but these switches can be built as in Section 5.3.2.

Listing 5.20 shows the same Verilog code in Listing 5.12, but `tranif1` and `tranif0` are used instead of NMOS and PMOS switches.

LISTING 5.20 HDL Code for the Logic $y = \overline{abc+de}$ ***: VHDL and Verilog***

```
module simple_logic (a, b, c, d, e, y);
input a, b, c, d, e;
output y;

supply1 vdd;
supply0 gnd;

tranif1 (s1, gnd, c);
tranif1 (s2, s1, b);
tranif1 (y, s2, a);
tranif1 (s3, gnd, e);
tranif1 (y, s3, d);

tranif0 (y, s4, a);
tranif0 (y, s4, b);
tranif0 (y, s4, c);

tranif0 (s4, vdd, d);
tranif0 (s4, vdd, e);

endmodule
```

5.8 Summary

In this chapter, HDL descriptions based on switches have been presented. The switches are built from perfect transistors. The transistor is either conducting to saturation or not conducting; this corresponds to two switch states, closed and open, respectively. Switch-level is the lowest level of HDL description. Verilog has an extensive switch-level description library. Standard VHDL does not have switch-level; if we use VHDL for switch-level description, packages have to built or imported from vendors.

VHDL switches have also been built as components (see Chapter 4). The power supply (vdd) and ground (gnd) in all systems covered in this chapter are the sources of the strongest 1s and 0s; nmos switches pass strong 0, and pmos switches pass strong 1. To produce strong output signals, nmos switches are employed as pull-down to ground networks, and pmos switches are employed as pull-up to vdd networks. Parallel and serial combinations of pmos and nmos switches have been implemented to describe combinational and sequential circuits. Other switches such as cmos, tran, tranif0, and tranif1, constructed from parallel combinations of NMOS and PMOS switches, have been also been discussed.

Many publications are available on the examples covered in this chapter. These publications may use innovative ways to reduce the number of transistors. The reader is encouraged to consult them (see References) if the main goal is to find a design with the minimum number of transistors.

5.9 Exercises

In all the following questions, unless otherwise mentioned, choose the design that yields strong outputs.

1. Derive the switch-level (transistor) logic of an XOR gate using a minimum number of transistors. Write and verify by simulation the VHDL code using PMOS and NMOS switches.

2. Without using the computer, inspect the Verilog code shown in Listing 5.21 and find the Boolean function of the output y. Translate the code to VHDL and verify your code by simulating it.

LISTING 5.21 Verilog Code for Exercise 5.2

```
module Problem_2 (a, b, c, y);
input a, b, c;
output y;

supply1 vdd;
supply0 gnd;

pmos (d, vdd, c);
nmos (d, gnd, c);
cmos (y, a, c, d);
```

```
cmos (y, b, d, c);

endmodule
```

3. For the XNOR gate discussed in Listing 5.13, use NAND gates and inverters to design the XNOR gate. Write the switch-level Verilog code and verify your design. Contrast this approach with that of Listing 5.13 in terms of the total number of transistors needed.

4. Referring to Listing 5.15, construct the gate level of the multiplexer using NOR gates and inverters. Write the switch-level VHDL code and verify your design. What is the total number of transistors used in this gate-level design?

5. Design the switch level for an SR-latch from the Boolean function using the minimum number of switches. Compare the number of switches used with that of Listing 5.15. Write the VHDL code and verify your design by simulation.

6. Write the VHDL code for Listing 5.18. Verify your code by simulation.

7. In Figure 5.22, the control E is active high. Modify the figure to show an active low enable.

8. Repeat Listing 5.15 using `tranif0` and `tranif1` instead of PMOS and NMOS switches.

9. Add active low clear signal to the D-Latch in Example 5.14 and rewrite the VHDL and Verilog codes.

5.10   References

Wang, J., S. Fang, and W. Feng. 1995. New Efficient Designs for XOR and XNOR Functions on the Transistor Level. *IEEE Journal of Solid State Circuits* 29:780–786.

Weste, Neil H. E. and D. Harris, *CMOS VLSI Design*, 3rd ed. Upper Saddle River, NJ: Addison-Wesley, 2004.

CHAPTER 6

PROCEDURES, TASKS, AND FUNCTIONS

Chapter Objectives

- Understand the concept of procedures (VHDL), tasks (Verilog), and functions (both VHDL and Verilog)
- Review and understand how to convert between different types of data
- Review signed vector multiplication
- Understand combinational arrays multiplier
- Review IEEE 754 representation of floating point
- Understand a simple enzyme mechanism

6.1 Highlights of Procedures, Tasks, and Functions

Facts

- Procedures, tasks, and functions are HDL tools to optimize the writing style of HDL code. They are implemented to instantiate a segment or a construct of code. Instead of writing the segment/construct every time it is needed, a single call statement to a function, task, or procedure that references the segment/construct is all that is needed.

- Procedures and tasks can have more than one input and more than one output. Functions have a single output, but they can have more than one input.

- Procedures and functions in VHDL can be called only from within `process`. Tasks and functions in Verilog can be called only from within `always` or `initial`.

6.2   Procedures and Tasks

Procedures (VHDL) and tasks (Verilog) are similar to subroutines in other software languages such as C. In many modules, a routine is repeatedly used, such as a multiplication algorithm, addition algorithm, or a conversion between two numbering systems. Instead of writing these routines every time they are needed, the routines' codes can be stored as the body of a `procedure` (VHDL) or as the body of a `task` (Verilog). Whenever the routine needs to be executed, the procedure (task) is called by writing just one call statement. Section 6.2.1 discusses procedures, and Section 6.2.2 discusses tasks.

6.2.1   Procedure (VHDL)

Procedure is a behavioral statement (see Chapter 3). A procedure has two parts: the declaration and the body. The declaration includes the name of the procedure, the inputs to the procedure and their types, and the outputs of the procedure and their types. For example, the declaration:

```
procedure Booth (X, Y : in signed (3 downto 0);
                 Z: out signed (7 downto 0)) is
```

declares a procedure by the name (identifier) `Booth`. The inputs are variables `X` and `Y`, each is four bits, and the type of the inputs is `signed`. The output is a four-bit variable `Z`, and its type is `signed`. In the declaration statement, `procedure` and `is` are predefined words and have to be inserted in the order shown. If the inputs or outputs are signals, they should be explicitly specified as follows:

```
procedure exmple (signal a : in std_logic ;
                  signal y: out std_logic) is
```

The body of the procedure contains the behavioral statements that describe the details of the procedure, mainly the relationship between the input(s) and the output(s). The body of the procedure cannot include the behavioral statement `process`. An example of a procedure is:

```
procedure exmple (signal a : in std_logic;
                  signal y : out std_logic) is
```

```
variable x : std_logic;
begin
x := a;
case x is
.....................
end case;
y <= x;
end exmple;
```

The procedure is called by a sequential statement that appears inside `process`. For example, the above procedure `exmple` is called as follows:

```
process (d, clk)
begin
......
exmple (d, z);
.........
end process
```

The input of the procedure `a` is linked to `d`, and accordingly, `d` assumes the value of `a`. The type of `a` should match the type of `d`. After execution of the procedure, the output of the procedure, `y`, is passed to `z`. The type of `z` should match the type of `y`. If a vector is an output or input of a procedure, it should not be constrained in length. Consider the declaration of procedure `Vect_constr`:

```
procedure Vect_constr (X : in std_logic_vector;
                       Y : out std_logic_vector) is
```

The length of vectors `X` and `Y` should not be constrained (i.e., they should not be specified). VHDL has a large number of built-in procedures in its standard package. Other procedures can be imported from external packages. An example of a built-in procedure is `open file` (see Chapter 8).

6.2.2 Task (Verilog)

Task is a Verilog subprogram. It can be implemented to execute specified routines repeatedly. The format in which the task is written can be divided into two parts: the declaration and the body of the task. In the declaration, the name of the task is specified, and the outputs and inputs of the task are listed. An example of a task declaration is:

```
task addr;
output cc, dd;
input aa, bb;
```

addr is the name (identifier) of the task. The outputs are cc and dd, and the inputs are aa and bb. task is a predefined word. The body of the task shows the relationship between the outputs and inputs. An example of the body of a task is:

```
begin
cc = aa ^ bb;
. . . . . . . . . . . .
end
endtask
```

The body of the task cannot include always or initial. A task must be called within the behavioral statement always or initial (see Chapter 3). An example of calling the task addr is:

```
. . . . . . . . . . . .
always @ (a, b)
begin
addr (c, d, a, b);
end
```

addr is the name of the task. Inputs a and b are passed to aa and bb. The outputs of the task cc and dd are passed, after execution, to c and d, respectively. Verilog has a large number of built-in tasks included in its package.

6.2.3 Examples: Procedures and Tasks

The following examples discuss procedures and tasks. Some of the examples have been covered in the previous chapters (2–4). Here, they are rewritten using procedure and task.

EXAMPLE 6.1 **HDL BEHAVIORAL DESCRIPTION OF A FULL ADDER USING PROCEDURE AND TASK**

In Chapter 4, a full adder was constructed from two half adders. Here, the same concept is used to design a full adder from two half adders using behavioral description. The code for a half adder is written using procedure in VHDL or task in Verilog to construct the full adder. Figure 6.1 shows a block diagram of the full adder.

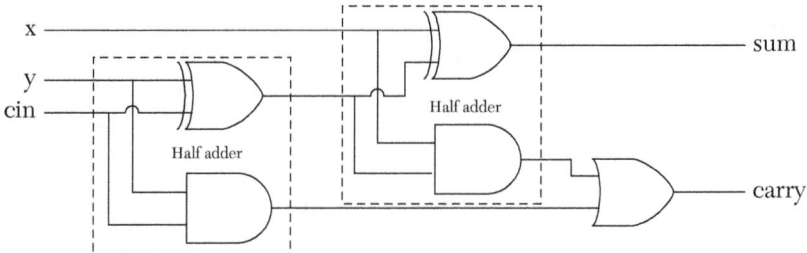

FIGURE 6.1 Block diagram of a full adder as two half adders.

Listing 6.1 shows the HDL code for the full adder using `procedure` (VHDL) and `task` (Verilog). Referring to the VHDL Listing, the code for the half adder is written as a procedure:

```
procedure Haddr(sh, ch : out std_logic;
ah, bh : in std_logic) is
begin
sh := ah xor bh;
ch := ah and bh;
end Haddr;
```

The name of the above procedure is `Haddr`; the inputs are `sh` and `ch`, and the outputs are `ah` and `bh`. The type of the outputs and inputs is `std_logic`. The code of the procedure is based on the Boolean functions of the half adder. To call the procedure, the call statement has to be inside `process`. To call the procedure `Haddr`:

```
Haddr (sum1, c1, y, cin);
```

where `Haddr` is the name of the procedure, and the values `y` and `cin` are passed to the inputs of the procedure `ah` and `bh`, respectively. After calculating the outputs (`sh` and `ch`), the procedure passes the value of those outputs to `sum1` and `c1`, respectively.

For the Verilog code, the task is:

```
task Haddr;
output sh, ch;
input ah, bh;
begin
    sh = ah ^ bh;
    ch = ah & bh;
end
endtask
```

The name of the task is `Haddr`; the inputs are `ah` and `bh`, and the outputs are `sh` and `ch`. The task code is based on the Boolean functions of the half adder. To call `task`, it has to be inside `always` or `initial` because `task` is a behavioral (sequential) statement. Therefore, to call the task `Haddr`:

```
Haddr (sum1, c1, y, cin);
```

where `haddr` is the name of the task. The values `y` and `cin` are passed to the inputs of the task `ah` and `bh`, respectively. After calculating the outputs (`sh` and `ch`), the task passes the value of the outputs to `sum1` and `c1`, respectively.

LISTING 6.1 HDL Description of a Full Adder Using `procedure` and `task`: VHDL and Verilog

VHDL Description

```
library IEEE;
use IEEE.STD_LOGIC_1164.ALL;

entity full_add is
port (x, y, cin : in std_logic; sum, cout : out std_logic);
end full_add;

architecture two_halfs of full_add is

-- The full adder is built from two half adders
procedure Haddr (sh, ch : out std_logic;
ah, bh : in std_logic) is
--This procedure describes a half adder
begin
sh := ah xor bh;
ch := ah and bh;
end Haddr;

begin

addfull : process (x, y, cin)
variable sum1, c1, c2, tem1, tem2 : std_logic;
begin
    Haddr (sum1, c1, y, cin);
    Haddr (tem1, c2, sum1, x);
    --The above two statements are calls to
    --the procedure Haddr
```

```
    tem2 := c1 or c2;
    sum <= tem1;
    cout <= tem2;
end process;

end two_halfs;
```

Verilog Description

```
module Full_add (x, y, cin, sum, cout);
//The full adder is built from two half adders
input x, y, cin;
output sum, cout;
reg sum, sum1, c1, c2, cout;
always @ (x, y, cin)
begin

Haddr (sum1, c1, y, cin);
Haddr (sum, c2, sum1, x);
//The above two statements are calls to the task Haddr.
cout = c1 | c2;
end

task Haddr;
//This task describes the half adder
output sh, ch;
input ah, bh;
begin
    sh = ah ^ bh;
    ch = ah & bh;
end
endtask
endmodule
```

Notice that the half adder procedure (task) was executed twice, but the body of the procedure (task) was only written once.

EXAMPLE 6.2 HDL DESCRIPTION OF AN *N*-BIT RIPPLE-CARRY ADDER USING PROCEDURE AND TASK

In Listing 4.14 of Chapter 4, the HDL structural description of a three-bit ripple-carry adder was introduced. Here, the behavioral code for an

N-bit ripple-carry adder is written using procedure (VHDL) and task (Verilog). Listing 6.2 shows the HDL code of the adder. The procedure (task) Faddr describes a one-bit full adder. To describe an N-bit adder, the procedure (task) is called N times.

LISTING 6.2 HDL Description of an N-Bit Ripple-Carry Adder Using procedure and task: VHDL and Verilog

VHDL N-Bit Ripple-Carry Adder Using procedure

```
library IEEE;
use IEEE.STD_LOGIC_1164.ALL;

entity adder_ripple is
generic (N : integer := 3);
    port (x, y : in std_logic_vector (N downto 0);
          cin : in std_logic;

          sum : out std_logic_vector (N downto 0);
          cout : out std_logic);
end adder_ripple;

architecture adder of adder_ripple is
procedure Faddr (sf, cof : out std_logic;
                 af, bf, cinf : in std_logic) is
--This procedure describes a full adder
begin
sf := af xor bf xor cinf;
cof := (af and bf) or (af and cinf) or (bf and cinf);
end Faddr;

begin
addrpl : process (x, y, cin)
variable c1, c2, tem1, tem2 : std_logic;
variable cint : std_logic_vector (N+1 downto 0);
variable sum1 : std_logic_vector (N downto 0);
begin
cint(0) := cin;
for i in 0 to N loop
    Faddr (sum1(i), cint(i+1), x(i), y(i), cint(i));
    --The above statement is a call to the procedure Faddr
end loop;
sum <= sum1;
```

```
cout <= cint(N+1);
end process;
end adder;
```

Verilog N-Bit Ripple-Carry Adder Using task
```
module adder_ripple (x, y, cin, sum, cout);
parameter N = 3;
input [N:0] x, y;
input cin;
output [N:0] sum;
output cout;

reg [N+1:0] cint;
reg [N:0] sum;
reg cout;
integer i;
always @ (x, y, cin)
begin
    cint[0] = cin;
    for (i = 0; i <= N; i = i + 1)
    begin
        Faddr (sum[i], cint[i+1], x[i], y[i], cint[i]);
        //The above statement is a call to task Faddr
end
cout = cint[N+1];
end
task Faddr;
//The task describes a full adder
output sf, cof;
input af, bf, cinf;
begin
    sf = af ^ bf ^ cinf;
    cof = (af & bf) | (af & cinf) | (bf & cinf);
end
endtask
endmodule
```

EXAMPLE 6.3 UNSIGNED BINARY-VECTOR-TO-INTEGER CONVERSION USING PROCEDURE AND TASK

Because VHDL is known to be a strict data-type-oriented language, conversion between data types, such as integer and binary, is important.

Verilog, however, is flexible when dealing with data types and allows computational operations between different types without any conversion. However, to understand `task`, Verilog code will be used to describe conversion examples.

In Chapter 3, behavioral code was written for conversions between binary and integer data. Here, `procedure (task)` is used to perform the conversion.

Listing 6.3 shows the HDL code for converting an unsigned binary vector to an integer. The conversion is based on accumulating the weighted sum of the binary bits. The code

```
result := result + 2**i; --VHDL
int = int + 2**i        //Verilog
```

performs the accumulation. To create a global constant N that represents the number of binary bits to be converted write the following code:

```
generic (N : integer := 3); --VHDL
parameter N = 3;             //Verilog
```

In Listing 6.3, $N = 3$ is used as an example; the number of bits can be changed just by changing the value of N. The statement

```
for i in bin'Range loop
```

has an index `i` with a range equal to that of `bin`. `Range` is a predefined attribute.

In addition to conversion, the procedure (task) outputs a flag (Z): Z = 1 if the value of the binary vector is zero. Otherwise, Z = 0.

LISTING 6.3 HDL Code for Converting an Unsigned Binary to an Integer Using `procedure` and `task`: VHDL and Verilog

```
VHDL: Converting an Unsigned Binary to an Integer Using proce-
dure
library ieee;
use ieee.std_logic_1164.all;
use ieee.numeric_std.all; --This Library is for
-- type "unsigned"

entity Bin_Int is
generic (N : natural := 3);
```

```
    port (X_bin : unsigned (N downto 0);
        Y_int : out natural; Z : out std_logic);
        --Y is always positive
end Bin_Int;

architecture convert of Bin_Int is

procedure bti (bin : in unsigned; int : out natural;
               signal Z : out std_logic) is

-- the procedure bti is to change binary to integer
-- Flag Z is chosen to be a signal rather than a variable
-- Since the binary vector is always positive,
-- use type natural for the output of the procedure.
variable result : natural;
begin

result := 0;
for i in bin'Range loop

--bin'Range represents the range of the unsigned vector bin
--Range is a predefined attribute
    if bin(i) = '1' then
    result := result + 2**i;
    end if;
end loop;
int := result;
if (result = 0) then
    Z <= '1';
    else
    Z <= '0';
    end if;
end bti;

begin
process (X_bin)
variable tem : natural;

begin
bti (X_bin, tem, Z);
Y_int <= tem;
end process;
end convert;
```

Verilog: Converting an Unsigned Binary to an Integer Using task
```verilog
module Bin_Int (X_bin, Y_int, Z);
parameter N = 3;
input [N:0] X_bin;
output integer Y_int;
output Z;
reg Z;
always @ (X_bin)
begin
    bti (Y_int, Z, N, X_bin);
end

task      bti;
parameter P = N;
output integer int;
output Z;
input N;
input [P:0] bin;
integer i, result;
begin
    int = 0;
//change binary to integer
    for (i = 0; i <= P; i = i + 1)
    begin
        if (bin[i] == 1)
        int = int + 2**i;
    end
    if (int == 0)
    Z = 1'b1;
else
Z = 1'b0;
end
endtask
endmodule
```

The simulation output for Listing 6.3 is shown in Figure 6.2

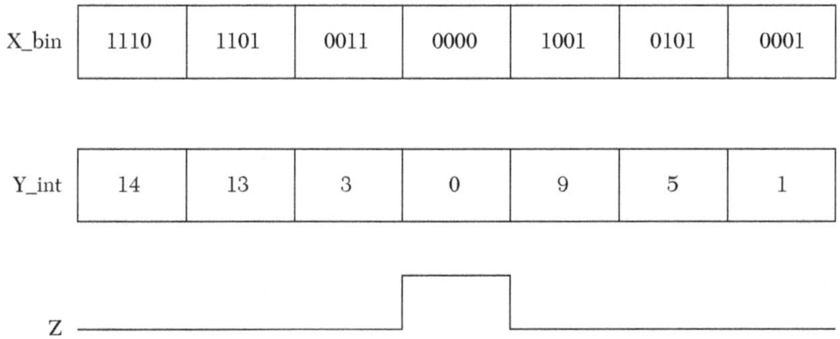

X_bin	1110	1101	0011	0000	1001	0101	0001

Y_int	14	13	3	0	9	5	1

Z

FIGURE 6.2 Simulation output for binary-to-integer conversion.

EXAMPLE 6.4 FRACTION-TO-REAL CONVERSION USING PROCEDURE AND TASK

Here, a fraction is represented as a fixed-point number where the binary point is at the left of the most significant bit. Examples of such binary numbers are 0.11 (equivalent to $2^{-1} + 2^{-2} = 0.5 + 0.25 = 0.75$) and 0.001 (equivalent to $2^{-3} = 0.125$). Listing 6.4 shows the HDL code for converting the binary using procedure and task by multiplying each bit by its weight. The first leftmost bit has a weight of 2^{-1}; the next bit to the right has a weight of 2^{-2}, and so on.

LISTING 6.4 HDL Code for Converting a Fraction Binary to Real Using procedure and task: VHDL and Verilog

```
VHDL: Converting a Fraction Binary to Real Using procedure
library IEEE;
use IEEE.STD_LOGIC_1164.ALL;
entity Bin_real is
generic (N : integer := 3);
port (X_bin : in std_logic_vector (0 to N); Y : out real);
end Bin_real;
architecture Bin_real of Bin_real is
procedure binfloat (a : in std_logic_vector (0 to 3);
float : out real) is

--This procedure converts fraction expressed
--in fixed-point binary to real
variable tem,j : real;
begin
    tem := 0.0;
```

```
      j := 1.0;
   for i in 0 to N loop
     j := j/ 2.0;

        if (a(i) = '1') then

        tem := tem + j;
        end if;
     end loop;
float := tem;
end binfloat;
begin


rel : process (X_bin)
variable temp : real;
begin
    binfloat (X_bin, temp);
    Y <= temp;
end process rel;
end Bin_real;
```

Verilog: Converting a Fraction Binary to Real Using task

```
module Bin_real (X_bin);
parameter N = 3;
input [N:0] X_bin;
real Z;

always @ (X_bin)
begin
    binfloat (X_bin, Z);
end

task binfloat;
parameter P = N;
input [0:P] a;
output real float;
integer i;
begin
    float = 0.0;
    for (i = 0; i <= P; i = i + 1)
    begin
        if (a[i] == 1)
        float = float + 1.0 / 2**(i+1);
```

```
// The above statement multiplies each bit by its weight.
    //
    end
end
endtask
endmodule
```
The simulation output is shown in Figure 6.3.

X_bin	1000	0100	0010	0001	0101	0001

Z	0.5	0.25	0.125	0.0625	0.3125	0.75

FIGURE 6.3 Simulation output for fraction binary conversion to real.

EXAMPLE 6.5 UNSIGNED INTEGER CONVERSION TO BINARY USING PROCEDURE AND TASK

In this example, integer-type data is converted to binary-type data. As was done in Chapter 3, the integer is successively divided by two to find the equivalent binary. The mod function is used to find the remainder of the division by two. Listing 6.5 shows the HDL code for converting an integer to binary using procedure and task. The code also checks to see if the integer is even or odd. The even_flag in Listing 6.5 equals one when the integer is even; if even _flag is zero, then the integer is odd.

LISTING 6.5 HDL Code for Converting an Unsigned Integer to Binary Using procedure and task: VHDL and Verilog

VHDL: Converting an Unsigned Integer to Binary Using procedure
```
library IEEE;
use IEEE.STD_LOGIC_1164.ALL;

entity Int_Bin is
generic (N : integer := 3);
port (X_bin : out std_logic_vector (N downto 0);
      Y_int : in integer;
    flag_even : out std_logic);
end Int_Bin;

architecture convert of Int_Bin is
```

```
procedure itb (bin : out std_logic_vector;
               signal flag : out std_logic;
               N : in integer; int : inout integer) is

-- The procedure itb is to convert the integer to binary
-- The dimension of bin does not have to be specified
-- at the above declaration statement; the procedure
-- can determine the dimension of bin later in its body.

begin
if (int MOD 2 = 0) then
--The above statement checks int to see if it is even.
    flag <= '1';
    else
    flag <= '0';
end if;
for i in 0 to N loop

    if (int MOD 2 = 1) then
        bin (i) := '1';
        else
        bin (i) := '0';
    end if;

-- perform integer division by 2
int := int/2;
end loop;
end itb;

begin
process (Y_int)
variable tem : std_logic_vector (N downto 0);
variable tem_int : integer ;
begin
    tem_int := Y_int;
    itb (tem, flag_even, N, tem_int);
    X_bin <= tem;
end process;
end convert;
```

Verilog: Converting an Unsigned Integer to Binary Using task

```
module Int_Bin (X_bin, flag_even, Y_int );
/*In general Verilog, in contrast to VHDL, does not
  strictly differentiate between integers and binaries;
```

```
    for example, if bin is declared as a binary of width 4,
    bin = bin/2 can be written, and the Verilog, but not
    VHDL, performs this division as if bin is integer.
    In the following, the corresponding VHDL program in
    Listing 6.5a is just translated to practice with
    the command task */
parameter N = 3;
output [N:0] X_bin;
output flag_even;
input [N:0] Y_int;
reg [N:0] X_bin;
reg flag_even;
always @ (Y_int)
begin
itb (Y_int, N, X_bin, flag_even);
end
task itb;
parameter P = N;
input integer int;
input N;
output [P:0] bin;
output flag;
integer j;
begin

if (int %2 == 0)
//The above statement checks int to see if it is even.
    flag = 1'b1;
    else
    flag = 1'b0;

for (j = 0; j <= P; j = j + 1)
    begin
        if (int %2 == 1)
        bin[j] = 1;
        else
        bin[j] = 0;
        int = int/2;
    end
end
endtask

endmodule
```

The simulation output of this conversion is shown in Figure 6.4.

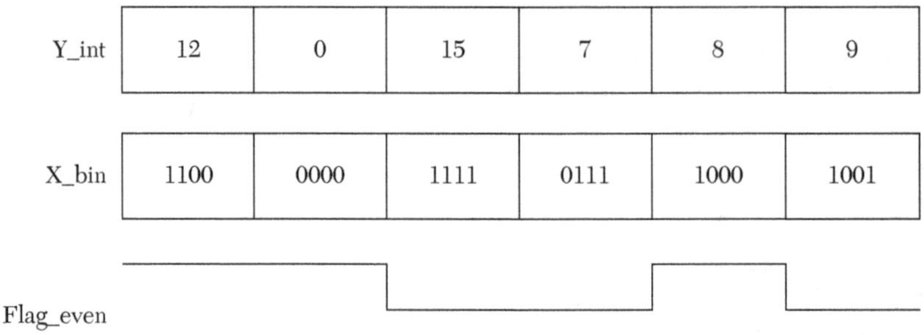

Y_int	12	0	15	7	8	9

X_bin	1100	0000	1111	0111	1000	1001

Flag_even

FIGURE 6.4 Simulation output for integer conversion to binary.

EXAMPLE 6.6   SIGNED BINARY-TO-INTEGER CONVERSION USING PROCEDURE AND TASK

Here, a signed binary is considered. The value of the binary data can be negative or positive. As is common, the negative data is represented by its 2s complement. The most significant bit of the data is the sign bit: if it is 0, the number is positive; otherwise, it is negative.

If the data is positive (the most significant bit is 0), then it is identical to unsigned data. For example, for four-bit data, 0101 is a positive number because its most significant bit is 0; the value of the number is +5. The data 1011 is a negative number; the value is −5. The decimal value of any negative number Y in the 2s-complement format can be written as $Y' - 2^N$, where N is the number of bits of Y, and Y' is equal to the decimal value of unsigned Y. For example, if Y = 1011, then N = 4. Y' = 1011, unsigned = 11_d. So, Y = 11 − 16 = −5.

Listing 6.5 shows the HDL code for converting from signed binary to integer using procedure (task). In addition to conversion, the procedure determines the parity (even or odd) of the input binary. Because a code has previously been written for conversion from binary to integer (Listing 6.3), it is used here. If the binary data is positive, it is the same code as in Listing 6.3. If the binary data is negative, its integer value is calculated as if it is unsigned, and this integer value is corrected by subtracting sixteen. Referring to the HDL code in Listing 6.6, the sign of the input date is tested for its most significant bit; if it is 1, the integer value (result) is corrected as follows:

VHDL

```
if (binsg(M) = '1') then
result := result - 2**(M+1);
end if;
```

Verilog
```
if (bin [P] == 1)
int = int - 2**(P+1);
```

To know whether the parity is odd or even, the 1s in the input data are counted and stored in the variable parity. Then, parity is divided by two: if there is a remainder, the parity is odd. Otherwise, it is even. The built-in function modulus, mod (VHDL) or % (Verilog), is used to determine whether parity is odd or even as follows:

VHDL
```
if (parity mod 2 = 1) then
--if parity is divisible by 2, then it is even,
--otherwise it is odd.
    even <= '0';
    else
    even <= '1';
end if;
```

Verilog
```
if ((parity % 2) == 1)
//if parity is divisible by 2, then it is even,
otherwise it is odd.

even = 0;
else
even = 1;
```

The signal even_parity in Listing 6.6 identifies the parity of the input data. If the parity is even, then even_parity is one, otherwise, it is zero.

LISTING 6.6 HDL Code for Converting a Signed Binary to an Integer Using procedure and task: VHDL and Verilog.

VHDL: Converting a Signed Binary to an Integer Using procedure
```
library ieee;
use ieee.std_logic_1164.all;
use ieee.numeric_std.all;

entity signed_btoIn is
```

```vhdl
generic (N : integer := 3);
port (X_bin : in signed (N downto 0); Y_int : out integer;
    even_parity : out std_logic );
end signed_btoIn;

architecture convert of signed_btoIn is

procedure sbti (binsg : in signed; M : in integer;
    int : out integer; signal even : out std_logic) is

--The procedure sbti is to change signed binary to integer
-- and also to find whether the parity of the binary
--is odd or even.
--The dimension of "sbin" does not have to be specified
--at the declaration statement; it can be declared later
--in the body of the procedure.

variable result, parity : integer;
begin

result := 0;
for i in 0 to M loop
if binsg(i) = '1' then
result := result + 2**i;
parity := parity + 1;
end if;
end loop;

if (binsg(M) = '1') then
result := result - 2**(M+1);
end if;
int := result;

if (parity mod 2 = 1) then
    even <= '0';
    else
    even <= '1';
end if;

end sbti;

begin
```

```
process (X_bin)
variable tem : integer;
begin
    sbti (X_bin, N, tem, even_parity);
    Y_int <= tem;
end process;
end convert;
```

Verilog: Converting a Signed Binary to an Integer Using task

```
module signed_btoIn(X_bin, Y_int, even_parity);
/*In general, Verilog (in contrast to VHDL) does not
  strictly differentiate between integers and binaries;
  for example if bin is declared as binary of width 4,
  write bin = bin/2, and the Verilog (but not VHDL) will
  perform this division. In the following, just translate
  the corresponding VHDL counterpart program. */

parameter N = 3;
input signed [N:0] X_bin;
output integer Y_int;
output even_parity;
reg even_parity;

always @ (X_bin)
begin
    sbti (Y_int, even_parity, N, X_bin);
end
task sbti;
parameter P = N;
output integer int;
output even;
input N;
input [P:0] bin;
integer i;
reg parity;

begin

int = 0;
parity = 0;
//change binary to integer
for (i = 0; i <= P; i = i + 1)
```

```
    begin
        if (bin[i] == 1)
        begin
            int = int + 2**i;
            parity = parity + 1;
        end
    end
    if ((parity % 2) == 1)
    even = 0;
    else
    even = 1;

    if (bin [P] == 1)
    int = int - 2**(P+1);
end
endtask

endmodule
```

The simulation output of the conversion is shown in Figure 6.5.

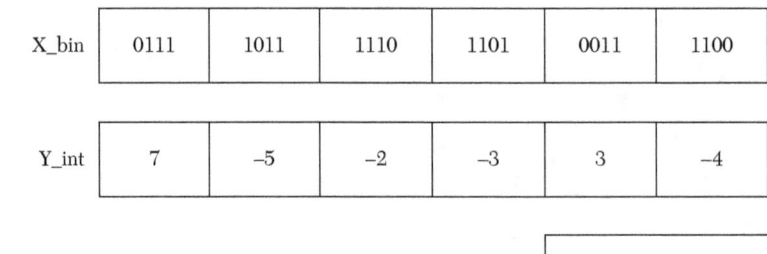

X_bin	0111	1011	1110	1101	0011	1100

Y_int	7	−5	−2	−3	3	−4

even_parity

FIGURE 6.5 Simulation output for converting a signed binary to an integer.

EXAMPLE 6.7   INTEGER-TO-SIGNED-BINARY CONVERSION USING PROCEDURE

In this example, an integer is converted to signed binary (see Listing 6.7). The sign of the integer is tested and negated only if it is negative. The same code from Listing 6.3 is applied. The outcome of the sign test is stored in the variable `flag`. After calculating the equivalent binary, its value is adjusted according to the flag. If the flag is one, this means that the integer has been negated, so the binary is negated. If the flag is zero, no action is taken. The conversion is written in procedure `sitb` and declared as:

```
procedure sitb (sbin : out signed; M, int : in integer) is
```

In the declaration statement above, the dimension of `sbin` does not have to be specified. The procedure can determine the dimension later in its body.

LISTING 6.7 VHDL Code for Converting an Integer to a Signed Binary Using **procedure**

```
library IEEE;
use IEEE.STD_LOGIC_1164.ALL;
use ieee.numeric_std.all;
entity signed_IntToBin is
generic (N : integer := 3);
port (X_bin : out signed (N downto 0); Y_int : in integer);
end signed_IntToBin;

architecture convert of signed_IntToBin is
procedure sitb (sbin : out signed; M, int : in integer) is

-- The procedure sitb is to convert integer into signed
-- binary. The dimension of "sbin" does not have to be
-- specified at the declaration statement; it can be
--declared later in the body of the procedure.

variable temp_int : integer;
variable flag : std_logic;
variable bin : signed (M downto 0);
begin

if (int < 0) then
    temp_int := - int;
    flag := '1';
    --if flag = 1, the number is negative
    else
temp_int := int;
end if;
for i in 0 to M loop
if (temp_int MOD 2 = 1) then
bin (i) := '1';
else
bin (i) := '0';
end if;
--integer division by 2
temp_int := temp_int/2;
```

```
end loop;
if (flag = '1') then
sbin := - bin;
else sbin := bin;
end if;
end sitb;
begin
process (Y_int)
variable tem : signed (N downto 0);
begin
sitb(tem, N, Y_int);
X_bin <= tem;
end process;
end convert;
```

EXAMPLE 6.8 SIGNED VECTOR MULTIPLICATION USING PROCEDURE AND TASK

In this example, vector multiplication is performed. Equation 6.1 shows vector multiplication

$$d = a \times b \tag{6.1}$$

where a is a row vector with three elements, and b is a column vector with three elements. Accordingly, d is a row vector with three elements. Equation 6.1 can be written as:

$$d = \begin{bmatrix} a0 & a1 & a2 \end{bmatrix} \times \begin{bmatrix} b0 \\ b1 \\ b2 \end{bmatrix} \tag{6.2}$$

From Equation 6.2,

$$d = a0\,b0 + a1\,b1 + a2\,b2 \tag{6.3}$$

In this example, all elements of Equation 6.3 are signed binary. To multiply two signed numbers, Booth algorithm (as discussed in Chapter 3) is implemented. Listing 6.8 shows the HDL code for signed vector multiplication using `procedure` (`task`). The inputs to the multiplication algorithm are written as:

```
port (a0, a1, a2, b0, b1, b2 : in signed (N downto 0);
d : out signed (3*N downto 0));
```

If a large number of elements are being multiplied, the above code may not be practical because a large number of ports must be listed. In Chapter 7, the ports are listed as an array; this will shorten the code. The Booth algorithm is written as a procedure or task with the declaration:

VHDL
```
procedure booth (X, Y : in signed (3 downto 0);
Z : out signed (7 downto 0));
```

Verilog
```
task booth;
input signed [3:0] X, Y;
output signed [7:0] Z;
```

where the inputs are x and y, and the output is z. The procedure (task) is restricted to 4x4 bits. The procedure (task) can be generalized to multiply any *NxN* bits (see Exercise 6.2). The three procedure (task) callings calculate the partial products a0b0, a1b1, and a2b2 as:

```
booth (a0, b0, tem0);
booth (a1, b1, tem1);
booth (a2, b2, tem2);
```

The partial products are stored in the eight-bit registers tem0, tem1, and tem2, respectively. To find the product d, add tem0 + tem1 + tem2, and, according to Listing 6.8, the product is stored in ten-bit register d. By choosing ten bits, any overflow is avoided that might occur after accumulating the partial products in register d. To calculate d in Verilog, simply write:

```
d = tem0 + tem1 + tem2;
```

In VHDL, the language is strictly type and size oriented, so the VHDL simulator may not perform the above operation because d has a different size than tem0, tem1, and tem2. Several approaches can be taken to adjust the size. The approach here is to convert tem0, tem1, and tem2 to integers by using the procedure sbti and add all integers and convert back to binary by using the procedure sitb. Another approach is to extend the sizes of tem0, tem1, and tem2 to ten bits and then add tem0+tem1+tem2 (see Exercise 6.1).

LISTING 6.8 *HDL Code for Signed-Vector Multiplication Using* procedure *and* task: *VHDL and Verilog*

VHDL: Signed-Vector Multiplication Using procedure
```
library IEEE;
use IEEE.STD_LOGIC_1164.ALL;
use ieee.numeric_std.all;
```

```
entity Vector_Booth is
generic (N : integer := 3);
port (a0, a1, a2, b0, b1, b2 : in signed (N downto 0);
    d : out signed (3*N downto 0));
end Vector_Booth;

architecture multiply of Vector_Booth is
procedure booth (X, Y : in signed (3 downto 0);
    Z : out signed (7 downto 0)) is
--Booth algorithm here is restricted to 4x4 bits.
--It can be adjusted to multiply any NxN bits.
variable temp : signed (1 downto 0);
    variable sum : signed (7 downto 0);
    variable E1 : unsigned (0 downto 0);
    variable Y1 : signed (3 downto 0);
begin

sum := "00000000"; E1 := "0";
    for i in 0 to 3 loop
    temp := X(i) & E1(0);
    Y1 := -Y;
    case temp is
        when "10" => sum (7 downto 4) :=
        sum (7 downto 4) + Y1;
        when "01" => sum (7 downto 4) :=
        sum (7 downto 4) + Y;
        when others => null;
    end case;
    sum := sum srl 1;
    sum (7) := sum(6);
    E1(0) := x(i);
    end loop;
    if (y = "1000") then

--If Y = 1000; then according to the code,
--Y1 = 1000 (-8 not 8 because Y1 is 4 bits only).
--The statement sum = -sum adjusts the answer.

    sum := -sum;
    end if;
    Z := sum;
    end booth;
```

```
procedure sitb (sbin : out signed; M, int : in integer) is
-- The procedure sitb is to convert integer into signed
-- binary. The dimension of "sbin" does not have to be
-- specified at the declaration statement; it can be
--declared later in the body of the procedure.
variable temp_int : integer;
variable flag : std_logic;
variable bin : signed (M downto 0);
begin

if (int < 0) then
temp_int := -int;
flag := '1';
else
temp_int := int;
end if;

for i in 0 to M loop
if (temp_int MOD 2 = 1) then
bin (i) := '1';
else
bin (i) := '0';
end if;
temp_int := temp_int/2;
end loop;
if (flag = '1') then
sbin := -bin;
else
sbin := bin;
end if;
end sitb;

procedure sbti (binsg : in signed; M : in integer;
                int : out integer) is

-- The procedure sbti is to change signed binary to
-- integer.No need to specify the dimension of "sbin"
-- at the declaration statement; it can be declared
-- later in the body of the procedure.

variable result : integer;

begin
```

```
result := 0;
for i in 0 to M loop
if binsg(i) = '1' then
result := result + 2**i;
end if;
end loop;

if (binsg(M) = '1') then
result := result - 2**(M+1);
end if;
int := result;
end sbti;

begin
process (a0, b0, a1, b1, a2, b2)
variable tem0, tem1, tem2 : signed ((2*N + 1) downto 0);
variable d_temp : signed (3*N downto 0);
variable temi0, temi1, temi2, temtotal : integer;

begin
--Find the partial products a0b0, a1b1, a2b2
booth (a0, b0, tem0);
booth (a1, b1, tem1);
booth (a2, b2, tem2);

-- Change the partial products to integers
sbti (tem0, (2*N+1), temi0);
sbti (tem1, (2*N+1), temi1);
sbti (tem2, (2*N+1), temi2);

-- Find the total integer sum of partial products
temtotal := temi0 + temi1 + temi2;

-- Change the integer to binary
sitb (d_temp, 3*N, temtotal);

d <= d_temp;
end process;
end multiply;
```

Verilog: Signed-Vector Multiplication Using task
```
module Vector_Booth (a0, a1, a2, b0, b1, b2, d);
```

```
parameter N = 3;
input signed [N:0] a0, a1, a2, b0, b1, b2;
output signed [3*N : 0] d;
reg signed [2*N+1 : 0] tem0, tem1, tem2;

reg signed [3*N : 0] d;

always @ (a0, b0, a1, b1, a2, b2)
begin
booth (a0, b0, tem0);
//booth is a task to multiply a0 x b0 = tem0

booth (a1, b1, tem1);
booth (a2, b2, tem2);
d = tem0 + tem1 + tem2;
end
task booth;
input signed [3:0] X, Y;
output signed [7:0] Z;
reg signed [7:0] Z;
reg [1:0] temp;
integer i;
reg E1;
reg [3:0] Y1;

begin
Z = 8'd0;
E1 = 1'd0;

for (i = 0; i < 4; i = i + 1)
begin
temp = {X[i], E1}; //This is catenation
Y1 = -Y; //Y1 is the 2' complement of Y
case (temp)
    2'd2 : Z [7:4] = Z [7:4] + Y1;
    2'd1 : Z [7:4] = Z [7:4] + Y;
    default : begin end
endcase
Z = Z >> 1; /*This is a logical shift of one position to
            the right*/
Z[7] = Z[6];
    /*The above two statements perform arithmetic shift
```

```
                where the sign of the number is preserved after
                the shift.*/
        E1 = X[i];

        end

        if (Y == 4'b1000) Z = -Z;

        /* If Y = 1000, then Y1 = 1000 (should be 8 not -8).
           This error is because Y1 is 4 bits only.
           The statement Z = -Z adjusts the value of Z. */

            end

        endtask
        endmodule
```

Figure 6.6 shows the output simulation of the vector multiplication.

a0	1001	1000	1000	1011
a1	0010	1000	1000	1111
a2	1011	1000	1000	0101
b0	0111	0111	1000	1110
b1	0010	0111	1000	1001
b2	0011	0111	1000	1000
d	1111000100	1101011000	0011000000	1111101001

FIGURE 6.6 Simulation output of vector multiplication.

EXAMPLE 6.9 SIGNED 3x3-BIT MULTIPLICATION USING COMBINATIONAL ARRAY

The Booth algorithm was introduced in Chapter 3. The algorithm compares two consecutive bits of the multiplicand $(a_i a_{i-1})$ and performs addition if $a_i a_{i-1} = 01$, subtraction if $a_i a_{i-1} = 10$, or nothing if $a_i a_{i-1} = 00$ or ai $a_{i-1} = 11$. After the comparison, a single arithmetic right shift is performed. The algorithm repeats the comparison-shift N times where N is the number of bits of the multiplicand. A clock is used to count the iterations and load the intermediate values. The multiplication algorithm, instead of using sequential approach, can be implemented using combinational circuits. Comparison, addition, and subtraction can be implemented by combinational circuits; the shift can be accomplished physically by shifting (placing) the combinational-circuit elements according to their placement in the partial products. For example, if the product is four bits $(P_3 P_2 P_1 P_0)$, then the circuit that generated P_1 is placed on the left-hand side of the circuit generating P_0. Figure 6.7 shows a combinational circuit that multiplies x × b = p. Each x and b is three-bit signed number. Figure 6.7a shows two types of cells: *comp*, which compares two bits, and *FAS*, which can add or subtract. In comp cells, the comparison between two consecutive bits x(i) and x(i-1) is done as follows:

$$C1 = x(i) \; xor \; x(i\text{-}1) \qquad C2 = x(i) \; and \; \overline{x(i-1)}$$

C1C2 indicates whether x(i) x(i-1) are equal to 11, 00, 01, or 10. Cell FAS is mainly a combinational adder/subtractor circuit. According to the output of the comp cell, the FAS cell will add if C1C2 = 10, will subtract if C1C2 = 11, and will be transparent (output = input) if C1 = 0. Table 6.1 shows the relationship between x(i) x(i-1) and the operation executed by the FAS cell.

TABLE 6.1 FAS Cell

x(i) x(i-1)	C2C1	FAS operation
00	00	Transparent: S = a
01	01	Addition: S1 S = a + b + c
10	11	Subtraction: S1 S = a - b - c
11	00	Transparent: S = a

From Table 6.1:

$$C1 = x(i) \; xor \; x(i\text{-}1)$$

$$C2 = x(i) \; and \; \overline{x(i-1)}$$

Figure 6.7b shows the multiplication of -3 × 2 where x = 2 and b = 3; initially, a is set to 0.

Listing 6.9 shows the HDL code for the multiplier. In Listing 6.9, procedure COMPR is generating C1 and C2. Procedure Fulladdr performs the addition if C2C1 = 01 and generates sum and carryout. Procedure Fullsub performs subtraction when C2C1 = 11 and generates difference (Diff) and Borrow. More details about the logic design of the array can be found in Hayes, 1998 [1].

LISTING 6.9 HDL Code for Signed 3x3-Bit Multiplication Using Combinational Array: VHDL and Verilog

VHDL
```
library IEEE;
use IEEE.STD_LOGIC_1164.ALL;
entity boothArrayProc is
port ( start: in std_logic;
        x : in std_logic_vector (2 downto 0);
y: in std_logic_vector (4 downto 0);
p:out std_logic_vector(4 downto 0));
--Input y is 3-bit plus two additional sign extension
-- bits.
end boothArrayProc;

architecture boothar of boothArrayProc is
procedure COMPR (n1 : in std_logic_vector;
i : in integer range 0 to 3; c1, c2: out std_logic) is
begin
c1 := n1(i+1) xor n1(i);
c2 := n1(i+1) and not n1(i);
end procedure COMPR;

procedure Fulladdr (a, b,c : in std_logic;
sum, Carryout : out std_logic) is
begin
Sum := (not a and not b and c) or
       (not a and b and not c) or
       (a and not b and not c) or
       (a and b and c);
Carryout := (a and b) or (a and c) or (b and c);
end procedure Fulladdr;
procedure Fullsub (a, b,c : in std_logic; Diff,
```

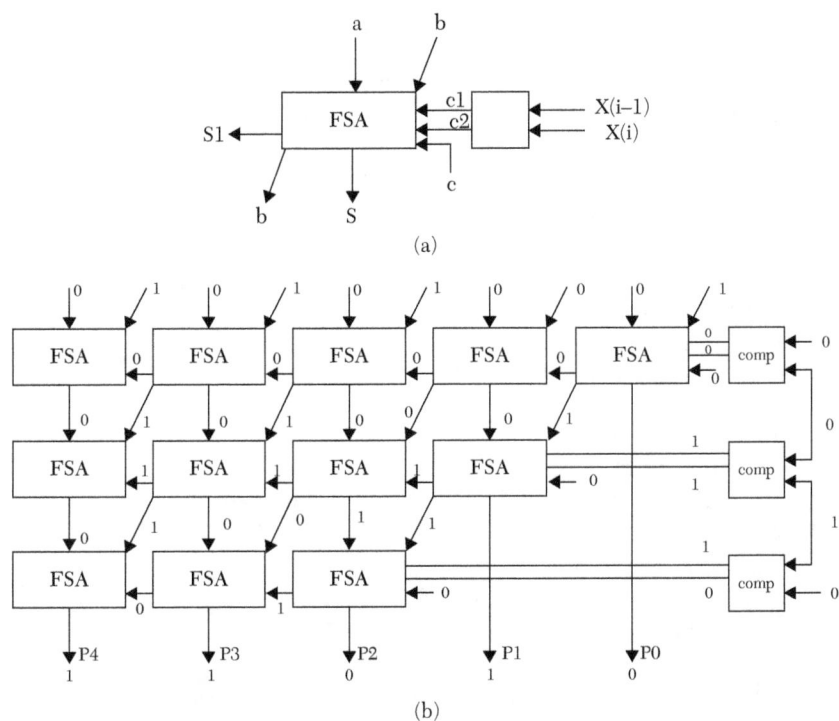

FIGURE 6.7 A combinational array multiplier a) Cells comp and FAS. b) A multiplier circuit that multiplies -3 × 2 where x = 2 and b = -3.

```
Borrow : out std_logic) is
begin
Diff := (not a and not b and c) or
        (not a and b and not c) or
        (a and not b and not c) or
        (a and b and c);
Borrow := ((not a) and c) or ((not a) and b) or (b and c);

end procedure Fullsub;


begin
B1: process (start,x,y)
variable s1, s2: std_logic;
variable T : std_logic_vector(3 downto 0);

variable f, E, z: std_logic_vector(4 downto 0):= "00000";
variable carry_temp: std_logic_vector(5 downto 0);
variable i,j,layer, j1 :integer range 0 to 6 ;
```

```
begin
if (start = '1') then
f := "00000"; E := "00000";z := "00000";
carry_temp := "000000";
T := (x & '0');-- T now represents all the inputs to
                --cells COMP including the initial
                --zero on the first COMP cell.
for layer in 0 to 2 loop

carry_temp(0) := '0';

COMPR (T,layer,s1,s2);

if (s1 = '0') then
for j in 0 to (4-layer) loop
z(j + layer) := f(j + layer);
end loop;

elsif (s2 = '0') then
for j in 0 to (4-layer) loop

Fulladdr (f(j+layer), y(j), carry_temp(j),
 z(j+ layer), carry_temp(j+1));
end loop;
else
for j in 0 to (4-layer) loop

Fullsub (f(j+layer), y(j),
carry_temp(j), z(j+layer), carry_temp(j+1));
end loop;
end if;
for j1 in 0 to (4-layer) loop

f(j1 + layer) := z(j1 + layer);end loop;

end loop;
end if;
p <= z;
end process B1;
end boothar;
```

Verilog
```
module bootharrayTask(start, x, y, p);
```

```verilog
input start;
input [2:0] x;
input [4:0]y;
output [4:0] p;
/*input y is 3-bit with additional 2 bits as sign extension*/

reg [4:0] f,E,p;
reg [3:0] T;
integer layer,i, j,j1;
reg [5:0] carry_temp;
reg s1, s2;
always @ (start,x,y)
begin

if (start == 1'b1)
begin
T = {x,1'b0};/* T now represents all the inputs to
                cells COMP including the initial
                zero on the first COMP cell.*/
f = 5'd0; E = 5'd0; p = 5'd0; carry_temp = 6'd0;
for (layer = 0; layer <= 2; layer = layer + 1)

begin
carry_temp [0] = 1'b0;

COMPR (T,layer,s1,s2);

if (s1 == 1'b0)
begin
for (j = 0; j <= (4-layer); j = j+1)
begin

p[j + layer] = f[j + layer];

end
end

else if (s2 == 1'b0)
begin
for (j = 0; j <= (4-layer); j = j+1)
begin
```

```
Fulladdr (f[j+layer], y[j], carry_temp[j],
 p[j+ layer], carry_temp[j+1]);
end
end
else
begin
for (j = 0; j <= (4-layer); j = j+1)
begin

Fullsub (f[j+layer], y[j], carry_temp[j],
 p[j+ layer], carry_temp[j+1]);
end
end

for (j1 = 0; j1 <= (4-layer); j1 = j1+1)
begin

f[j1 + layer] = p[j1 + layer];
end

end
end
end
task COMPR;
input [3:0]n1;
input integer i;
output c1, c2;
begin
c1 = n1[i+1] ^ n1[i];
c2 = n1[i+1] & ( ~ n1[i]);
end
endtask
task Fulladdr;
input a, b,c;
output sum, Carryout;
reg sum, Carryout;
begin
sum = (~a & ~b & c) |
      (~a & b & ~ c) |
      (a & ~ b & ~ c) |
      (a & b & c);
Carryout = (a & b) | (a & c) | (b & c);
end
```

```
endtask
task Fullsub;
input a, b,c;
output Diff, Borrow;
begin
Diff = (~a & ~b & c) |
       (~a & b & ~ c) |
       (a & ~ b & ~ c) |
       (a & b & c);
Borrow = ((~ a) & c) | ((~ a) & b) | (b & c);
end
endtask

endmodule
```

EXAMPLE 6.10   **DESCRIPTION OF ENZYME-SUBSTRATE ACTIVITY USING PROCEDURE AND TASK**

Enzymes are molecules (generally proteins) that increase the speed of a chemical reaction. The human body uses a large number of different enzymes to speed up various types of chemical reactions such as those involved in metabolism. Each enzyme is specific for a certain reactant, called a *substrate*. For the substrate-enzyme complex to work, the enzyme must be capable of binding to the substrate; if the enzyme cannot bind to the substrate, the enzyme will not be active. The activity of the enzyme increases with the strength of binding. There are several theories that explain this binding such as the key-lock mechanism. In this mechanism, the physical shape of the enzyme matches a groove on the substrate where it can bind. Figure 6.8a illustrates a potentially strong bond between the enzyme and substrate. Figure 6.8b illustrates a case where binding is almost impossible between the substrate and enzyme.

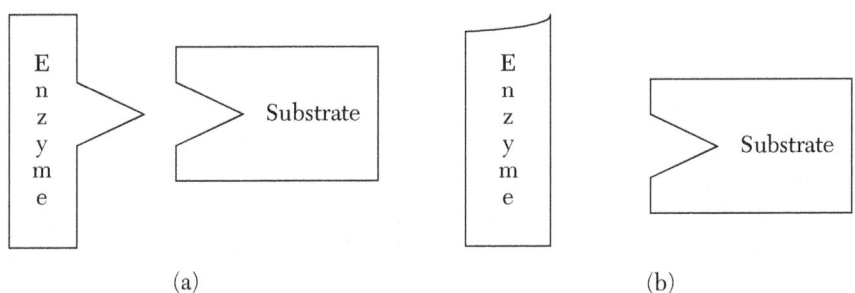

(a)   (b)

FIGURE 6.8 Binding between substrate and enzyme. (a) Strong. (b) Week.

The binding strength between the substrate and the enzyme is measured by a parameter called the *dissociation constant* (*M*). If M is large, the binding or affinity between substrate and enzyme is weak and vice versa. The rate of reaction between an enzyme with dissociation constant, M, and a substrate with concentration, S, is represented by Equation 6.4

$$V = V_{max}\frac{S}{S + M}$$
(6.4)

where S is the concentration of the substance and V_{max} is the maximum possible rate of reaction when S >> M. Usually, V_{max} is assigned the value of 1 (100%), and accordingly, V is measured as a fraction or percentage. Figure 6.9 shows a graphical representation of Equation 6.4 for a dissociation constant of three units. Notice that if S = M, then V = 0.5V_{max}, so M can be viewed as the concentration of substrate at which the rate of reaction is 50% of V_{max}. Figure 6.9 shows the relationship between the substrate concentration, S, and rate of reaction, V.

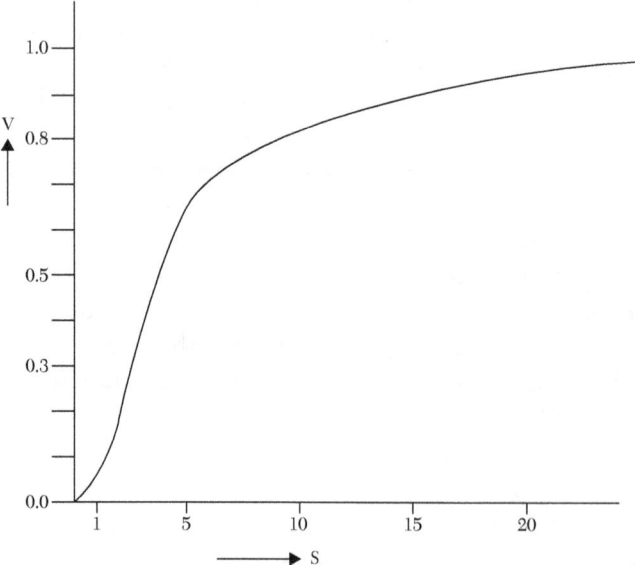

FIGURE 6.9 Relationship between the substrate concentration, S, and rate of reaction, V.

Listing 6.10 shows the HDL code describing the enzyme-substrate activity. The challenge here is to write HDL code that can operate on real numbers. Recall that V is a fraction if V_{max} is taken as 1.

Several basic VHDL simulators, in contrast to Verilog, will not accept a mix of integer and real numbers in the same assignment statement. Here, we assume that the user does not have an external library or packages that convert between integer and real. Also, in Listing 6.10, the VHDL code assumes that S and M are fractions; if not, the user can scale down both S and M to be fractions. In VHDL code, S and M are represented by fractions in the Q4 (four bits) format. The four bits are converted to real fraction by the procedure flt. The procedure multiplies bit, i, with its weight, 2^{-i}, and accumulates the product into a real number float. The rate of reaction (V) is calculated using Equation 6.4. To display the rate of reaction in binary, the real number is converted to an integer by the procedure rltointg.

On the other hand, Verilog code is very easy to write because Verilog allows for the mixing of integer and real.

Listing 6.10 simulates Equation 6.4 (the relationship between the rate of reaction and concentrate of substrate). This same approach can be used to simulate other equations similar to Equation 6.4 such as the output of certain filters as a function of frequency or a transistor collector's voltage as a function of collector current.

LISTING 6.10 HDL Description for Enzyme Activity Using procedure and task: VHDL and Verilog

```
VHDL: Enzyme Activity Using procedure
library ieee;
use ieee.std_logic_1164.all;
use ieee.numeric_std.all;
use ieee.std_logic_arith.all;
-- A description of enzyme-substrate binding mechanism.
-- S= (Vmax *S)/(S + M) where S is the substrate
-- concentration, M is the dissociation constant, and Vmax
-- is the maximum rate of reaction. In this example,
-- Vmax = 1. The inputs are S and M in binary (fraction);
-- the output v is in Q4 format. This means that v is
-- always less than one, with the binary point placed to
-- the left of the most significant bit. For example,
-- if v = 1010, the decimal equivalent is .5 + .125 =
-- 0.625. To calculate v, convert S and M to real fraction,
-- find the real value of (S/(S + M)), convert this real
-- value to Q4 by multiplying it with 2**4 = 16, and
```

```
-- convert the integer to binary.
library ieee;
use ieee.std_logic_1164.all;
use ieee.numeric_std.all;
use ieee.std_logic_arith.all;

entity enzyme_beh is
port (S : in std_logic_vector (0 to 3);
      M : in std_logic_vector(0 to 3);
          v : out real ; intg_bin : out integer;
          start: in bit);
end enzyme_beh;

architecture enzyme of enzyme_beh is
procedure flt (a : in std_logic_vector (0 to 3);
                   float : out real) is

--This procedure converts fraction expressed
--in fixed-point binary (Q4) to real
variable tem,j : real;
begin
    tem := 0.0;
      j := 1.0;

    for i in 0 to 3 loop
       j := j/ 2.0;
         if (a(i) = '1') then

         tem := tem + j;
         end if;
    end loop;
float := tem;
end flt;
procedure rltointg (a1 : in real ; Bin1 : out integer ) is

--This procedure converts real to integer
--The procedure does not round off
variable tem,j : real := 0.0;
variable tmp : integer := 0;
begin

while (tmp <= 256) loop
```

```
--Assume the maximum integer value is 256; if not,
--adjust accordingly
if (a1 > j) then
tmp := tmp + 1;
    j := j + 1.0;
    else
    exit;
        end if;
    end loop;
        Bin1 := tmp;
end rltointg;

begin
P1 : process(S, M, start)
variable temp1, temp11, vmax : real;
variable temp2: integer := 0;

begin
if (start = '1') then
vmax := 1.0;
flt( S,temp1);
flt( M,temp11);
temp1 := Vmax * ( temp1/(temp1 + temp11));
v <= temp1;
temp1 := temp1 *16.0;
rltointg (temp1, temp2);
intg_bin <= temp2;
end if;
end process P1;
end enzyme;
```

Verilog: Enzyme Activity Using task
```
/* A description of enzyme-substrate binding mechanism.
   S= (Vmax *S)/(S + M), where S is the substrate
   concentration, M is the dissociation constant,
   and Vmax is the maximum rate of reaction. In this
   example, Vmax =1. The inputs are S and M in binary
   (integer);
   the output v is in Q4 format. This means that v is
   always less than 1, with the binary point placed to the
   left of the most significant bit. For example, if
   v = 1010, the
```

decimal equivalent is .5 + .125 = 0.625. To calculate V, find the real value of (S /(S + M)), convert this real value to Q4 by multiplying it with 2**4 =16, and convert the integer to binary. */

```verilog
module enzyme_beh (S, M, V);
input [3:0] S, M;
output [3:0] V;
integer vmax;
reg [3:0] V;
real vr;
always @ (S, M)
begin
vmax = 1;
vr = vmax * (1.0 * S) / (S * 1.0 + M * 1.0);
vr = vr * 2**4;
rti (vr, V);
end
task rti;
/* This task can be replaced by just one statement, v1= r.
   Verilog, in contrast to VHDL, can handle different
   types of the assignment statement. Verilog finds the
   equivalent integer value v1 for the real r. The task has
   been designed here only to match the same steps done in
   VHDL. */

input real r;
output [3:0] v1;
real temp;
begin
temp = r;
v1 = 4'b0000;
while (temp >= 0.5)
begin
    v1 = v1 + 1;
    temp = r - 1.0 * v1;
end
end
endtask
endmodule
```

Figure 6.10 shows the simulation output of the relationship between substrate concentration, S, and rate of reaction, V, for M = 3 units.

M	0011	0011	0011	0011	0011	0011

S	0001	0011	0111	1001	1011	1111

V	0100	1000	1011	1100	1101	1101

Vr	0.2500	0.500	0.700	0.75	0.785714	0.8333

FIGURE 6.10 Simulation output of the relationship between substrate concentration, S, and rate of reaction, V, for M = 3 units.

6.3   Functions

Functions are behavioral statements. As is the case when calling `procedure` or `task`, functions must be called within `process` (VHDL) or `always` or `initial` (Verilog). Functions take one or more inputs, and, in contrast to `procedure` or `task`, they return only a single output value.

6.3.1   VHDL Functions

As in `procedure`, functions have a declaration and a body. An example of a function declaration is:

```
function exp (a, b : in std_logic) return std_logic is
```

where `function` is a predefined word, `exp` is the user-selected name of the function, and `a` and `b` are the inputs. Only inputs are allowed in the function declaration. The function returns a single output by the use of the predefined word `return`. The function `exp` returns a variable of type `std_logic`, and `is` is a predefined word that has to be at the end of the declaration statement. The name of the output is not listed in the declaration; it is listed in the body of the function. The body of the function lists the relationship between the inputs and the output to be returned. All statements in the body of the function should be behavioral (sequential) statements, and `return` is used to point to the output of the function. An example of a function's declaration and body (VHDL) is shown in Listing 6.11.

LISTING 6.11 Example of a VHDL Function

```
library IEEE;
use IEEE.STD_LOGIC_1164.ALL;
```

```
entity Func_exm is
port (a1, b1 : in std_logic; d1 : out std_logic);
end Func_exm;

architecture Behavioral of Func_exm is
function exp (a, b : in std_logic) return std_logic is
variable d : std_logic;
begin
d := a xor b;
return d;
end function exp;

begin
process (a1, b1)
begin
d1 <= exp (a1, b1);
--The above statement is a function call
end process;
end Behavioral;
```

In Listing 6.11, the name of the function is `exp`; it has two inputs, `a` and `b`, of type `std_logic`. The type of the output to be returned is `std_logic`. The output to be returned is `d`. The function, as seen from its body, is performing a *xor function* on the inputs `a` and `b`. To call the function, it should be written inside a `process`. The function is called by the following statement:

```
d1 <= exp (a1, b1);
```

The function call passes `a1` and `b1` to `a` and `b`, respectively, then calculates `a1 XOR b1` and passes the output of the `XOR` to `d1`.

The standard VHDL package has many built-in functions; other functions can be imported from packages attached to the VHDL module. Some examples of built-in functions are:

`mod`: finds the modulo of x mod y

`abs`: finds the absolute value of a signed number

`To_INTEGER`: returns an integer value of a signed input

`TO_SIGNED`: takes an integer and returns its signed binary equivalent

The package ieee.numeric_std.all has a large number of built-in functions.

6.3.2 Verilog Functions

Functions in Verilog have a declaration statement and a body. In the declaration, the size (dimension), type, and name of the output are specified, as well as the names and sizes (dimensions) of the inputs. For example, the declaration statement

```
function exp;
input a, b;
```

declares a function with the name (identifier) exp. The function has two inputs, a and b, and one output, exp. All inputs are one-bit data, and the output is also one-bit data. The inputs and output can take 0, 1, x ("don't care"), or Z (high impedance). The body of the function follows the declaration in which the relationship between the output and the inputs is stated. An example of a function and its call is shown in Listing 6.12. The function calculates exp = a XOR b.

LISTING 6.12 Verilog Function That Calculates exp = a XOR b

```
module Func_exm (a1, b1, d1);
input a1, b1;
output d1;
reg d1;

always @ (a1, b1)
begin

/*The following statement calls the function exp
and stores the output in d1.*/

d1 = exp (a1, b1);
end

function exp ;
input a, b;
begin

exp = a ^ b;
end
endfunction
endmodule
```

In addition to user-defined functions, the standard Verilog package includes a large number of built-in functions such as `modulus %`.

6.3.3 Function Examples

EXAMPLE 6.11 **FUNCTION TO FIND THE GREATER OF TWO SIGNED NUMBERS**

In this example, the greater of two signed numbers, x and y, is determined. Each number is a signed binary of four bits, and `function` is called in the main module to find the greater of the two input numbers. The result is stored in z. Listing 6.13 shows the HDL code of this example.

LISTING 6.13 HDL Function to Find the Greater of Two Signed Numbers: VHDL and Verilog

```
VHDL Function to Find the Greater of Two Signed Numbers
library IEEE;
use IEEE.STD_LOGIC_1164.ALL;
use ieee.numeric_std.all;

entity greater_2 is
port (x, y : in signed (3 downto 0);
      z : out signed (3 downto 0));
end greater_2;

architecture greater_2 of greater_2 is
function grt (a, b : signed (3 downto 0)) return signed is
-- The above statement declares a function by the name grt.
-- The inputs are 4-bit signed numbers.

variable temp : signed (3 downto 0);
begin
    if (a >= b) then
        temp := a;
        else
        temp := b;
    end if;
return temp;
end grt;

begin
```

```
process (x, y)
begin
    z <= grt (x, y); --This is a function call.
end process;
end greater_2;
```

Verilog Function to Find the Greater of Two Signed Numbers
```
module greater_2 (x, y, z);
input signed [3:0] x;
input signed [3:0] y;
output signed [3:0] z;
reg signed [3:0] z;
always @ (x, y)
begin
z = grt (x, y); //This is a function call.
end

function [3:0] grt;

/*The above statement declares a function by the name grt;
grt is also the output of the function*/

input signed [3:0] a, b;
/*The above statement declares two inputs to the function;
both are 4-bit signed numbers.*/

begin
if (a >= b)
grt = a;
else
grt = b;
end
endfunction

endmodule
```

EXAMPLE 6.12 **FUNCTION TO FIND THE FLOATING SUM** $Y = \sum_{i=0}^{3}(-1)^{i}(x)^{i}$,
$0 < X < 1$

In this example, a function is written that accumulates the polynomial summation of x. The polynomial in this example is of the third degree. The input number x is a positive fraction, and it is represented as a fixed-point

Q4 format. This means that the binary point is at the left of the most significant bit of x, and the total number of bits is four. For example, if the number is 1010, then its decimal value is $2^{-1} + 0 + 2^{-3} + 0 = 0.5 + 0.125 = 0.625$. The output y in this example is assigned a Q8 format. To calculate y, first convert x to real, then calculate the real sum of $\sum_{i=0}^{3}(-1)^i(x)^i = 1 - x + x^2 - x^3$. To convert the sum to Q8, multiply the real sum by 28; this generates a real number. This real number is converted to an integer, and finally, the integer is converted to binary. For example, if x = 1011, the following steps are executed:

1. Convert x to real: 1011 is converted to 0.5 + 0.125 + 0.0625 = 0.6875

2. Multiply the real number in Step 1 by 28: 0.6875 × 28 = 176.0

3. Convert the real number in Step 2 to an integer: 176.0 is converted to 176

4. Convert the integer in Step 3 to an eight-bit binary:
176 is converted to B0 (hex)

Listing 6.14 shows the HDL code for calculating y. Referring to the VHDL, three procedures are built: `flt`, `rltointg`, and `itb`. The procedure `flt` converts `std_logic` to real. We need this procedure to convert the input x. The procedure `rltointg` converts the real value to an integer, and itb converts the integer to `std_logic`. The function `exp` implements the three procedures to calculate y. In VHDL, procedures are allowed to be written in the body of the function.

The Verilog code in Listing 6.14 consists of three functions: `float`, `rti`, and `exp`. In contrast to VHDL, Verilog does not allow tasks to be written in the body of the function. The function `float` converts binary numbers that represent fractions (Q4) to real numbers. The function `rti` converts real numbers to integers.

Because Verilog is not a very strict type-oriented language, we can rewrite `function rti` as:

```
function [15:0] rti;
input real r;

begin
```

```
        rti = r;
end
endfunction
```

The statement `rti = r;` has a left-hand `rti` of type integer and a right-hand side of `r` (real). Verilog allows this mixing of two types; it calculates the right-hand side as real, and when assigned to the left-hand side, the type is converted from real to integer.

LISTING 6.14 HDL Code for $y = \sum_{i=0}^{3} (-1)^i (x)^i$, ***0 < x < 1: VHDL and Verilog***

```
VHDL Floating Sum Description
library IEEE;
use IEEE.STD_LOGIC_1164.ALL;
use IEEE.STD_LOGIC_ARITH.ALL;
use IEEE.STD_LOGIC_UNSIGNED.ALL;
entity segma is
port (x : in std_logic_vector (0 to 3);
      y: out std_logic_vector (7 downto 0));
end segma;

architecture segm_beh of segma is

procedure flt (a : in std_logic_vector (0 to 3);
          float : out real) is

--This procedure converts fraction expressed
-- in fixed-point binary to real
variable tem,j : real;
begin
    tem := 0.0;
      j := 1.0;

    for i in 0 to 3 loop
      j := j/ 2.0;

        if (a(i) = '1') then

      tem := tem + j;
        end if;
    end loop;
float := tem;
end flt;
```

```
procedure rltointg (a1 : in real ; Bin1 : out integer ) is

--This procedure converts real to integer
--The procedure does not round off
variable tem,j : real := 0.0;
variable tmp : integer := 0;
begin

while (tmp <= 256) loop
--Assume the maximum integer value is 256;
--if not, adjust accordingly
if (a1 > j) then
tmp := tmp + 1;
    j := j + 1.0;
    else
    exit;
        end if;
      end loop;
        Bin1 := tmp;
end rltointg;

procedure itb (bin : out std_logic_vector;
N : in integer; int : in integer) is
--This procedure is to convert integer to binary
variable temp_int : integer := int;
begin

    for i in 0 to N loop
        if (temp_int MOD 2 = 1) then
            bin(i) := '1';
            else bin(i) := '0';
        end if;
        temp_int := temp_int/2;
        end loop;
    end itb;
function exp (a : in std_logic_vector (0 to 3))
    return std_logic_vector is

variable z1 : real;
variable intgr : integer;
variable tem : std_logic_vector (7 downto 0);
```

```
begin
    flt (a, z1);
        z1 := 1.0 - z1 + z1 * z1 - z1 * z1 * z1;
        z1 := z1 * 256.0; -- 256 is for 8 bits
        rltointg (z1, intgr);
        itb (tem, 7, intgr);
        return tem;

    end exp;
begin
sg1 : process (x)
variable tem1 : std_logic_vector (7 downto 0);
variable tem2: integer;
begin

tem1 := exp(x);
y <= tem1;

end process sg1;

end segm_beh;
```

Verilog Floating Sum Description

```
module segma1(x,y);
input [0:3] x;
// x is a fraction in Q4 format, 0 < x < 1.
output [7:0] y;
reg [7:0] y;
always @ (x)
begin
    y = exp (x);
end
function [7:0] rti;
//This function convers real to integer with rounding off
input real r;

begin
    rti = r;
end
endfunction
function [7:0] exp;
input [0:3] a;
```

```
real z1;

integer i;
begin

    z1 = 0.0;
    for (i = 0; i <= 3; i = i + 1)
    begin
        if (a[i] == 1)

        z1 = z1 + 1.0 / 2**(i+1);
/*The above statement multiplies
each bit by its weight*/
end

    z1 = 1.0 - z1 + z1**2 - z1**3;
    z1 = z1 * 2**8;
    exp = rti(z1);

end
endfunction
endmodule
```

The simulation output of Listing 6.14 is shown in Figure 6.11.

x	1000	1111	0100

y	10100000	00011110	11001100

FIGURE 6.11 Simulation output of Listing 6.14.

EXAMPLE 6.13   IMPLEMENTATION OF IEEE 754 FLOATING-POINT REPRESENTATION

In Example 6.12, the real number was represented by four bits in Q4 format (fixed-point representation); the binary point is located just to the left of the most signifact bit of the number. For example, if the number is 1000_2, the binary point is located to the left of the most significant bit, in this case, the bit with value 1, and the value of the number is 8/16 = 0.5. Fixed-point representation is not used in computers due to its limited accuaracy. For four bits, the lowest number that can be represented with full accuracy is 1/16; any number less than 1/16 will not be represented with 100% accuracy and may be considered (depending on the rounding systems used) as zero. On the other hand, floating-point representation, which represents the number using exponent and mantessa fields, are more accurate and is the common representation for real numbers in computers. Because any number can be represented with unlimited variations of exponent and mantessa, the Institute of Electrical and Eleconics Engineers (IEEE) has established a standard format (IEEE 754) for the representation of floating-point numbers. According to this format, any floating-point number is represented by 32 bits for single precision and 64 bits for double precision, as shown in Figure 6.12. The value of the number N is:

```
N=(-1)ˢ x (1 + Fraction) x 2^(Exponent - Bias)
```

The fraction is less than one and is represented by 23 bits for single precision and 53 for double precision. The "1" that is added to the fraction is hidden and does not appear in the format (see Figure 6.12). The bias is 127 for single precision and 1023 for double precision. For example, the number 0.5×2^{-10} is represented in single precision as

```
00111010110000000000000000000000
```

where the sign is positive (0); the exponent is 127 + (-10) = 117_{10} = 011101012; the fraction is

$$10000000000000000000000_2$$

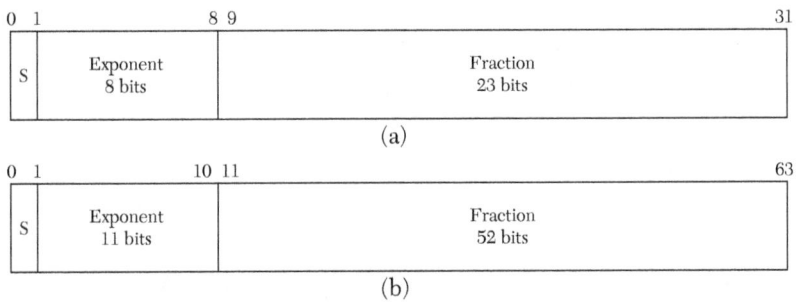

FIGURE 6.12 IEEE 754 floating-point representation. a) Single precision. b) Double precision.

Listing 6.15 shows a Verilog code for the conversion of any positive number to the IEEE 754 single-precision floating-point representation. For example, if the number is x = 24.0, it is converted first to (1 + fraction) $2^{(Exponent - Bias)}$ by dividing x by 2^4 which yields to x = 1.5 × 2^4; the fraction is 0.5, and the exponent is 127 + 4 = 131. The output IEEE_flt for x = 24.0 is:

```
01000001110000000000000000000000
```

LISTING 6.15 Verilog Code for Conversion to Single-Precision IEEE 754 Floating-Point Representation

```
module IEEEflt( start, IEEE_flt);
input start;
output [0:31] IEEE_flt;//This is the 32-bit IEEE
reg [7:0] j;
reg [0:31] IEEE_flt;
real x = 24.0; /* x is the iput number in decimal
  format; assume all numbers entered here are
  positive; see Exercise at the end of the Chapter*/

real Mant;integer i;
always @ (start,x,j)
begin
if (start == 1'b1)
begin

j =8'd127;//The exponent is in excess 127
if (x != 0.0)
begin
while (x >= 2.0)
begin
x = x/ 2.0;
j = j +1;
end

while (x < 1.0)
begin
x = x* 2.0;
j = j -1;
end

end

end
Mant = x-1.0;

for (i =0 ; i <= 22; i = i + 1)
begin
IEEE_flt [9+i] = 1'b0;
Mant = Mant * 2.0;
if ( Mant >= 1.0)
```

```
begin
IEEE_flt [9+i] = 1'b1;
Mant = Mant-1.0;
end
end
IEEE_flt [0] = 1'b0;
IEEE_flt [1:8] = j[7:0];
end

endmodule
```

6.4 Summary

In this chapter, `procedure` (VHDL), `task` (Verilog), and `function` (both VHDL and Verilog) have been covered. Procedures, tasks, and functions can optimize the style of writing HDL code; they shorten the code. The procedure/task has a declaration statement and a body, and it can have more than one input and more than one output. On the other hand, a function can have more than one input but only one output. VHDL allows `procedure` calls to be written inside functions; Verilog does not allow such calls.

6.5 Exercises

1. In Listing 6.6, negative binary numbers were converted to integers by reverse-negating them. Another approach is to find the integer value of any twoscomplement number by detecting the beginning and end of strings of ones in the number (see Case Study 3.1). Apply this approach and write both the VHDL and Verilog codes for such a conversion. Verify your code by simulation and compare the two approaches.

2. In Listing 6.8, a VHDL Booth procedure was written that multiplies 4x4 bits. Modify the VHDL procedure so it can multiply any *NxN*. Verify your answer by simulation.

3. In Listing 6.8 (VHDL), `temtotal = tem0 + tem1 + tem2` was added by converting to integer, adding, and then changing back to binary. An alternate approach is to adjust the width of all partial products (after they are calculated) to be the same as `temtotal` and then add. Perform this alternate approach and verify your results by simulation.

4. Derive the HDL code (both VHDL and Verilog) for the function y = Ln x; 0 < x < 1. Express y in Q15 format. Hint: use polynomial representation for Ln (x).

5. Write a function to calculate the area of a sphere, given the radius.

6. Rewrite the code for the Booth array (Listing 6.9) to simulate an *NxN array multiplier. Hint: For N layers, the bottom layer contains N cells, and the number of cells increases by one.*

7. In Listings 6.10 and 6.14, the VHDL code for the procedure `rlto-intg` does not round off. For example, if the real value is 215.3175, the procedure will output 216 for the integer equivalent. On the other hand, the Verilog code of the procedure is rounding off. Rewrite the VHDL procedure so it will round off. Adjust the VHDL code to output the same value as the Verilog.

8. Repeat Listing 6.15 but for positive or negative numbers and use double precision.

6.6 Reference

Hayes, J., *Computer Architecture and Organization*, 3rd ed. McGraw Hill, Boston, Massachusetts, USA, 1998.

MIXED-TYPE DESCRIPTION

Chapter Objectives

- Learn how to use different types (styles) of descriptions to write HDL modules
- Learn which type or style of description to use for optimal writing style
- Understand the concept of packages in VHDL and how to use them
- Practice with single and multidimensional arrays
- Practice with real (floating) numbering systems
- Practice user-defined types
- Practice using finite sequential-state machines
- Review and understand the steps needed to design and describe a basic computer

7.1 Why Mixed-Type Description?

Our definition of mixed-type description is an HDL code that mixes different types (styles) of descriptions within the same module. In previous chapters, description codes consisted mainly of one type such as data flow (see Chapter 2), behavioral (see Chapter 3), structural (see Chapter 4), or switch level (see Chapter 5). Here, the code is written using more than one type of description in the same module.

In fact, it is very common to write mixed descriptions because each part of the selected system may be written best by a certain type of description. For example, consider a system that performs two operations: addition

($Z = x + y$) and division ($Z = x / y$). The code can be written using a few styles. The first style is to use behavioral statements to model the addition and the division. This style is somewhat easy to write because HDL has built-in addition and division functions. The behavioral statements are written inside `process` (VHDL) or `always` (Verilog) as:

```
Z := a + b; --VHDL    or    Z := x / y; --VHDL
Z = a + b; //Verilog     or    Z = x / y; //Verilog
```

The ultimate goal of VHDL or Verilog description is to synthesize the description on electronic chips (see Chapter 10). If the behavioral description is used, the author has no control over selecting the components or the methods used to implement the addition and division. The HDL package may contain addition or division algorithms not suitable for the current needs of the user. For example, the addition algorithm might need to be as fast as possible; to achieve this fast addition, adders such as carry lookahead or carry-save should be used. There is no guarantee, however, that behavioral description will implement those adders in its addition function. A second option is to use data-flow or structural description. These descriptions can be implemented to describe the specific adder. It is, however, hard to implement these descriptions in complex algorithms such as division. The third option is to use a mixture of two types (styles) of descriptions: structural or data-flow for addition and behavioral for division. This description is referred to here as a mixed type.

Before considering examples of mixed-type description, some tools and commands that could be used to write more complex codes are discussed. Section 7.2 discusses user-defined types, and Section 7.3 discusses packages and arrays.

7.2 VHDL User-Defined Types

VHDL has an extensive set of predefined data types such as `bit`, `std_logic`, `array`, and `natural` (see Chapter 1). In some applications, other data types that are not included in the basic HDL package are needed. Examples of such types are weekdays, weather, or grades. These are user-defined types. To instantiate a user-defined type, the predefined word `type` is used. An example of instantiating a user-defined type is:

```
type week_days is (mon, tues, wed, th, fr, sat, sun);
```

This statement declares a user-defined type by the name of `week_days`, and `type` is a predefined word; the elements or members of `week_days` are `mon`, `tues`, `wed`, `th`, `fr`, `sat`, and `sun`. Another example of a user-defined type is:

```
type states is (S0, S1, S2, S3);
```

This statement declares a user-defined type by the name of `states`. The elements (members) of `states` are `s0`, `s1`, `s2`, and `s3`. Another example of a user-defined type is:

```
type weather is (sunny, cloudy, rain, snow);
```

This statement declares a user-defined type by the name of `weather`, and the elements (members) of `weather` are `sunny`, `cloudy`, `rain`, and `snow`. Another example of a user-defined type is:

```
type decimal_numbers is ('0', '1', '2', '3', '4', '5');
```

This statement declares a user-defined type by the name of `decimal_numbers`; the elements (members) of `decimal_numbers` are the integers: `'0'`, `'1'`, `'2'`, `'3'`, `'4'`, and `'5'`. If the members of a type are digits, they should be written between two apostrophes, such as '5.' Another example of a user-defined type is:

```
type grades is (A, B, C, D, F, I);
```

This statement declares a user-defined type by the name of `grades`; the elements (members) of `grades` are `A`, `B`, `C`, `D`, `F`, and `I`. The statement:

```
signal scores : grades;
```

declares `scores` as the of type `grades`. This means that `scores` can be assigned a value of `A`, `B`, `C`, `D`, `F`, or `I`.

A subtype of a type can be declared by using the predefined word `subtype`, as shown below:

```
subtype failed is grades range D to I;
signal scores : failed;
```

where `failed` is a subtype of `grades` and has a range from `D` to `I`, `range` is a predefined attribute, so `scores` can be assigned a value of `D`, `F`, or `I`. Another example is:

```
subtype values is integer range 10 to 100;
signal x : values;
```

Signal `x` can be assigned an integer value from 10 to 100. Remember from Chapter 1 that `integer` is a predefined type.

7.3   VHDL Packages

Packages constitute an essential part of VHDL description. Packages allow the user to access built-in constructs. Packages may include type and subtype declarations, constant definitions, function and procedure, and component declarations. VHDL has default built-in packages that include predefined words such as bit, bit_vector, and integer. In addition to the defaults, the user can attach a variety of packages to the VHDL module.

Several packages have been implemented in previous chapters; examples include packages authored by **IEEE**: IEEE.STD_LOGIC_1164, IEEE.STD_LOGIC_ARITH, IEEE.NUMERIC_STD, IEEE.STD_LOGIC_UNSIGNED, and IEEE.STD_LOGIC_SIGNED. In addition to such built-in packages, the user can attach other packages to the VHDL module. A package consists of a declaration and a body. The declaration states the name (identifier) of the package and the names (identifiers) of types, procedures, functions, and components. The body of the package contains the code for all the identifiers listed in the declaration. Listing 7.1 shows an example of a user-defined package.

LISTING 7.1 An Example of a VHDL Package

```
package conversions is
    type wkdays is (mon, tue, wed, th, fr);
    procedure convert (a : in bit; b : out integer);
    function incr (b : std_logic_vector) return std_logic_vec-
tor;
    end conversions;
```

package body of conversions is written as:

```
procedure convert (a : in bit; b : out integer) is
begin
.......
end convert;
function incr (b : std_logic_vector) return std_logic_vector is
begin
...
end incr;
end conversions;
```

As shown in Listing 7.1, the name of the package is conversions; the package contains Type wkdays, Procedure convert, and function incr. The package body lists the code of the procedure convert and function incr. Listing 7.2 shows another package example. The name of the package is codes; the members are add, mul, divide, and none.

LISTING 7.2 An Example of a VHDL Package

```
library ieee;
use ieee.std_logic_1164.all;
package codes is
type op is (add, mul, divide, none);
end;
use work.codes;

entity ALUS2 is
    port (a, b : in std_logic_vector (3 downto 0);
        cin : in std_logic; opc : in op;
        z : out std_logic_vector (7 downto 0);
        cout : buffer std_logic);
end ALUS2;
```

To use this package in a VHDL module, the statement use work. Codes; is entered. Notice that in the entity ALUS2, opc is declared as a of type op; this means opc can be assigned a value of add, mul, divide, or none.

7.3.1 Implementations of Arrays

As discussed in Chapter 1, arrays are a data type; all elements of the array should have the same type. The array can be single-dimensional or multidimensional. HDL allows for multidimensional arrays. Arrays can be composed of signals, constants, or variables. This section covers arrays in detail, as well as several implementations.

7.3.1.1 Single-Dimensional Arrays

Single-dimensional arrays have single index. They are declared as follows:

VHDL Single-Dimensional Array

The two statements

```
type datavector is array (3 downto 0) of wordarray;
subtype wordarray is std_logic_vector (1 downto 0);
```

declare an array by the name of `datavector`; it has four elements, and each element is two bits. An example of this array is:

("11", "10", "10", "01")

The value of each element of the array in decimal is:

```
datavector(0) = 1, datavector(1) = 2, datavector(2) = 2,
datavector(3) = 3.
```

Verilog Single-Dimensional Array

In Verilog, arrays are declared using the predefined word `reg`. An example of array declaration in Verilog is:

```
reg [1:0] datavector[0:3];
```

This declares an array by the name of `datavector`; it has four elements, and each element is two bits. An example of this array is:

```
datavector[0] = 2'b01;
datavector[1] = 2'b10;
datavector[2] = 2'b10;
datavector[3] = 2'b11;
```

The following examples cover array implementations.

EXAMPLE 7.1   FIND THE GREATEST AMONG N ELEMENTS OF AN ARRAY

Listing 7.3 shows the HDL code for finding the greatest element (`grtst`) of array `a`. First, initialize `grtst` with 0. Then, `grtst` is compared with the first element of array `a`. If the first element is greater than `grtst`, then set `grtst` to be equal to the first element; otherwise, `grtst` is left unchanged. The same is done with the other elements.

LISTING 7.3 HDL Code for Finding the Greatest Element of an Array: VHDL and Verilog

```
VHDL: Finding the Greatest Element of an Array
library IEEE;
use IEEE.STD_LOGIC_1164.all;

--Build a package for an array
package array_pkg is
constant N : integer := 4;
--N+1 is the number of elements in the array.

constant M : integer := 3;
```

```
--M+1 is the number of bits of each element
--of the array.
subtype wordN is std_logic_vector (M downto 0);
type strng is array (N downto 0) of wordN;

end array_pkg;
library IEEE;
use IEEE.STD_LOGIC_1164.ALL;
use work.array_pkg.all;
-- The above statement makes the package array_pkg visible
-- in this module.

entity array1 is
    generic (N : integer :=4; M : integer := 3);

--N + 1 is the number of elements in the array; M = 1 is
-- the number of bits of each element.
    Port (a : inout strng;
          z : out std_logic_vector (M downto 0));
end array1;

architecture max of array1 is

begin
com: process (a)
variable grtst : wordN;
begin

--enter the data of the array.
    a <= ("0110", "0111", "0010", "0011", "0001");

    grtst := "0000";

    lop1 : for i in 0 to N loop

    if (grtst <= a(i)) then
        grtst := a(i);
        report " grtst is less or equal than a";
    -- use the above report statement if you want to
    -- monitor the progress of the program.
        else
        report "grtst is greater than a";
```

```
         -- Use the above report statement to monitor the
         -- progress of the program
            end if;
            end loop lop1;
     z <= grtst;


end process com;


end max;
```

Verilog: Finding the Greatest Element of an Array

```
module array1 (start, grtst);
parameter N = 4;
parameter M = 3;
input start;
output [3:0] grtst;
reg [M:0] a[0:N];

/*The above statement is declaring an array of N + 1 elements;
each element is M bits. */

reg [3:0] grtst;
integer i;
always @ (start)
begin
a[0] = 4'b0110;
a[1] = 4'b0111;
a[2] = 4'b0010;
a[3] = 4'b0011;
a[4] = 4'b0001;
grtst = 4'b0000;
for (i = 0; i <= N; i= i +1)
    begin
        if (grtst <= a[i])
        begin
            grtst = a[i];
            $display (" grtst is less or equal than a");
// use the above statement to monitor the program
        end
        else
        $display (" grtst is greater than a");
```

```
// use the above statement to monitor the program
    end
end
endmodule
```

EXAMPLE 7.2 MULTIPLICATION OF TWO SIGNED *N*-ELEMENT VECTORS USING ARRAYS

This example describes the multiplication of two signed vectors. The two vectors have the dimension of $1{\times}(N{+}1)$ and $(N{+}1){\times}1$. Chapter 6 covered the multiplication of two three-element vectors; here, arrays are used to expand the multiplication to N elements. Listing 7.4 shows the description of two signed vectors of N elements. A Booth algorithm is implemented, (see Chapter 3), and code from Chapter 6 is used to multiply signed numbers in twos-complement format. The algorithm is written as `procedure` in VHDL or `task` in Verilog.

In VHDL, the procedure `booth` is included in a package. The package `booth_pkg` is declared as:

```
package booth_pkg is
constant N : integer := 4;

constant M : integer := 3;

subtype wordN is signed (M downto 0);
type strng is array (N downto 0) of wordN;
procedure booth (X, Y : in signed (3 downto 0);
    Z : out signed (7 downto 0));
end booth_pkg;
```

The package `booth_pkg` includes the procedure `booth` and an array declaration. The array is declared as a user-defined type, `strng`, and a user-defined subtype, `wordN`. It has $N + 1$ elements; each element is $M + 1$ bits. In our example, N is selected as 4, and $M = 3$, so the array has five elements, and each element is four bits in signed (twos complement) format.

In Verilog, the array is declared as:

```
reg signed [M:0] b[0:N];
```

which is an array of $N + 1$ elements. Each element is $M + 1$ bits.

LISTING 7.4 Multiplication of Two Signed N-Element Vectors: VHDL and Verilog

VHDL: Multiplication of Two Signed N-Element Vectors

```vhdl
library IEEE;
use IEEE.STD_LOGIC_1164.all;
use ieee.numeric_std.all;

package booth_pkg is
constant N : integer := 4;
--N + 1 is the number of elements in the array.

constant M: integer := 3;
--M + 1 is the number of bits of each element
--of the array.

subtype wordN is signed (M downto 0);
type strng is array (N downto 0) of wordN;
procedure booth (X, Y : in signed (3 downto 0);
    Z : out signed (7 downto 0));

end booth_pkg;

package body booth_pkg is
procedure booth (X, Y : in signed (3 downto 0);
    Z : out signed (7 downto 0)) is
--Booth algorithm here is restricted to 4x4 bits.
--It can be adjusted to multiply any NxN bits.
variable temp : signed (1 downto 0);
    variable sum : signed (7 downto 0);
    variable E1 : unsigned (0 downto 0);
    variable Y1 : signed (3 downto 0);
begin

sum := "00000000"; E1 := "0";
    for i in 0 to 3 loop
    temp := X(i) & E1(0);
    Y1 := -Y;
    case temp is
        when "10" => sum (7 downto 4) :=
        sum (7 downto 4) + Y1;
        when "01" => sum (7 downto 4) :=
```

```
            sum (7 downto 4) + Y;
            when others => null;
    end case;
    sum := sum srl 1;
    sum(7) := sum(6);
    E1(0) := x(i);
    end loop;
    if (y = "1000") then

            sum := -sum;
            --If Y = 1000; then Y1 is calculated as 1000;
            --that is -8, not 8 as expected. This is because Y1 is
            --4 bits only. The statement sum = -sum corrects
            --this error.
            end if;
    Z := sum;
    end booth;
end booth_pkg;
-- We start writing the multiplication algorithm using
-- the package booth_pkg
library IEEE;
use IEEE.STD_LOGIC_1164.ALL;
use ieee.numeric_std.all;
use work.booth_pkg.all;

entity vecor_multply is
generic (N : integer := 4; M : integer := 3);
--N + 1 is the number of elements in the array; M + 1 is
-- the number of bits of each element.

    Port (a, b : in strng; d : out signed (3*N downto 0));
end vecor_multply;
architecture multply of vecor_multply is

begin
process (a, b)
variable temp : signed (7 downto 0);
variable temp5 : signed (3*N downto 0) := "0000000000000";

begin
```

```
for i in 0 to 4 loop
booth(a(i), b(i), temp);

--accumulate the partial products in the product temp5
temp5 := temp5 + temp;
end loop;
d <= temp5;
end process;

end multply;
```

Verilog: Multiplication of Two Signed N-Element Vectors

```verilog
module vecor_multply (start, d);
parameter N = 4;
parameter M = 3;
input start;
output signed [3*N:0] d;
reg signed [M:0] a[0:N];
reg signed [M:0] b[0:N];
reg signed [3*N:0] d;
reg signed [3*N:0] temp;
integer i;

always @ (start)
begin
    a[0] = 4'b1100;
    a[1] = 4'b0000;
    a[2] = 4'b1001;
    a[3] = 4'b0011;
    a[4] = 4'b1111;

    b[0] = 4'b1010;
    b[1] = 4'b0011;
    b[2] = 4'b0111;
    b[3] = 4'b1000;
    b[4] = 4'b1000;
    d = 0;
    for (i = 0; i <= N; i = i + 1)
        begin
            booth (a[i], b[i], temp);
            d = d + temp;
        end
end
```

```
task booth;
input signed [3:0] X, Y;
output signed [7:0] Z;
reg signed [7:0] Z;
reg [1:0] temp;
integer i;
reg E1;
reg [3:0] Y1;

begin
Z = 8'd0;
E1 = 1'd0;

for (i = 0; i < 4; i = i + 1)
    begin
        temp = {X[i], E1}; //This is catenation
        Y1 = -Y; //Y1 is the 2'complement of Y
            case (temp)
            2'd2 : Z[7:4] = Z[7:4] + Y1;
            2'd1 : Z[7:4] = Z[7:4] + Y;
            default : begin end

            endcase
        Z = Z >> 1;
        /*The above statement is a logical shift of
        one position to the right*/

        Z[7] = Z[6];
/*The above two statements perform arithmetic shift where the
sign of the number is preserved after the shift. */
    E1 = X[i];

    end
if (Y == 4'b1000)

/* If Y = 1000, then Y1 = 1000 (should be 8 not -8).
This error is because Y1 is 4 bits only.
The statement Z = -Z adjusts the value of Z. */

Z = -Z;
end
```

```
endtask
endmodule
```

Figure 7.1 shows the simulation output of the vector multiplication. Array a is written here in integer format for convenience:

$$a = \{-1 \quad 3 \quad -7 \quad 0 \quad -4\}$$
$$b = \{-8 \quad -8 \quad 7 \quad 3 \quad -6\}$$

multiplying a × b = 8 − 24 − 49 + 0 + 24 = −41 =d

As shown in Figure 7.1, d has the correct value of −41.

a   1111 0011 1001 0000 1100

b   1000 1000 0111 0011 1010

d   1111111010111

FIGURE 7.1 Simulation output of vector multiplication.

7.3.1.2 Two-Dimensional Arrays

VHDL and Verilog (after 2003) allow for multidimensional arrays. In VHDL, two-dimensional arrays are described by using `type` statements. For example, the statements

```
subtype wordg is integer;
type singl is array (2 downto 0) of wordg;
type doubl is array (1 downto 0) of singl;
```

describe a two-dimensional array. Each single-dimensional array has three elements, and each element is an integer. An example of a two-dimensional array is the array y:

$$y = ((10 \quad 5 \quad 6), (3 \quad -2 \quad 7))$$

The elements of the array y are:

$$y(0)(0) = 7 \text{ refers to element 0 of array 0}$$
$$y(1)(1) = 5 \text{ refers to element 1 of array 1}$$
$$y(2)(0) = 3 \text{ refers to element 2 of array 0}$$
$$y(2)(1) = 10 \text{ refers to element 2 of array 0}$$

In Verilog, the statement

```
reg [5:0] Y [0:4] [0:4];
```

represents a two-dimensional array (a matrix) with five rows and five columns; each element of the matrix is six bits. For example, such an array can be:

$$[25,24,23,22,21], [20,19,18,17,16], [15,14,13,12,11],$$
$$[10,9,8,7,6], [5,4,3,2,1]$$

with Y[0][0]=1,Y[0][1]=2, Y[4]4]=25

Another two-dimensional array statement

```
reg [3:0] Y [0:5] [0:3];
```

represents a two-dimensional array (a matrix) with six rows and four columns; each element of the matrix is four bits.

EXAMPLE 7.3   TWO-DIMENSIONAL ARRAYS

This example considers a two-dimensional array. Listing 7.5 shows the HDL description of a two-dimensional array. In VHDL, the package twodm_array is used to declare a two- dimensional array with five single arrays; each single array has five elements. The elements are of type integer.

In Verilog, five single arrays of five elements where each element is six bits has been created. The loop assigns a value (K) to each elemt Y[i][j]; K is incremented by one starting from Y[0][0] and continuing until Y[4][4]. To find the value of any element, the user enters N (row) and M(column).

LISTING 7.5 HDL Code for a Two-Dimensional Array

VHDL Two-Dimensional Array
```
library IEEE;
use IEEE.STD_LOGIC_1164.all;

--Build a package to declare the array
package twodm_array is

constant N : integer := 4;
-- N+1 is the number of elements in the array.
-- this is [N+1,N+1] matrix with N+1 rows and N+1 columns
```

```
subtype wordg is integer;
type strng1 is array (N downto 0) of wordg;
type strng2 is array (N downto 0) of strng1;
end twodm_array;
--use the package to describe a two-dimensional array

library IEEE;
use IEEE.STD_LOGIC_1164.ALL;
use work.twodm_array.all;

-- The above statement instantiates the package twodm_array

entity two_array is
    Port (N, M : integer; z : out integer);
end two_array;

architecture Behavioral of two_array is

begin
com : process (N, M)
variable t : integer;
constant y : strng2 := ((7, 6, 5, 4, 3), (6, 7, 8, 9, 10),
            (30, 31, 32, 33, 34), (40, 41, 42, 43, 44),
            (50, 51, 52, 53, 54));
begin

t := y (N)(M);

--Look at the simulation output to identify the elements of the
--array
z <= t;
end process com;
end Behavioral;
```

Verilog Two-Dimensional Array
```
module twodmarrays(start,N,M,Z );
parameter N1 = 4;
parameter M1 = 4;
input start;
input [2:0] N,M;
output integer Z;
```

```
reg [5:0] Y [0:4] [0:4];
//The following statements generate the array as
//[25,24,23,22,21], [20,19,18,17,16], [15,14,13,12,11],
// [10,9,8,7,6], [5,4,3,2,1] with Y[0][0]=1,Y[0][1]=2

integer i,j,K = 0;
always @ ((start == 1'b1),N,M)
begin
K = 0;
for (i = 0; i <= N1; i= i +1)
begin
for (j = 0; j <= M1; j= j +1)

    begin
K= K +1;
    Y[i][j]= K;

end
    end
Z = Y[N][M];
end
endmodule
```

Figure 7.2 shows the VHDL simulation output of Listing 7.5. From the simulation:

y[0][0], the first element in the first array = 54

y[0][3], the fourth element of the first array = 51

y[2][4], the fifth element of the third array = 30

N	0	0	0	2	4	4
M	0	3	4	4	4	3
Z	54	51	50	30	7	6

FIGURE 7.2 VHDL simulation output of the array in Listing 7.5.

EXAMPLE 7.4 MATRIX ADDITION

Here, an HDL code is written to add two matrices. The matrices must have the same dimensions. The addition of the two matrices yields a matrix with the same dimension as the two matrices. Consider the addition of the two matrices:

$$
\begin{bmatrix}
1 & 2 & 3 & 4 & 5 \\
6 & 7 & 8 & 9 & 10 \\
11 & 12 & 13 & 14 & 15 \\
16 & 17 & 18 & 19 & 20 \\
21 & 22 & 23 & 24 & 25
\end{bmatrix}
+
\begin{bmatrix}
4 & 5 & 6 & 7 & 8 \\
9 & 10 & 11 & 12 & 13 \\
14 & 15 & 16 & 17 & 18 \\
19 & 20 & 21 & 22 & 23 \\
24 & 25 & 26 & 27 & 28
\end{bmatrix}
$$

The addition is done by adding row by row. Listing 7.6 shows the HDL description of the addition of two [5×5] matrices. In VHDL, the two matrices are entered as inputs. In Verilog, the two matrices are generated by two loops.

LISTING 7.6 VHDL Description: Addition of Two [5×5] Matrices

```
-- First, write a package to declare a two-dimensional
--array with five elements
library IEEE;
use IEEE.STD_LOGIC_1164.all;

package twodm_array is

constant N : integer := 4;
-- N+1 is the number of elements in the array.
-- This is an NxN matrix with N rows and N columns.
subtype wordg is integer;
type strng1 is array (N downto 0) of wordg;
type strng2 is array (N downto 0) of strng1;
end twodm_array;

--Second, write the code for addition
library IEEE;
use IEEE.STD_LOGIC_1164.ALL;
use work.twodm_array.all;
entity matrices is
    Port (x, y : strng2; z : out strng2);
```

```
--strng2 type is 5x5 matrix
end matrices;

architecture sum of matrices is

begin
com : process (x, y)
variable t : integer := 0;
begin

for i in 0 to 4 loop
for j in 0 to 4 loop
t := x(i)(j) + y(i)(j);
z(i)(j) <= t;

end loop;

end loop;
end process com;
end sum;
```

Verilog Description: Addition of Two [5×5] Matrices
```
module sumMatrices(start,N,M,Z );
//The program generates two matrices
//X,Y (two dimensional arrays)
//and add them up and store the sum
//in matrix sum
parameter N1 = 4;
parameter M1 = 4;
input start;
input [2:0] N,M;
output integer Z;
reg [6:0] sum [0:4] [0:4];
reg [5:0] Y [0:4] [0:4];
reg [5:0] X [0:4] [0:4];
//initial values in the array are generated by the loop
//statements as:
//[25,24,23,22,21], [20,19,18,17,16], [15,14,13,12,11],
//[10,9,8,7,6], [5,4,3,2,1}with Y[0][0]=1,Y[0][1]=2
integer i,j,K = 0;
always @ ((start == 1'b1),N,M)
```

```
begin
K =0;
for (i = 0; i <= N1; i= i +1)
begin
for (j = 0; j <= M1; j= j +1)
    begin
K= K +1;
   Y[i][j]= K;
    X[i][j] = K + 3;

end
    end
Z = Y[N][M];

for (i = 0; i <= N1; i= i +1)
begin
for (j = 0; j <= M1; j= j +1)
    begin
sum[i][j]= X[i][j] + Y[i][j];

end
end
end
endmodule
```

After simulation of the above code, the sum matrix is displayed as:

$$\begin{bmatrix} 5 & 7 & 9 & 11 & 13 \\ 15 & 17 & 19 & 21 & 23 \\ 25 & 27 & 29 & 31 & 33 \\ 35 & 37 & 39 & 41 & 43 \\ 45 & 47 & 49 & 51 & 53 \end{bmatrix}$$

7.4   Mixed-Type Description Examples

This section presents some examples of mixed-type descriptions. The strategy is to use the type or style of description that best fits the needs of the system (or parts of the system) to be described. Structural or data-flow description may be the best fit for any part of the system that needs specific hardware architecture. Behavioral description is best used when describing, for example, a complex arithmetic operation with no specific hardware architecture is desired. If the system to be described consists of transistors or transistor-based circuits, then switch-level description may be the best fit.

EXAMPLE 7.5   HDL DESCRIPTION OF AN ARITHMETIC-LOGIC UNIT

The arithmetic-logic unit (ALU) is one of the major units in a computer. The unit performs arithmetic operations such as addition, subtraction, and division, and logical operations such as AND, OR, and INVERT. The ALU in this example has three inputs (see Figure 7.3) a, b, and cin. Inputs a and b are four bits, and cin is one bit. The output z is six bits. The unit can perform addition, multiplication, integer division, or no operation. To select one operation out of the available four, a two-bit signal opc is implemented to select the desired operation. The selection is shown in Table 7.1.

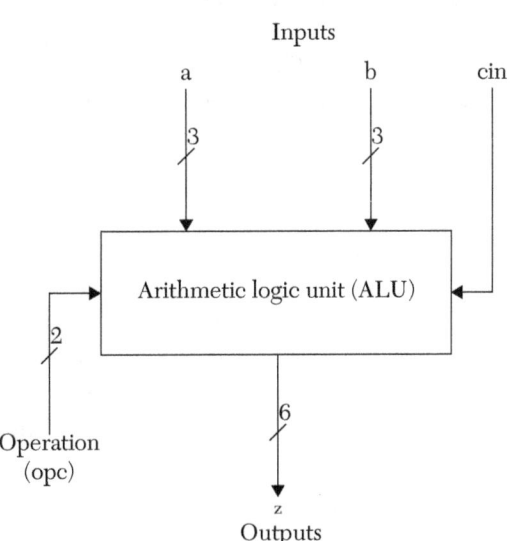

FIGURE 7.3 Block diagram of the arithmetic-logical unit.

TABLE 7.1 Operation Selection of the ALU

Operation Code (opc)	Operation
00	Addition
01	Multiplication
10	Integer division
11	No operation

In this example, the implementation of carry-lookahead adders is desired. Because the adders are specified, the most convenient style of description is structural or data flow. Chapter 2 described these adders using data-flow description, so it is repeated here. Recall that data-flow description is usually implemented by writing the Boolean functions of the system. Because no specific hardware structure is specified, multiplication and division can be described by behavioral statements within a process. The multiplication operator (∗) and the division operator (/) are used to perform the multiplication and division.

Listing 7.7 shows the VHDL code for the ALU. The package codes_Arithm declares a user-defined type op; the elements of op are the operation codes for addition (add), multiplication (mul), division (divide), and no operation (none). The package also includes a user-defined function, TO_UNSIGN. This function converts from integer to unsigned. This function (or a similar one) may be built in to some vendors' packages. Converting between integer and unsigned is needed because many VHDL simulators cannot perform the unsigned division z = a / b. VHDL can perform integer division.

Listing 7.7 also shows the Verilog code for the ALU. In Verilog, it is easy to perform addition, multiplication, and division on unsigned numbers; no conversion to integer is needed. On the other hand, Verilog does not have as extensive user-defined type statements as does VHDL. The parameter statement is used to assign values to ALU operations. For example, to assign 00 to the addition operation code add, write:

```
parameter add = 0;
```

LISTING 7.7 HDL Description of an ALU: VHDL and Verilog

VHDL ALU Description
```
--Here the code for a package for user-defined
--type and function is written.
library ieee;
use ieee.std_logic_1164.all;
use IEEE.STD_LOGIC_1164.ALL, IEEE.NUMERIC_STD.ALL;
package codes_Arithm is
type op is (add, mul, divide, none);
-- type op is for the operation codes for the ALU.
--The operations are: addition, multiplication,
--division, and no operation

function TO_UNSIGN (b : integer) return unsigned;
end;

package body codes_Arithm is
function TO_UNSIGN (b : integer) return unsigned is
```

```
--The function converts integers to unsigned. This function
--can be omitted if it is included in the vendor's package;
--the vendor's package, if available, should be attached.

variable temp : integer;
variable bin : unsigned (5 downto 0);
begin
temp := b;
for j in 0 to 5 loop

    if (temp MOD 2 = 1) then
    bin (j) := '1';
    else bin (j) := '0';
    end if;
    temp := temp/2;
    end loop;
    return bin;

end TO_UNSIGN;
end codes_Arithm;

--Now we write the code for the ALU
library IEEE;
use IEEE.STD_LOGIC_1164.ALL;
use ieee.numeric_std.all;
use work.codes_arithm.all;

--The above use statement is to set the user-
--defined package "codes_arithm.all" visible to this
-- module.

entity ALU_mixed is

port (a, b : in unsigned (2 downto 0);
      cin : in std_logic;
     opc : in op; z : out unsigned (5 downto 0));
--opc is of type "op"; type op is defined in the
--user-defined package "codes_arithm"
end ALU_mixed;
architecture ALU_mixed of ALU_mixed is
    signal c0, c1 : std_logic;
    signal p, g : unsigned (2 downto 0);
```

```
    signal temp1 : unsigned (5 downto 0);

begin

--The following is a data flow-description of a 3-bit
-- lookahead adder. The sum is stored in the three least
-- significant bits of temp1.
adder.
-- The carry out is stored in temp1(3).

g(0) <= a(0) and b(0);
g(1) <= a(1) and b(1);
g(2) <= a(2) and b(2);
p(0) <= a(0) or b(0);
p(1) <= a(1) or b(1);
p(2) <= a(2) or b(2);
c0 <= g(0) or (p(0) and cin);
c1 <= g(1) or (p(1) and g(0)) or
    (p(1) and p(0) and cin);
temp1(3) <= g(2) or (p(2) and g(1)) or (p(2) and p(1)
    and g(0)) or (p(2) and p(1) and p(0) and cin);

--temp1(3) is the final carryout of the adders
    temp1(0) <= (p(0) xor g(0)) xor cin;
    temp1(1) <= (p(1) xor g(1)) xor c0;
    temp1(2) <= (p(2) xor g(2)) xor c1;
    temp1 (5 downto 4) <= «00»;

process (a, b, cin, opc, temp1)
--The following is a behavioral description for the
-- multiplication and division functions of the ALU.
    variable temp : unsigned (5 downto 0);
    variable a1, a2, a3 : integer;
begin
    a1 := TO_INTEGER (a);
    a2 := TO_INTEGER (b);
--The predefined function «TO_INTEGER»
--converts unsigned to integer.
--The function is a member of the VHDL package
-- IEEE.numeric.
    case opc is
        when mul =>
```

```
        a3 := a1 * a2;
        temp := TO_UNSIGN(a3);
--The function «TO_UNSIGN» is a user-defined function
--written in the user-defined package «codes_arithm.»
      when divide =>
        a3 := a1 / a2;
        temp := TO_UNSIGN(a3);

    when add =>
        temp := temp1;
    when none =>
        null;

end case;

z <= temp;
end process;

end ALU_mixed;
```

Verilog ALU Description

```
module ALU_mixed (a, b, cin, opc, z);
parameter add = 0;
parameter mul = 1;
parameter divide = 2;
parameter nop = 3;
input [2:0] a, b;
input cin;
input [1:0] opc;
output [5:0] z;
reg [5:0] z;
wire [5:0] temp1;
wire [2:0] g, p;
wire c0, c1;

// The following is data-flow description
// for 3-bit lookahead adder
    assign g[0] = a[0] & b[0];
    assign g[1] = a[1] & b[1];
    assign g[2] = a[2] & b[2];
    assign p[0] = a[0] | b[0];
```

```
            assign p[1] = a[1] | b[1];
            assign p[2] = a[2] | b[2];
            assign c0 = g[0] | (p[0] & cin);
            assign c1 = g[1] | (p[1] & g[0]) | (p[1] & p[0] & cin);
            assign temp1[3] = g[2] | (p[2] & g[1]) | (p[2] & p[1]
                & g[0]) | (p[2] & p[1] & p[0] & cin);
            // temp1[3] is the final carryout of the adders
            assign temp1[0] = (p[0] ^ g[0]) ^ cin;
            assign temp1[1] = (p[1] ^ g[1]) ^ c0;
            assign temp1[2] =. (p[2] ^ g[2]) ^ c1;
            assign temp1[5:4] = 2'b00;

            //The following is behavioral description

    always @ (a, b, cin, opc, temp1)
    begin
        case (opc)
            mul : z = a * b;
            add : z = temp1;
            divide : z = a / b;
            nop : z = z;
        endcase
    end
    endmodule
```

Figure 7.4 shows the simulation output of the ALU. Notice the integer division of $5 / 7 = 0$.

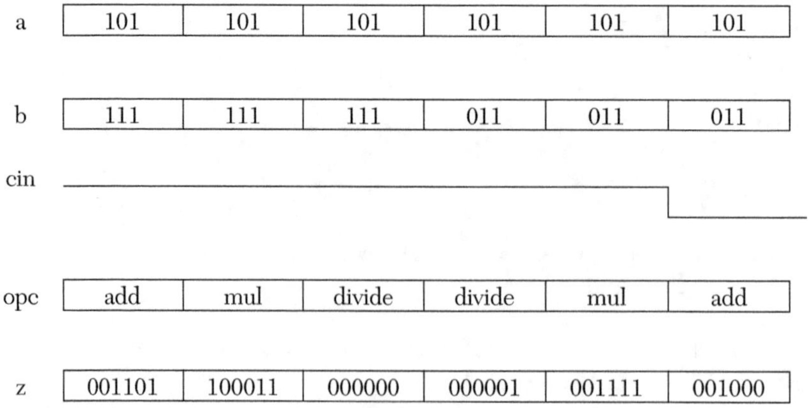

a	101	101	101	101	101	101

b	111	111	111	011	011	011

cin

opc	add	mul	divide	divide	mul	add

z	001101	100011	000000	000001	001111	001000

FIGURE 7.4 Simulation output of the ALU.

EXAMPLE 7.6   HDL DESCRIPTION OF A 16×8 SRAM

In Chapter 4, a static memory cell using structural description was described. In this example, a 16×8 SRAM is described. Because the description of this memory in structural style would be huge, and no specific logic is required, behavioral statements to describe the memory will be implemented. Figure 7.5 shows a block diagram of the memory. The memory has eight-bit input data (Data_in), eight-bit output data (Data_out), four-bit address bus (ABUS), a chip select (CS), and read/write signal (R_ WR).

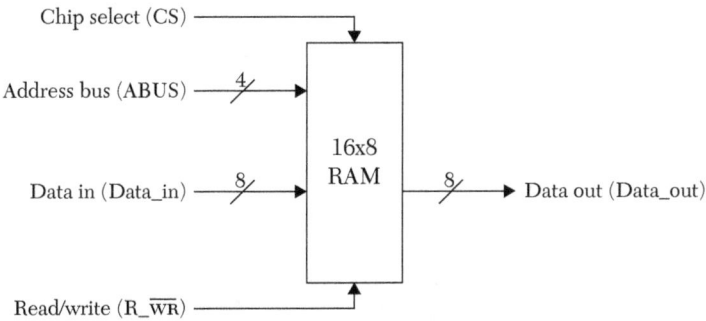

FIGURE 7.5 A block diagram of 16×8 static memory.

The function table of the memory is shown in Table 7.2. Listing 7.8 shows the HDL code for the RAM.

TABLE 7.2 Function Table of SRAM

CS	R_ WR	Data_out	Memory Function
0	x	Z (high impedance)	The memory is deselected
1	1	M (ABUS)	This is a read; M refers to memory locations, and contents of a memory location pointed to by ABUS are placed in the output data
1	0		This is a write cycle; data in the Data_in are stored in M (ABUS)

Referring to Listing 7.8 VHDL, an array to represent the memory is implemented. Because the memory is 16×8 bits, an array of sixteen elements is used, and each element is eight bits. The array is written in a package array_pkg. Because the index of the array should be an integer, and the ABUS in the entity memory16x8 is declared unsigned, the ABUS is converted from unsigned to integer using the predefined function TO_IN-TEGER.

In the Verilog version of Listing 7.8, an array is used to represent the memory. The array is instantiated by the statement:

```
reg [7:0] Memory [0:15];
```

that describes an array by the name Memory; it has sixteen words, and each word is eight bits.

LISTING 7.8 HDL Description of 16×8 SRAM: VHDL and Verilog

VHDL 16×8 SRAM Description
```
library IEEE;
use IEEE.STD_LOGIC_1164.all;

--Build a package for an array
package array_pkg is
constant N : integer := 15;
--N+1 is the number of elements in the array.

constant M : integer := 7;
--M+1 is the number of bits of each element
--of the array.
subtype wordN is std_logic_vector (M downto 0);
type strng is array (N downto 0) of wordN;

end array_pkg;
library IEEE;
use IEEE.STD_LOGIC_1164.ALL;
use ieee.numeric_std.all;
use work.array_pkg.all;
entity memory16x8 is
generic (N : integer := 15; M : integer := 7);
--N+1 is the number of words in the memory; M+1 is the
--number of bits of each word.
    Port (Memory : inout strng; CS : in std_logic;
          ABUS: in unsigned (3 downto 0);
        Data_in : in std_logic_vector (7 downto 0);
        R_WRbar : in std_logic;
        Data_out : out std_logic_vector (7 downto 0));
end memory16x8;

architecture SRAM of memory16x8 is
begin
com : process (CS, ABUS, Data_in, R_WRbar)
```

```
variable A : integer range 0 to 15;
begin

if (CS = '1') then

A := TO_INTEGER (ABUS);
-- TO_INTEGER is a built-in function

if (R_WRbar = '0') then

Memory (A) <= Data_in;
else

Data_out <= Memory(A);

end if;
else
Data_out <= "ZZZZZZZZ";
--The above statement describes high impedance.
end if;
end process com;
end SRAM;
```

Verilog 16×8 SRAM Description
```
module memory16x8 (CS, ABUS, Data_in, R_WRbar, Data_out);
input CS, R_WRbar;
input [3:0] ABUS;
input [7:0] Data_in;
output [7:0] Data_out;
reg [7:0] Data_out;
reg [7:0] Memory [0:15];

always @ (CS, ABUS, Data_in, R_WRbar)
begin

if (CS == 1'b1)
    begin
        if (R_WRbar == 1'b0)
        begin
            Memory [ABUS] = Data_in;
            end
            else
```

```
            Data_out = Memory [ABUS];
        end
    else
    Data_out = 8'bZZZZZZZZ;

//The above statement describes high impedance
end
endmodule
```

The simulation output of Listing 7.8 is shown in Figure 7.6. Data are written in memory locations 0, 14, 15, and 8, and the contents of two memory locations, 0 and 15, are read; the read data match the written. The memory is deselected by setting CS to zero, and consequently, the memory Data_out, as expected, goes on high impedance.

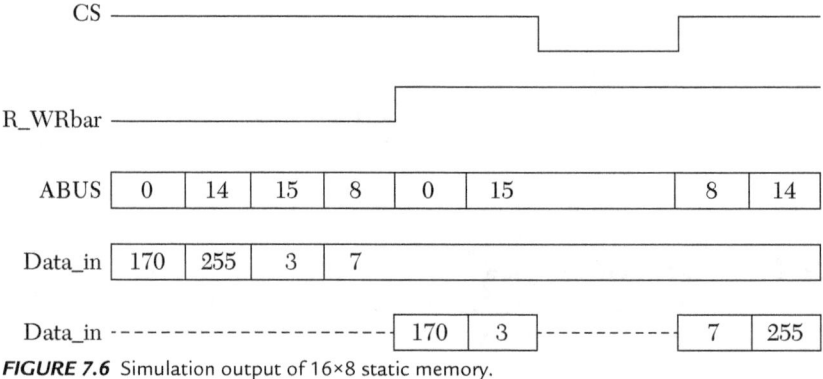

FIGURE 7.6 Simulation output of 16×8 static memory.

EXAMPLE 7.7   DESCRIPTION OF A FINITE SEQUENTIAL-STATE MACHINE

State machines are very useful tools for designing systems because their operation can be described in time events or steps. The control unit of a computer is an example of such a system. (See Case Study 7.1, which will include information from this example to write a mixed-type description of a basic computer.) The control unit generates different signals at certain time events. For example, when it boots up, a reset signal is needed to initialize components or registers in the computer. The control unit should generate this reset signal at the right time, that is, when the operation starts.

In this example, the control unit will be designed as a finite state machine. In Chapter 4, finite state machines were designed using structural description. Here, the machine is designed by using behavioral description. The state diagram of the machine shows the signals that need to be generated at each step, and it also shows the next step to which the machine has to go. The term *states* will be used here to refer to steps. Consider the state diagram shown in Figure 7.7.

The state machine in Figure 7.7 shows that, for example, if the machine is in state0 and the input is 0, the machine stays in state0 and generates a signal of 1 at the output. If the input is 1, the machine gener-

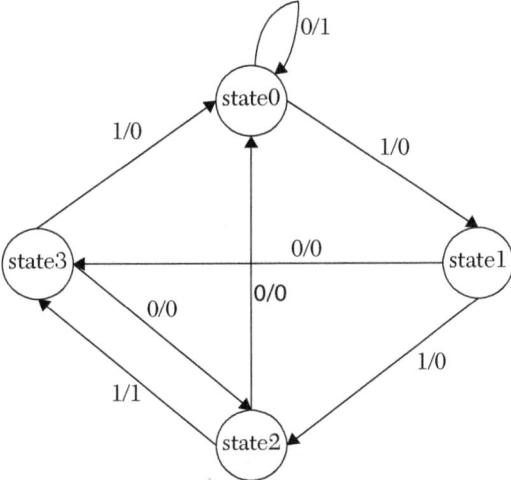

FIGURE 7.7 State diagram of a finite sequential-state machine.

ates a signal of 0 at the output, and transits to state1. Listing 7.9 lists the HDL code for the finite sequential-state machine shown in Figure 7.7.

LISTING 7.9 HDL Code for the State Machine in Figure 7.7: VHDL and Verilog

VHDL State-Machine Description

```
library IEEE;
use IEEE.STD_LOGIC_1164.all;

--First we write a package that includes type "states."
package types is
type op is (add, mul, divide, none);
type states is (state0, state1, state2, state3);
end;
    -- Now we use the package to write the code for the
    -- state machine.
library IEEE;
use IEEE.STD_LOGIC_1164.ALL;

use work.types.all;
entity state_machine is
    port (A, clk : in std_logic; pres_st : buffer states;
          Z : out std_logic);
end state_machine;
```

```
architecture st_behavioral of state_machine is

begin

FM : process (clk, pres_st, A)
variable present : states := state0;
begin
if (clk = '1' and clk'event) then
case pres_st is
    when state0 =>
        if A ='1' then
        present := state1;
        Z <= '0';
        else
        present := state0;
        Z <= '1';
        end if;
when state1 =>
if A ='1' then
        present := state2;
        Z <= '0';
        else
        present := state3;
        Z <= '0';
        end if;

when state2 =>
    if A ='1' then
    present := state3;
    Z <= '1';
    else
    present := state0;
    Z <= '0';
end if;

when state3 =>
    if A ='1' then
    present := state0;
    Z <= '0';
    else
    present := state2;
    Z <= '0';
```

```
    end if;
end case;
pres_st <= present;
end if;
end process FM;
end st_behavioral;
```

Verilog State-Machine Description

```
`define state0 2'b00
`define state1 2'b01
`define state2 2'b10
`define state3 2'b11
// We could have declared these states as parameters.
// See Listing 7.7.

module state_machine (A, clk, pres_st, Z);

input A, clk;
output [1:0] pres_st;
output Z;
reg Z;

reg [1:0] present;
reg [1:0] pres_st;

initial
begin
    pres_st = 2'b00;
end
always @ (posedge clk)
begin
        case (pres_st)
        `state0 :
            begin
                if (A == 1)
                    begin
                        present = `state1;
                        Z = 1'b0;
                    end
                    else
                    begin
```

```
                        present = `state0;
                        Z = 1'b1;
                    end
        end
    `state1 :
        begin
            if (A == 1)
            begin
                present = `state2;
                Z = 1'b0;
                end
            else
            begin
                present = `state3;
                Z = 1'b0;
            end
        end
        `state2 :
        begin
            if (A == 1)
            begin
                present = `state3;
                Z = 1'b1;
                end
            else
            begin
                present = `state0;
                Z = 1'b0;
                end
        end
        `state3 :
            begin
                if (A == 1)
                begin
                    present = `state0;
                    Z = 1'b0;
                    end
                else
                begin
                    present = `state2;
                    Z = 1'b0;
                    end
                end
```

```
endcase
pres_st = present;
end
endmodule
```

The simulation waveform is shown in Figure 7.8.

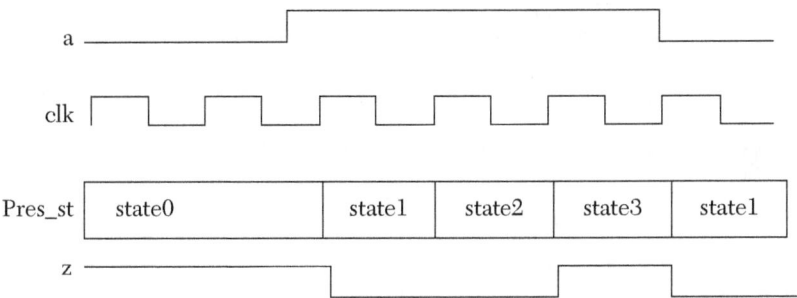

FIGURE 7.8 Simulation waveform of the state machine shown in Figure 7.7.

CASE STUDY 7.1   HDL DESCRIPTION OF A BASIC COMPUTER

In this case study, the HDL description for a basic computer will be written. In this computer, the CPU consists of ALU, registers, and a control unit. The ALU performs all arithmetic and logic operations (see Table 7.3). The registers inside the CPU store data, and communicate with the ALU and the memory. The memory here is 16×8 bits. Figure 7.9 shows the different computer registers.

Listed below are definitions of the components shown in Figure 7.9:

- **Program Counter (PC):** Stores the address of the instruction to be executed. It is four bits wide because the memory has sixteen words.

- **Address Register (AR):** Connected to the address bus of the memory, it supplies addresses to the memory. It is four bits wide because the memory has sixteen words. In this computer, AR is the only register that can provide addresses to the memory.

- **Data Register (DR):** Connected to the data bus of the memory, it receives and stores data from the memory. It is eight bits wide because the width of the memory word is eight bits. In our computer, DR is the only register that can communicate with memory data bus.

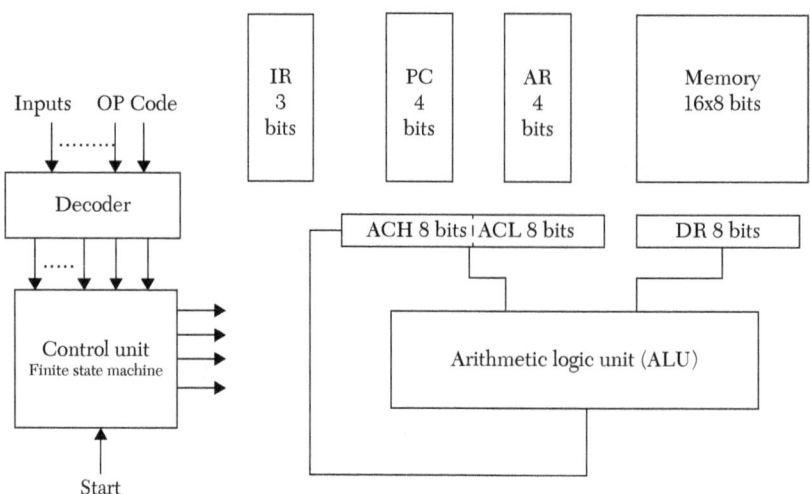

FIGURE 7.9 Registers in the basic computer.

- **Accumulator (AC):** A general register that stores data. This register has two equal halves, low (ACL) and high (ACH), and each is eight bits. The AC is sixteen bits wide.

- **Instruction Register (IR):** Stores three-bit operation code (op code).

The control unit supervises all other units in the computer, providing timing and control signals. In our basic computer, all programs are stored in the memory. A program is a group of instructions and the data it is processing. The instruction is eight bits wide (see Figure 7.10) and has two fields: operation code (op code) and address.

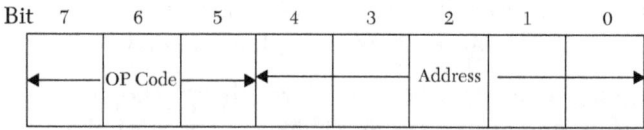

FIGURE 7.10 Basic computer instruction format.

The op code field determines the type of operation the computer should perform. The address determines the location of the operand in memory, and the operand is the data on which the operation is performed. The computer has eight different instructions, so three bits are needed to decode the instruction operations. Table 7.3 shows a possible decoding for these operations.

TABLE 7.3 Operation Codes

Operation in Mnemonic	OP Code
HALT	000
ADD	001
MULT	010
DIVID	011
XOR	100
PRITY	101
NAND	110
CLA	111

The memory used here is 16×8 bits. To access this memory, a four-bit address is needed. We will use five bits for the address; the extra bit is for any future expansion of the memory. Therefore, the instruction is eight bits wide with three bits for the op code and five bits for the address. The following is a description of the instructions shown in Table 7.3:

- **HALT:** Halts the computer by deactivating the master clock; all registers retain their current data.

- **ADD:** This is an addition operation. The contents of the lower half of the accumulator register (ACL) are added to the contents of a memory location; the result is stored in ACL.

- **MULT:** Multiplies the contents of the lower half of the AC with an operand in the memory and stores the result in AC (both halves).

- **DIVID:** This is integer division. It divides the contents of the lower half of the AC by the contents of the memory location; the result is stored in ACL.

- **XOR:** Performs the logical operation EXCLUSIVE-OR between the contents of ACL and a memory location; the result is stored in ACL.

- **PRITY:** This is an even parity generator. The parity bit for the least significant seven bits of ACL is calculated, and the parity bit is inserted in the most significant bit of ACL.

- **NAND:** Performs the logical operation NAND between the contents of ACL and a memory location; the result is stored in ACL.

- **CLA:** Clears the contents of the ACL.

The memory location in all of the above instructions is determined by the address provided by the instruction and is stored in the address register (AR). A couple of detailed instruction explanations are:

- **ADD 7:** This instruction adds the contents of the lower half of the accumulator to the contents of memory location 7; the result is stored in ACL.

- **DIVID 5:** This instruction divides the contents of the lower half of the accumulator by the contents of memory location 5; the result is stored in ACL.

The computer executes the instructions in two cycles: fetch and execute (see Figure 7.11). The control unit supplies all required signals necessary for operation of the two cycles. In the fetch cycle, the instruction is moved from the memory to the DR. The lower four bits (0 to 3) of DR are stored in AR; bits 5–7 of the DR are stored in IR. The PC is incremented to point at the next instruction to be fetched. The three bits of the IR are decoded into eight outputs by a 3×8 decoder. The output of the decoder determines the type of operation requested by the instruction. For example, if the least significant output of the decoder is active, then the operation requested belongs to the op code 000, which is HALT.

In the execute cycle, the computer executes the instruction that has been fetched. For example, if the instruction is ADD, the execute cycle issues a memory read, DR M [AR], to move the operand from the memory to the DR. M stands for memory. This movement is necessary because the ALU can operate only on DR and AC, but not on data stored in memory. After moving the operand to DR, an ADD operation in the ALU is selected. Different ALU operations are selected according to control signals supplied by the control unit. Accordingly, the ALU executes the microoperation AC AC + DR. For the instruction PRITY, the execute cycle calculates the parity bit (bit 7 of the accumulator) as:

Parity (ACL(7)) =ACL(6) XOR ACL(5) XOR ACL(4) XOR ACL(3)

XOR ACL(2) XOR ACL(1) XOR ACL(0)

The control unit oversees the fetch-and-execute cycle. The control unit here is designed as a finite sequential-state machine. Figure 7.12 shows the state diagram of the machine. The figure only shows transitions between states; it does not show outputs. The states state0, state1, and state2 are

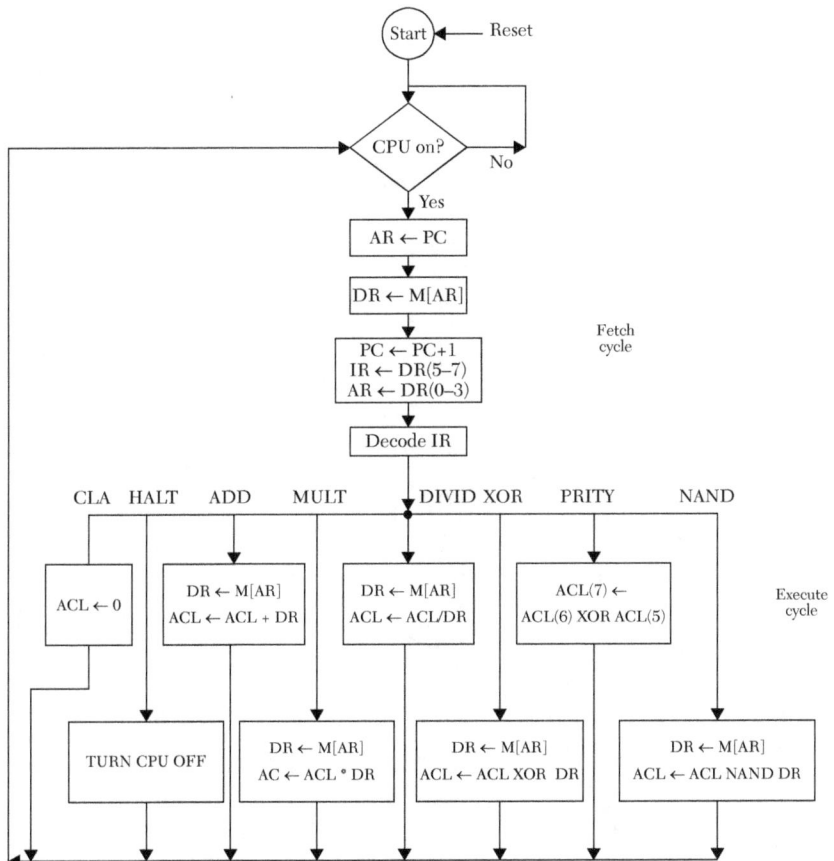

FIGURE 7.11 Fetch-and-execute cycles of the basic computer.

performing the three steps of the fetch cycle, while state3 performs the execute cycle (see Figure 7.11).

To define the basic computer's operation, we store a program in the memory. Table 7.4 shows the instructions of the program with the op code written in mnemonic and the instructions written in hexadecimal. For example, the instruction:

1   ADD   9

is stored in memory location 1, the op code is ADD, and the address is 9. The instruction adds the contents of the accumulator (AC) to the contents

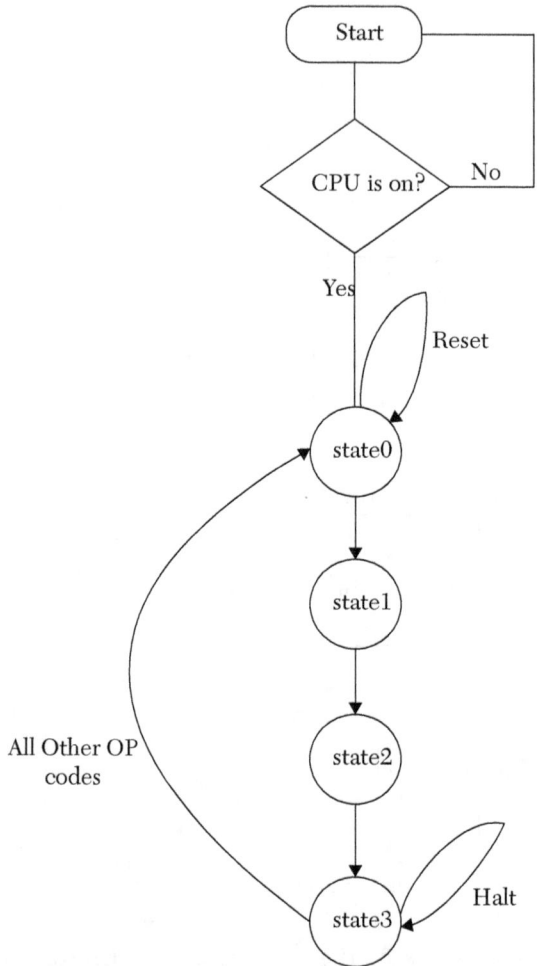

FIGURE 7.12 State diagram of the finite sequential-state machine.

of memory location 9; the result of addition is stored in the accumulator. The accumulator in our computer is always the default register. The binary value of the op code ADD is 001 (see Table 7.3). The instruction is eight bits wide, so the binary representation of the instruction is 00101001, which is 29 in hexadecimal.

TABLE 7.4 Contents of Memory of the Basic Computer

Instruction in Mnemonic or Memory Contents (Eight Bits)

Location	Instruction		Data in Hex
0	CLA		E0
1	ADD	9	29
2	XOR	A	8A
3	MULT	B	4B
4	DIVID	C	6C
5	XOR	D	8D
6	NAND	E	CE
7	PRITY		A0
8	HALT		00
9	C		0C
A	5		05
B	4		04
C	9		09
D	3		03
E	9		09
F	7		07

Listing 7.10 shows the HDL code for the basic computer program shown in Table 7.4. Referring to the VHDL Listing, the package `Comp_Pkg` declares a one-dimensional array with sixteen elements; each element is eight bits. This array represents the memory of the computer. In the entity `computer_basic`, the signal `clk_master` simulates the master clock of the computer. The signal `ON_OFF` simulates an on/off switch. The data-flow statement

```
clk <= clk_master and ON_OFF;
```

simulates an AND gate. The signal `clk` simulates the clock signal of the CPU; if the switch is off, the clock signal to the CPU is inactive, and accordingly, the CPU is inactive. The statement

```
z(0) <= ACL(6) xor ACL (5) xor ACL (4)xor
ACL (3)xor ACL (2)xor ACL (1)xor ACL (0);
```

generates an even-parity bit. We selected to write this statement as data flow (outside `always`), because it is easier to write the Boolean as data-flow description, rather than the behavioral of this parity generator circuit. The statements

```
ARI := TO_INTEGER(AR);
DR := Memory (ARI);
```

convert AR from unsigned to integer by the built-in function TO_INTEGER. This function is part of the package ieee.numeric_std. We convert to integer because the index of the array ARI has to be of type integer in VHDL.

Referring to the Verilog description, the memory is simulated by the statement:

```
reg [7:0] Memory [0:15];
```

which describes an array of fifteen elements (words); each element is eight bits. In contrast to VHDL, Verilog can accept an index of an array declared as bit_vector. For example, we can write

```
DR = Memory [AR];
```

without specifying AR to be of type integer.

LISTING 7.10 HDL Code for the Basic-Computer Memory Program (Table 7.4): VHDL and Verilog

VHDL Basic-Computer Memory Program
```
--Write the code for Package Comp_Pkg
library IEEE;
use IEEE.STD_LOGIC_1164.all;
use ieee.numeric_std.all;

package Comp_Pkg is
constant N: integer := 15;
--N+1 is the number of elements in the array.
constant M : integer := 7;
--M+1 is the number of bits of each element
--of the array.

subtype wordN is unsigned (M downto 0);
type strng is array (N downto 0) of wordN;
type states is (state0, state1, state2, state3);

end Comp_Pkg;
```

```
--Now write the code for the control unit
library IEEE;
use IEEE.STD_LOGIC_1164.ALL;
use IEEE.STD_LOGIC_UNSIGNED.ALL;
use ieee.numeric_std.all;
use work.Comp_Pkg. all;

entity computer_basic is

generic (N : integer := 15; M : integer := 7);
--N+1 is the number of words in the memory; M+1 is the
--number of bits of each word.

Port (Memory : inout strng;
    PC : buffer unsigned (3 downto 0);
   clk_master : std_logic;
   ACH : buffer unsigned (7 downto 0);
   ACL : buffer unsigned (7 downto 0);
   Reset : buffer std_logic; ON_OFF : in std_logic);

end computer_basic;

architecture Behavioral_comp of computer_basic is
signal z : unsigned (0 downto 0);
signal clk : std_logic;

begin

z(0) <= ACL(6) xor ACL(5) xor ACL(4) xor
   ACL(3) xor ACL(2) xor ACL(1) xor ACL(0);
--Z has to be in vector form to match ACL

clk <= clk_master and ON_OFF;

--The above two statements are data-flow description.
--The following is behavioral description.

cpu : process (Reset, PC, ACL, ACH, clk, Memory, z(0))
variable AR : unsigned (3 downto 0);
variable DR : unsigned (7 downto 0);
variable pres_st,     next_st : states;
variable ARI : integer range 0 to 16;
```

```
variable IR : unsigned (2 downto 0);
variable PR : unsigned (15 downto 0);

begin

if rising_edge (clk) then
if Reset = '1' then
pres_st := state0;
Reset <= '0';
PC <= «0000»;
end if;

case pres_st is
when state0 =>
next_st := state1;
--This is fetch cycle
AR := PC;

when state1 =>
next_st := state2;
ARI := TO_INTEGER(AR);
--This is fetch cycle
DR := Memory (ARI);
when state2 =>
next_st := state3;
--This is fetch cycle
PC <= PC + 1;
IR := DR (7 downto 5);
AR := DR (3 downto 0);
when state3 =>
--This is execute cycle

case IR is
    when «111» =>
    --The op code is CLA
    ACL <= «00000000»;
    next_st := state0;

    when «001» =>
    --The op code is ADD
    ARI := TO_INTEGER(AR);
    DR := memory (ARI);
    ACL <= ACL + DR;
```

```
next_st := state0;

when «010» =>
--The op code is MULT
ARI := TO_INTEGER(AR);
DR := memory (ARI);
PR := ACL * DR;
ACL <= PR (7 downto 0);
ACH <= PR (15 downto 8);
next_st := state0;

when «011» =>
--The op code is DIVID
ARI := TO_INTEGER(AR);
DR := memory (ARI);
ACL <= ACL / DR;
next_st := state0;

when «100» =>
--The op code is XOR
ARI := TO_INTEGER(AR);
DR := memory (ARI);
ACL <= ACL XOR DR;
next_st := state0;

when «110» =>
--The op code is NAND
ARI := TO_INTEGER(AR);
DR := memory (ARI);
ACL <= ACL NAND DR;
next_st := state0;

when «101» =>
--The op code is PRITY
ACL(7) <= z(0);
next_st := state0;

when «000» => null;
-- The op code is HALT
next_st := state3;

when others => null;
end case;
```

```
when others => null;
end case;
pres_st := next_st;
end if;
end process cpu;
end Behavioral_comp;
```

Verilog Basic-Computer Memory Program

```
module computer_basic (PC, clk_master, ACH, ACL,
                       Reset, ON_OFF);
parameter state0 = 2'b00;
parameter state1 = 2'b01;
parameter state2 = 2'b10;
parameter state3 = 2'b11;

output [3:0] PC;
input clk_master;
output Reset;
input ON_OFF;
output [7:0] ACH;
output [7:0] ACL;
reg [1:0] pres_st;
reg [1:0] next_st;
reg Reset;
reg [3:0] PC;
reg [3:0] AR;
reg [7:0] DR;
reg [2:0] IR;
reg [7:0] ACH;
reg [7:0] ACL;
reg [15:0] PR;
reg [7:0] Memory [0:15];

assign z = ACL[6] ^ ACL[5] ^ ACL[4]^
    ACL[3]^ ACL[2]^ ACL[1]^ ACL[0];

/*The above statement can be written using the reduction
    XOR as: assign z = ^ ACL[6:0];*/
//

assign clk = clk_master & ON_OFF;
always @ (Reset, PC, ACL, ACH, posedge(clk), z, pres_st)
```

```
begin

    if (Reset == 1'b1)
        begin
        pres_st = state0;
        Reset = 1'b0;
        PC = 4'd0;
        Memory [0] = 8'hE0; Memory [1] = 8'h29;
        Memory [2] = 8'h8A; Memory [3] = 8'h4B;
        Memory [4] = 8'h6C; Memory [5] = 8'h8D;
        Memory [6] = 8'hCE; Memory [7] = 8'hA0;
        Memory [8] = 8'h00; Memory [9] = 8'h0C;
        Memory [10] = 8'h05; Memory [11] = 8'h04;
        Memory [12] = 8'h09; Memory [13] = 8'h03;
        Memory [14] = 8'h09;
        Memory [15] = 8'h07;
        end

    case (pres_st)

    state0 :
    begin
        next_st = state1;
        AR = PC;
        end
    state1 :
    //This is fetch cycle
    begin
        next_st = state2;
        DR = Memory [AR];
        end

    state2 :
    //This is fetch cycle
    begin
        next_st = state3;
        PC = PC + 1;
        IR = DR [7:5];
        AR = DR [3:0];
        end

    state3 :
    //This is execute cycle
```

```
begin
    case (IR)
        3'd7 :
        //The op code is CLA
        begin
            ACL = 8'd0;
            next_st = state0;
            end
        3'd1 :
        //The op code is ADD
        begin
            DR = Memory [AR];
            ACL = ACL + DR;
            next_st = state0;
            end
        3'd2 :
        //The op code is MULT
        begin
            DR = Memory [AR];
            PR = ACL * DR;
            ACL = PR [7:0];
            ACH = PR [15:8];
            next_st = state0;
            end
        3'd3 :
        //The op code is DIVID
        begin
            DR = Memory [AR];
            ACL = ACL / DR;
            next_st = state0;
        end

        3'd4 :
        //The op code is XOR
        begin
            DR = Memory [AR];
            ACL = ACL ^ DR;
            next_st = state0;
            end
        3'd6 :
        //The op code is NAND
        begin
```

```
                    DR = Memory [AR];
                    ACL = ~(ACL & DR);
                    next_st = state0;
                    end
                3'd5 :
                //The op code is PRITY
                begin
                    ACL[7] = z;
                    next_st = state0;
                    end
                3'd0 :
                //The op code is HALT
                begin
                    next_st = state3;
                    end
                default :
                begin
                end
                endcase
            end
        default :
        begin
        end
        endcase
        pres_st = next_st;
    end
endmodule
```

Figure 7.13 shows the simulation output of the accumulator register. To start simulation, reset is forced high and then unforced.

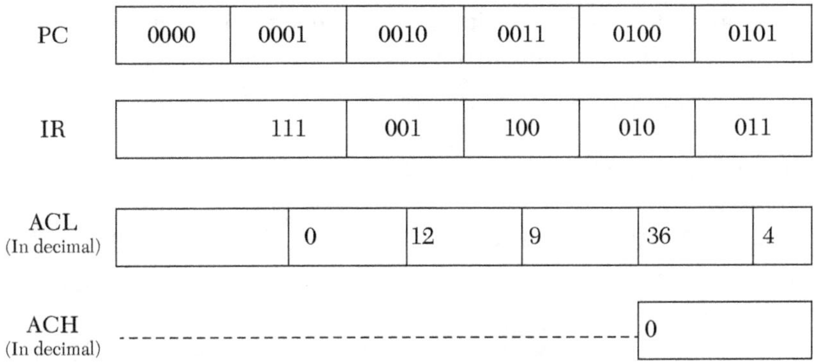

PC	0000	0001	0010	0011	0100	0101

IR		111	001	100	010	011

ACL (In decimal)		0	12	9	36	4

ACH (In decimal)					0	

FIGURE 7.13 Simulation output of the accumulator register.

7.5   Summary

This chapter covered mixed-type descriptions (HDL code that includes more than one style of description in the same module). An example of mixed description is when we write a module using behavioral and data-flow statements. In some systems, one part can be best described by behavioral statements, and other parts of the system can be best described by data-flow description. Instead of writing a module with only behavioral or only data flow, we write the module using both behavioral and data-flow descriptions; that is what is defined as mixed-type description.

An example to illustrate the mixed-type description is the ALU of a computer. Some operations of the ALU, such as division, are usually described by behavioral statements because it is not easy to find the Boolean function or the hardware logic for division. Other operations, such as addition, may be described by data-flow or structural description because it is usually easy to find the logic diagram of adders. In addition to mixed-type descriptions, packages and single/multidimensional arrays were covered. Packages are an essential construct in VHDL code. User-defined types, components, functions, and procedures can be written in a package and made visible to a VHDL module by attaching (including) the package with the module.

7.6   Exercises

1. Write the HDL code to find the value and the order of the smallest element in an array. The elements are four-bit signed numbers. Simulate and verify your code.

2. Given an array of N elements, write the HDL code to organize the elements of the array in ascending order. All elements are integers.

3. Consider the code shown in Listing 7.11.

LISTING 7.11 Exercise 7.3

```
library IEEE;
use IEEE.STD_LOGIC_1164.all;

package arrypack is
```

```
constant N : integer := 2;
constant M : integer := 1;

subtype wordg is integer;
type singl1 is array (N downto 0) of wordg;
type singl2 is array (N downto 0) of singl1;
type arry3 is array (M downto 0) of singl2;
end arrypack;

library IEEE;
use IEEE.STD_LOGIC_1164.ALL;
use work.arrypack.all;

entity exercise is
    Port(N, M, P : integer; z : out integer);
end exercise;

architecture exercise of exercise is

begin
com : process (N, M, P)
variable t : integer;
constant y : arry3 := (((5, 4, 3), (8, 9, 10), (32, 33, 34)),
               ((42, 43, 44), (52, 53, 54), (-10, -7, -5))));

begin

t := y (N)(M)(P);
z <= t;
end process com;

end exercise;
```

a) What is the value of the following elements of y?

y (0,0,0), y (0,0,1), y (0,0,2), y (0,1,2), y (1,1,2), y (1,2,2)

b) If we change all (N downto 0) and (M downto 0) in package arrypack to (0 to N) and (0 to M), what will be the values of the elements in part a?

4. Repeat Listing 7.8 but for a memory of 128×16. Store the following data in the corresponding memory locations:

Memory Location in Decimal	Contents in Decimal
0	123
127	1025
55	35
105	801
99	0

Verify your storage by reading the data from the above locations.

5. For the state diagram shown in Figure 7.14, write a behavioral HDL program to simulate the state machine. Verify your answer by simulation.

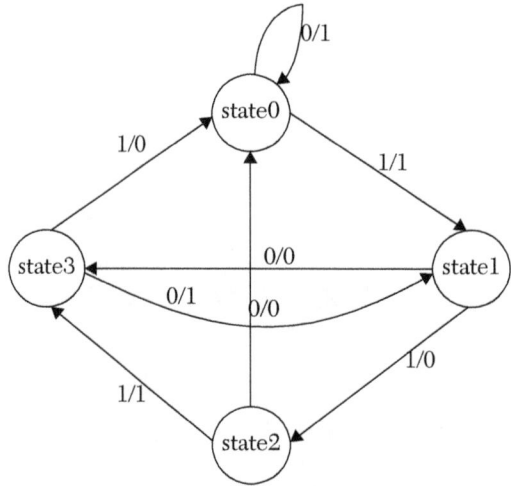

FIGURE 7.14 State-machine instruction format for Exercise 7.5.

6. Write a Verilog code to perform the following:

$$[C] = [A] - [B]$$

All matrices are 6x6. Matrix [A] has the following elements: $a(i+1,j) = a(i,j) + 4$ with $a(0,0) = 0$.

Matrix [B] has the following elements: $b(i+1,j) = b(i,j) * 2$ with $b(0,0) = 1$.

In your code, first generate the matrices and then perform the subtraction.

7. For Case Study 7.1, increase the memory size to 32×16. Also, change all instructions so the result of each instruction is stored in a memory location rather than the accumulator. The address of this memory location is the same as the address provided in the instruction. For example, the instruction ADD 9 would mean the addition of the contents of AC to the contents of memory location 9 with storage of the result in memory location 9. Keep the size of all registers at eight bits. Rewrite the VHDL code shown in Listing 7.10 and verify by simulation.

ADVANCED *HDL* DESCRIPTION

Chapter Objectives

- Explore several advanced topics in HDL Description, such as file processing, character and string implementation, and the type `record`
- Understand VHDL `Assert` and `Block` statements
- Understand Verilog user-defined primitives
- Acquire a basic knowledge of cache memories
- Acquire a basic knowledge of artificial neural networks

8.1   File Processing

Files are implemented when dealing with a large amount of data that need to be stored and accessed. Also, files can be used to display formatted output, such as reports. Files can be read or written. To read from or write to a file, it must be opened, and after reading or writing is finished, the file must be closed. A closed file cannot be accessed unless it is opened. In Section 8.1.1, the VHDL file description is explored, and Section 8.1.2 discusses Verilog file descriptions.

8.1.1 VHDL File Processing

File processing can be slightly different from one HDL simulator to another. Appropriate packages have to be attached to the VHDL module. The reader is advised to consult his or her VHDL package and simulator for files-handling capability. This section will present complete examples of file description with the names of the appropriate packages. Files have to be declared by the predefined object type `file`. File declaration includes the predefined word `file` followed by (in this order) the port direction or mode of the file (`infile` or `outfile`), a colon, and the subtype of the file. An example of file declaration is:

```
file infile : text;
```

The above statement declares a file with mode `infile`, and the subtype of the file is `text`. The IEEE package `textio` should be attached (see examples in the following sections). VHDL has built-in procedures for file handling. These procedures include `file_open`, `readline`, `writeline`, `read`, `write`, and `file_close`. In the following sections, each of these procedures is briefly discussed.

8.1.1.1 `File_open`

The `file_open` procedure opens the file; files cannot be accessed if not opened. This procedure has the following declaration:

```
Procedure file_open (status : file_open_status, infile : file
type, external_name : in string,
open_kind : file_open_kind) is
```

The statement `status` enables the VHDL to keep track of the activities of the file (e.g., open, close, read, etc.); `infile` is the type (mode) of the file. The `infile` is used for input files (their contents will be read), and `outfile` is used for output files (they will be written into). The `external_name` is the name of the file to be opened; the name has to be in `string` form such as "`rprt.txt`" or "`testfile.txt`." The `open_kind` is the mode of opening the `read_mode` or `write_mode`. An example of implementing `file_open` is:

```
file_open (fstatus, infile, "testfile.txt", read_mode);
```

The above procedure opens a file by the name of `testfile.txt` for reading. The file is an input file (`infile`) with type `txt`. It is located in the same path as the procedure. For example, if the procedure is written in a module stored in directory C under a subdirectory VHDL_files, then `testfile` should be stored in the subdirectory VHDL_files; otherwise, the path of `testfile` should be explicitly stated in the declaration of `file_open`. The file is then opened for reading.

The following procedure opens a text `outfile` by the name of `store.txt` for writing:

```
file_open (fstatus, outfile, "store.txt", write_mode);
```

8.1.1.2 `File_close`

The procedure `file_close` is used to close an open file. For example:

```
file_close (infile);
```

closes the open file `infile`. The name and path of `infile` are specified in the procedure `file_open`. The following statement closes `outfile`:

```
file_close (outfile);
```

8.1.1.3 `Readline`

The predefined procedure `readline` reads a line from the file opened in read mode. An example of implementing this procedure is:

```
readline (infile, temp);
```

The above statement reads a line from `infile` and stores the line in variable temp. Variable temp has to be of predefined type `line`. The name and type of `infile` should have been stated in the procedure `file_open`. Inside the file specified by `infile`, a carriage return is the separator between the lines. If `readline` is repeated before closing the file, another line is read. A carriage return indicates a new line.

8.1.1.4 `Writeline`

The predefined procedure `writeline` writes a line into an `outfile` that is open for write mode. An example of implementing this procedure is:

```
writeline (outfile, temp);
```

The above statement writes a line stored in the variable temp into the file `outfile`. Variable temp has to be of type `line`. The name and path of `outfile` should be specified in the procedure `file_open`. Only integers, real values, or characters can be written into `outfile`. If `writeline` is repeated before closing `outfile`, a new line is stored in `outfile`.

8.1.1.5 `Read`

To read an integer, a character, or a real value from a line in an `infile` that is open for read mode, the procedure `read` is used. For example, if `intg1` has been declared as of type integer, the statement

```
read (temp, intg1);
```

performs a single read of an integer from line `temp` of the open file (for read mode) and stores the value of this integer in `intg1`. If a character or a real value is to be read, the variable `intg1` should be of type `character` or `real`, respectively. If `intg1` is a single value (not an array), each time the read operation is executed, a single word of the line is read and stored in `intg1`. If the read statement is repeated before closing the file, the next word in the line is read and stored in `intg1`.

8.1.1.6 `Write`

The procedure `write` stores an integer, a character, or a real value from a line to an `outfile` that is open for `write_mode`. For example, if `intg1` has been declared as type `integer`, the statement

```
write (temp, intg1);
```

stores the integer `intg1` in the line `temp` of the open `outfile`, which is in `write` mode. If a character or a real value is to be written, the variable `intg1` should be of type `character` or `real`, respectively. Each time the `write` operation is executed, a single word is stored in the line. If the `write` statement is repeated before closing the file, a new value of `intg1` is stored in the line.

8.1.2   Verilog File Processing

Standard Verilog can handle several file operations. As in VHDL, before accessing a file, it must be opened. If the file is not open, it cannot be read from or written to. Accessing a file is accomplished through built-in tasks such as `$fopen`, `$fdisplay`, `$fmonitor`, and `$fclose`. More tasks are being introduced in newer Verilog packages. Let us briefly investigate each of these tasks.

8.1.2.1 `$fopen`

The task `$fopen` is used to open files. It is the counterpart of the VHDL procedure `file_open`. The format for opening a file is:

```
Channel = $fopen ("name of the file");
```

where `Channel` is a variable of type integer; it indicates the channel number. Verilog uses this channel number to track and identify which files are open. Verilog automatically assigns an integer value to each channel. For example, to open a text file named `testfile`:

```
ch1 = $fopen ("testfile.txt");
```

and `ch1` becomes the indicator (identifier) of file `testfile.txt`.

8.1.2.2 $fclose

The task `$fclose` closes a file indicated by the channel number. For example the task

```
$fclose (ch1);
```

closes the file `testfile.txt`.

8.1.2.3 $fdisplay

The task `$fdisplay` is the counterpart of `write` in VHDL. It can write variables, signals, or quoted strings. The format of `$fdisplay` is as follows:

```
$fdisplay (channel, V1, V2, V3, ....);
```

where V1, V2, V3, and so on are variables, signals, or quoted strings. For example, consider the following `$fdisplay` task:

```
$fdisplay (ch1, "item    description   quantity");
```

After executing the task, the file `testfile.txt` displays:

```
item    description    quantity
```

The number of spaces displayed in the file between each string is the same number of spaces inside the quotations.

8.1.2.4 $fmonitor

The task `$fmonitor` has the following format:

```
$fmonitor (channel, V1, V2, V3,.....)
```

The task monitors and records the values of V1, V2, V3, and so on. For example, consider the following `$fmonitor` task:

```
$fmonitor (ch1, "        %b", quantity);
```

The above task monitors the variable `quantity` and records its value in binary in the file `testfile.txt` indicated by `ch1`, and `%b` indicates binary format. If `quantity` = 7 in decimal, after execution of the above task, the file `testfile.txt` displays:

```
item    description       quantity
                          111
```

Different formats can be selected such as:

`%d`	Display in decimal
`%s`	Display strings
`%h`	Display in hex
`%o`	Display in octal
`%c`	Display in ASCII character
`%f`	Display real numbers in decimal format

Escape characters may also used; some of these characters are:

\n	Insert a blank line
\t	Insert tab
\\	Insert the character \
\"	Insert the character "
\	Insert the character %

8.2   Examples of File Processing

The following sections present and discuss some examples of file processing. Because VHDL and Verilog file processing are not very similar, their examples are discussed separately.

8.2.1   Examples of VHDL File Processing

The following examples cover file processing in VHDL.

EXAMPLE 8.1   READING A FILE CONTAINING INTEGER NUMBERS

Consider a text file (written by a Notepad, for example) by the name of `file_int.txt` in the same path as the VHDL module that accesses it (see Listing 8.1). The contents of the file are integers written in two lines (see Figure 8.1). The two lines are separated by a carriage return, and the integers are separated by a space band (the number of space bands can be one or more than one).

In this example, the first integer is to be multiplied by two, the second by five, the third by

```
12      −3      5
20
```

FIGURE 8.1 File `file_int.txt`.

three, and the fourth by four. The products are stored in the integer variables z, z1, z2, and z3, respectively.

To calculate the products, open the file, read its contents, perform the multiplication, and close the file. Referring to Listing 8.1, the statement

```
file_open (fstatus, infile, "file_int.txt", read_mode);
```

opens the infile file_int.txt for reading. The statement

```
readline (infile, temp);
```

reads a line from the file file_int.txt and stores this line in the variable temp of type line. If the statement is repeated, temp acquires the next line. The statement

```
read (temp, count);
```

reads a single integer from the line temp and stores the integer in the variable count. If the statement is repeated, count will acquire the next integer from the same line. The statement

```
file_close (infile);
```

closes the file. No operation can be performed on the file as long as it is closed. If the file is opened again, readline reads the first line of the file. To repeat the code for another file, be sure to create an event in the process by turning off (START = 0) and turning on (START = 1).

LISTING 8.1 VHDL Code for Reading and Processing a Text File Containing Integers

```
library ieee;
use ieee.std_logic_1164.all;
use std.textio.all;

entity FREAD_INTG is
port (START : in std_logic;
z, z1, z2, z3 : out integer);
end FREAD_INTG;

architecture FILE_BEHAVIOR of FREAD_INTG is
begin

process (START)
```

```
-- declare the infile as a text file
file infile : text;

--declare the variable fstatus (or any other variable name)
--as of type file_open_status
variable fstatus : file_open_status;
variable count : integer;

--declare variable temp as of type line
variable temp : line;

begin
if (START = '1') then
--open the file file_int.txt in read mode
file_open (fstatus, infile, "file_int.txt", read_mode);

--Read the first line of the file and store the line in temp
readline (infile, temp);
-- temp now has the data: 12 -3 5

-- Read the first integer (12) from the
--line temp and store it in the integer variable count.

    read (temp, count);

--count has the value of 12. Multiply by 2 and store in z
    z <= 2 * count;

-- Read the second integer from the line temp and
-- store it in count
    read (temp, count);
--count now has the value of -3

--Multiply by 5 and store in z1
    z1 <= 5 * count;

-- read the third integer in line temp
--and store it in count.
    read (temp, count);

--Multiply by 3 and store in z2
    z2 <= 3 * count;
```

```
--Read the second line and store it in temp
    readline (infile, temp);
--temp has only the second line

--Read the first integer of the second line
--and store it in count.
    read (temp, count);

--Multiply by 4 and store in z3
    z3 <= 4 * count;

--Close the infile
file_close (infile);
end if;
end process;
end FILE_BEHAVIOR;
```

After the code in Listing 8.1 executes, z, z1, z2, and z3 take the following values:

z = 24, z1 = −15, z2 = 15, z3 = 80

EXAMPLE 8.2   READING A FILE CONTAINING REAL NUMBERS

In this example, a file by the name of file_real.txt is read. The contents of this file are real numbers (containing fractions) written in decimal format such as 50.3 (see Figure 8.2). The contents are written in two lines separated by a carriage return. The numbers are separated by one or more spaces. Listing 8.2 shows the code for reading the file; it is very similar to Listing 8.1. Open the file with file_open and read a line from the file using the procedure readline. After reading a line, one word is read at a time by invoking the procedure read. Each word is a real number; spaces are not read but are recognized as separators between words.

| −13.4 | −5.654 | .023 |
| −55.32 | | |

FIGURE 8.2 File file_real.txt.

LISTING 8.2 VHDL Code for Reading a Text File Containing Real Numbers

```
library ieee;
use ieee.std_logic_1164.all;
use std.textio.all;
```

```
entity FREAD_REAL is
port (START : in std_logic;
z, z1, z2, z3 : out real);
end FREAD_REAL;

architecture FILE_BEHAVIOR of FREAD_REAL is
begin

process (START)
file infile : text;
variable fstatus : file_open_status;
variable count : real;
--Variable count has to be of type real
variable temp : line;

begin
if (START = '1') then
--Open the file
    file_open (fstatus, infile,
        "file_real.txt", read_mode);

-- Read a line
    readline (infile, temp);

--Read one number and store it in real variable count
    read (temp, count);
--multiply by 2
z <= 2.0 * count;

--read another number
    read (temp, count);
--multiply by 5
  z1 <= 5.0 * count;
  --read another number
     read (temp, count);
  --multiply by 3
     z2 <= 3.0 * count;

  --read another line
     readline (infile, temp);
     read (temp, count);
```

```
    --multiply by 4
        z3 <= 4.0 * count;
        file_close (infile);
            end if;
end process;
end FILE_BEHAVIOR;
```

After executing Listing 8.2, z, z1, z2, and z3 take the following values:

z = −26.8, z1 = −28.27, z2 = 0.069, z3 = −221.28

EXAMPLE 8.3   READING A FILE CONTAINING ASCII CHARACTERS

Figure 8.3 shows the file to be read, named file_chr.txt. The contents of this file are ASCII characters. ASCII characters can be digits (e.g., 0, 1, 2), letters of the alphabet (e.g., A, B, C), or special characters (e.g., ; *

A5B
M

FIGURE 8.3 File file_chr.txt.

& #). The space band is an ASCII character and is read as a character. Listing 8.3 shows the code for reading the file file_chr.txt. The file has two lines (see Figure 8.3) separated by a carriage return. The first line has three characters, A5B, and the second line has one character, M. If the first line contains A B instead of A5B, it is still read as three characters: A, space band, and B.

Listing 8.3 shows the VHDL code for reading an ASCII file. The file is opened with file_open. A line from the file is read using the procedure readline. After reading a line, one word at a time is read by invoking the procedure read; each word is a character, including spaces. The character is then stored in the variable count; this variable has to be of type character.

LISTING 8.3 VHDL Code for Reading an ASCII File

```
use ieee.std_logic_1164.all;
use std.textio.all;

entity FREAD_character is
port (START : in std_logic;
        z, z1, z2, z3 : out character);
end FREAD_character;
```

```
architecture FILE_BEHAVIOR of FREAD_character is
begin

process (START)
file infile : text;
variable fstatus : file_open_status;
variable count : character;
-- Variable count has to be of type character
variable temp : line;

begin

if(START = '1') then
    file_open (fstatus, infile, "file_chr.txt", read_mode);

--read a line from the file
    readline (infile, temp);
--read a character from the line into count.
--Count has to be of type character.
--

    read (temp, count);

--store the character in z
    z <= count;
    read (temp, count);
    z1 <= count;
    read (temp, count);
    z2 <= count;
    readline (infile, temp);
    read (temp, count);
    z3 <= count;
    file_close (infile);
        end if;
end process;
end FILE_BEHAVIOR;
```

After the code in Listing 8.3 executes, z, z1, z2, and z3 take the following values:

z = A, z1 = 5, z2 = B, z3 = M

Reading files has been covered in the previous examples. The following examples cover writing into files. As mentioned, VHDL files can store integers, real values, and characters.

EXAMPLE 8.4   WRITING INTEGERS TO A FILE

In this example, writing into the text file `Wfile_int.txt` is considered. Assume that the file is located in the same path as the VHDL module that accesses it (see Listing 8.4). Integers will be written into the file. Start by opening the file using the procedure `file_open`. Assemble the line to be stored using the procedure `write`, as follows:

```
write (temp, z);
```

The above statement stores the integer `z` into the line `temp`. Quoted characters can be stored in `temp` as follows:

```
write (temp, "This is an integer file");
```

Executing this statement results in storing the message "This is an integer file" in `temp`. If the statement is repeated, another integer or character is stored into `temp`. A space is to be stored between each two integers. After all integers and characters have been stored in the line, the line is written to the file using the procedure:

```
writeline (outfile, temp);
```

The above procedure, `writeline`, writes the line `temp` into the `outfile`, `Wfile_int.txt`.

LISTING 8.4 VHDL Code for Writing Integers to a File

```
library ieee;
use ieee.std_logic_1164.all;
use std.textio.all;

entity FWRITE_INT is
port (START : in std_logic);

end FWRITE_INT;

architecture FILE_BEHAVIOR of FWRITE_INT is
begin

process (START)
file outfile : text;
variable fstatus : file_open_status;
--declare temp as of type line
variable temp : line;
variable z,z1,z2,z3 : integer := 6;
```

```
begin
    if(START = '1' ) then
      z := 12; z1 := 23; z2 := -56; z3 := -45;
      file_open (fstatus, outfile,
                 "Wfile_int.txt", write_mode);
    --The generated file "Wfile_int.txt" is in
    --the same directory as this VHDL module
    --Insert the title of the file Wfile_int.txt.
    --Your simulator should support formatted text;
    --if not, remove all formatted statements " ".

    write (temp, "This is an integer file");

    --Write the line temp into the file
    writeline (outfile, temp);
    --store the first integer in line temp
    write (temp, z);

     -- leave space between the integer numbers.
    write (temp, " ");
    write (temp, z1);

    -- leave another space between the integer numbers.
    write (temp, " ");
    write (temp, z2);
    write (temp, " ");

    writeline (outfile, temp);
    --Insert the fourth integer value on a new line
    write (temp, z3);
    writeline (outfile, temp);
file_close(outfile);
  end if;
end process;
end FILE_BEHAVIOR;
```

After executing the code above, the outfile Wfile_int.txt appears as shown in Figure 8.4.

```
This is an integer file
12    23    -56
-45
```

FIGURE 8.4 File Wfile_int.txt.

In the same way as was done in Listing 8.4, characters or real numbers can be written into an outfile.

EXAMPLE 8.5 **READING A STRING OF CHARACTERS AND STORING THEM INTO AN ARRAY**

In previous examples, a single character from the file was read and stored in a single variable count. Here, a string of characters are read and stored in an array. To handle arrays, a package is built that contains an array of characters. The package array_pkg is shown in Listing 8.5. Subtype wordchr of type character is used. The array is written as type string_chr, which is an array of the subtype wordchr. The array consists of $N + 1$ elements, and each element is type character.

LISTING 8.5 VHDL Code for Writing a Package Containing a String of Five Characters

```
library IEEE;
use IEEE.STD_LOGIC_1164.all;

package array_pkg is
constant N : integer := 4;
--N+1 is the number of elements in the array.
subtype wordChr is character;
type string_chr is array (N downto 0) of wordChr;

end array_pkg;
```

Now, a string of characters need to be read from the file string_chr and stored in an array. Listing 8.6 shows the code for reading a string from the file. A single word composed of five characters, "STORE," is stored in the file. The package written in Listing 8.5 is used here to instantiate the array. In Listing 8.6, z is declared as type string_chr. This means that z is an array of five elements, (N down to 0) where N = 4; each element is a single character. The file is opened, the string is read and then stored in array z.

LISTING 8.6 VHDL Code for Reading a String of Characters into an Array

```
library ieee;
use ieee.std_logic_1164.all;
use std.textio.all;
```

```
--include the package with this module
use work.array_pkg.all;

entity FILE_CHARCTR is
port (START : in std_logic; z : out string_chr);

--string_char is included in the package array_pkg;
--z is a 5-character array

end FILE_CHARCTR;

architecture FILE_BEHAVIOR of FILE_CHARCTR is
begin

process (START)
file infile : text;
variable fstatus : file_open_status;
variable count : string_chr;
variable temp : line;

begin
file_open (fstatus, infile, "myfile1.txt", read_mode);
readline (infile, temp);

read (temp, count);
--Variable count has been declared as an array of five
-- elements, each element is a single character.
--
z <= count;

file_close (infile);
end process;

end FILE_BEHAVIOR;
```

After the code in Listing 8.6 is executed, the signal z contains "S" "T" "O" "R" "E."

EXAMPLE 8.6   FINDING THE WORD IN A FILE WITH THE SMALLEST ASCII VALUE

When an ASCII character is read, the VHDL package assigns the unique hexadecimal (hex) value for that character. Table 8.1 shows the hexadecimal values for several characters. Notice that A has the lowest hex value among the letters, while Z has the highest. In this example, we want to find the word that has the lowest ASCII hex value.

TABLE 8.1 ASCII Character Hexadecimal Values

Character	Hex Value	Character	Hex Value
A	41	U	55
B	42	V	56
C	43	W	57
D	44	X	58
E	45	Y	59
F	46	Z	5A
G	47	0	30
H	48	1	31
I	49	2	32
J	4A	3	33
K	4B	4	34
L	4C	5	35
M	4D	6	36
N	4E	7	37
O	4F	8	38
P	50	9	39
Q	51	CARRIAGE RET	0D
R	52	SPACE	20
S	53	)	29
T	54	=	3D

The file that contains the word to be found, the word with the smallest ASCII value (f_smallest), is shown in Figure 8.5. The file consists of eleven words; each word has a maximum of five characters and is followed by a carriage return. The file can have any number of words, but the last word must be "END."

Listing 8.7 shows the VHDL code for finding the word with the lowest ASCII value. The smallest value will be stored in a character-type variable, smallest, and the variable is initialized with the highest possible ASCIII value (in our example, "ZZZZZ"). Compare the value of smallest with

```
STORE
STOP
ADD
ADA
SUB
MTPLY
LOAD
JUMP
HLT
COMPR
END
```

FIGURE 8.5 File f_smallest.

each word. If the value of the word is less than the value of smallest, then smallest assumes the value of this word. Otherwise, smallest retains its value. Continue this comparison until the last word in the file is encountered. The code tests each word to see if it is "END." If it is, then the program stops; if not, the program continues. The statement that checks for the word "END" in Listing 8.7 is a while-loop:

```
while (count /= ('E', 'N', 'D', ' ', ' ')) loop
```

The operator /= is the logic NOT EQUAL. The variable count has to be declared as type character. The above loop will continue running until the variable count is equal to END. The statement

```
read (temp, count);
```

reads a character word from the line temp. Because count is declared as an array of characters (string_chr), each time a word is read, the ASCII value corresponding to the characters of the word (see Table 8.1) is computed and stored in the variable count. This is how the VHDL determines that "ADD" is less than "AND."

Listing 8.7 VHDL Code for Finding the Smallest ASCII Value

```
--The following package needs to be attached
--to the main module.

library IEEE;
use IEEE.STD_LOGIC_1164.all;

package array_pkg is
constant N : integer := 4;
--N+1 is the number of elements in the array.
subtype wordChr is character;
type string_chr is array (N downto 0) of wordChr;

end array_pkg;

library ieee;
use ieee.std_logic_1164.all;
use std.textio.all;
use work.array_pkg.all;

--Now start writing the code to find the smallest
entity SMALLEST_CHRCTR is
```

```
     port (START : in std_logic; z : out string_chr);

end SMALLEST_CHRCTR;

architecture BEHAVIOR_SMALLEST of SMALLEST_CHRCTR is

begin
process (START)
file infile : text;
variable fstatus : file_open_status;
variable count, smallest :
    string_chr := ('z ', 'z ', 'z ', 'z ', 'z ');

-- The above statement assigns initial values (Z's) to
-- count and smallest.

variable temp : line;

begin
    file_open (fstatus, infile,
                "f_smallest.txt", read_mode);
    while (count /= ('E', 'N', 'D', ' ', ' ')) loop
    readline (infile, temp);
    read (temp, count);
        if (count < smallest) then
            smallest := count;
        end if;
    end loop;
z <= smallest;
file_close (infile);
end process;
end BEHAVIOR_SMALLEST;
```

After execution, the output z is equal to "ADA."

EXAMPLE 8.7 **IDENTIFYING A MNEMONIC CODE AND ITS INTEGER EQUIVALENT FROM A FILE**

In many programming applications, the user writes the source code in mnemonic. The computer, if not equipped with the appropriate assembler or compiler, understands only machine language, which consists of zeroes and ones. Assemblers and compilers translate from mnemonic to machine

Mnemonic code	User-assigned integer code
HALT	0
ADD	1
XOR	4
MULT	2
DIVID	3
NAND	6
PRITY	5
CLA	7

FIGURE 8.6 File cods.txt.

language. In this example, the code is written for a simple assembler. An integer code is assigned to each mnemonic code. This assignment is user selected. The mnemonic code and its integer value are stored in the file cods.txt (see Figure 8.6).

Listing 8.8 is the VHDL code to find the integer code for each mnemonic. Referring to, the statement

```
if (temp = assmbly_code) then
```

The statement tests whether temp is equal to assmbly_code. This comparison can be done because temp and assmbly_code have been declared with the same type of arrays of characters.

LISTING 8.8 VHDL Code for Finding the Integer Code for a Mnemonic Code

```
--The following package needs to be
--attached to the main module

library IEEE;
use IEEE.STD_LOGIC_1164.all;
package array_pkg is
constant N : integer := 4;
--N+1 is the number of elements in the array.
subtype wordChr is character;
type string_chr is array (N downto 0) of wordChr;

end array_pkg;

--Start writing the code to find the assigned integer value
library ieee;
use ieee.std_logic_1164.all;
use std.textio.all;
use work.array_pkg.all;

entity OPCODES is
    port (assmbly_code : in string_chr; z : out string_chr;
        z1 : out integer);
```

```
end OPCODES;

architecture BEHAVIOR of OPCODES is

begin

process (assmbly_code)
file infile : text;

variable fstatus : file_open_status;
variable temp : string_chr := (' ', ' ', ' ', ' ', ' ');
variable tem_bin : integer;
variable regstr : line;

begin
file_open (fstatus, infile, "cods.txt", read_mode);
   for i in 0 to 8 loop
    -- while loop could have been used instead
    -- of for loop. See Exercise 8.3.

    readline (infile, regstr);

    read (regstr, temp);
        if (temp = assmbly_code) then
        z <= temp;

        read (regstr, tem_bin);
        z1 <= tem_bin;
    exit;
        else if (i > 7)then
        report ("ERROR: CODE COULD NOT BE FOUND");
        z <= ('E', 'R', 'R', 'O', 'R');

        -- assign -1 to z1 if an error occurs
            z1 <= -1;
            end if;
    end if;
end loop;

file_close(infile);

end process;
end BEHAVIOR;
```

EXAMPLE 8.8   VHDL CODE OF AN ASSEMBLER

An assembly program is a group of instructions written in mnemonic code. The instructions usually contain four fields: label, operation code (op-code), address, and comments. In this example, the instruction will have only two fields: opcode and address. The opcode determines the type of operation, such as addition, subtraction, or data movement.

Because the opcode in an assembly program is written in mnemonics, the operation for addition could be written, for example, as ADD. The address field determines the memory address of the operand. For example, the assembly code ADD 9 means the operation is addition, and the addition operation is adding the data (operand) in memory location address 9 to the contents of a CPU register (usually the accumulator). The result of the addition is stored in the accumulator. For the CPU to understand the assembly instruction, the contents of the instruction have to be translated into machine language code, which consists of zeroes and ones. The program that translates assembly code to machine code is called an *assembler* (see Figure 8.7).

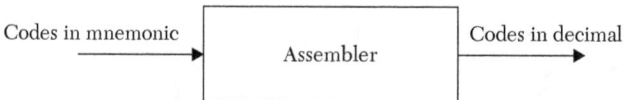

Codes in mnemonic → | Assembler | → Codes in decimal

FIGURE 8.7 The input and output of an assembler.

Listing 8.9 shows the code for an assembler. The assembly program to be translated is written in the file `asm.txt` (see Figure 8.8). Integer op-codes are assigned to the mnemonic codes, as shown in Table 8.2. This assignment is arbitrary; the programmer can assign any pattern of code to the mnemonic code as long as each code has a unique integer value. In this example, the same code pattern is followed as in Figure 8.6. The mnemonic codes ORIG and END have no integer codes; they are called *pseudo codes*. ORIG tells the assembler the starting memory location where the output of the assembler is stored. END tells the assembler where the last line of the assembly program is. Figure 8.9 shows the flowchart of our assembler.

ORG	200
CLA	0
ADD	9
XOR	10
MULT	11
DIVID	12
XOR	13
NAND	14
PRITY	0
HALT	0
HLT	5
END	

FIGURE 8.8 File `asm.txt`.

TABLE 8.2 Integer Codes Assigned to Mnemonic Codes

Mnemonic Code	Assigned Integer Code in Decimal
CLA	7
ADD	1
XOR	4
MULT	2
DIVID	3
NAND	6
PRITY	5
HALT	0

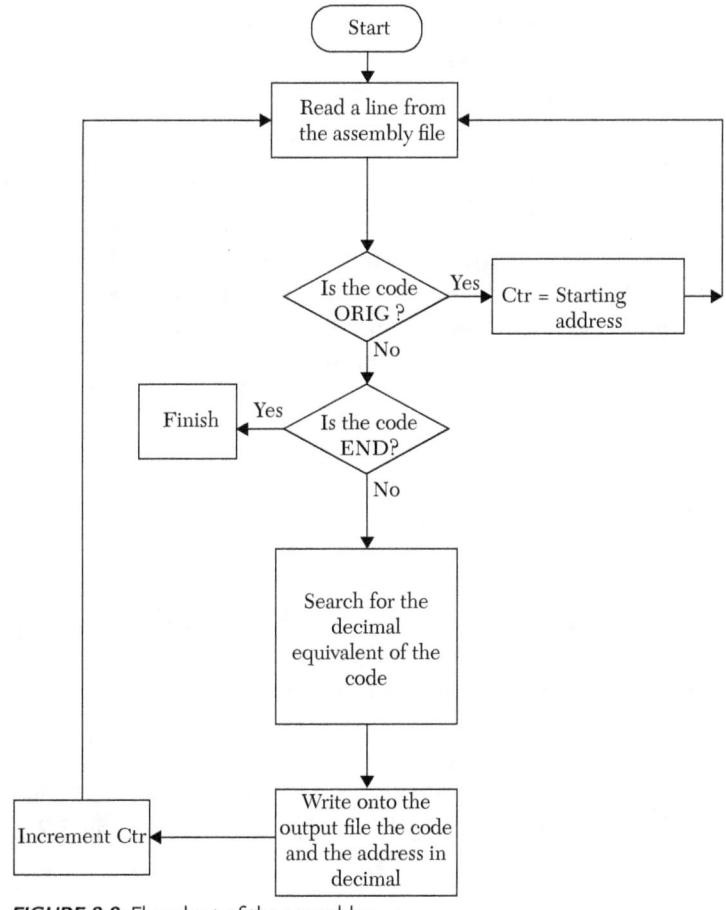

FIGURE 8.9 Flowchart of the assembler.

The assembler first reads a line from the assembly file `asm.txt`. In Listing 8.9, the line is read by the read procedure `readline`:

```
readline (infile, regstr);
```

The infile is the file `asm.txt`. If the first line is read, the contents of `regstr` would be:

```
ORIG    200
```

and `regstr` is read using the procedure `read`:

```
read (regstr, temp);
```

The `read` procedure above stores one word (an array of five characters) in `temp`. If this is the first line of `asm.txt`, then `temp` = "ORIG." As shown in Figure 8.9, the assembler tests the code to see what type it is. In the case of ORIG, the `if` statement is used as follows:

```
if (temp = ('O','R','I','G ',' ')) then
read (regstr, ctr);
```

If the code is "ORIG," the same line is read again, which results in storing the value 200 in `ctr`. If the code of the first line is not "ORIG," an error is reported. After the first line is finished (see Figure 8.9), the subsequent lines are read, and the `case` statement is used to determine the codes and the addresses. For example, the statements

```
when ('M','U','L','T',' ') => code := 2;
write (regstw, code);
write (regstw, " ");
write (regstw, addr);
writeline (outfile, regstw);
```

test the code to see if it is "MULT." If the code is "MULT," an integer of value 2 is assigned to MULT (see Table 8.2), and the address part of the code is written into the `outfile`. In Listing 8.9, a for-loop is implemented to test all lines of the `infile`. Using the for-loop means that the exact number of lines in the `infile` is known; if the exact number is not known, while-loop could have been implemented to test all the lines, regardless of the number of lines. This can be done by specifying an end-of-file word, such as "END," as the condition for terminating the while-loop (see Listing 8.7).

LISTING 8.9 VHDL Assembler Code

```vhdl
--The following package needs to be attached
--to the main module
library IEEE;
use IEEE.STD_LOGIC_1164.all;

package array_pkg is
constant N : integer := 4;
--N+1 is the number of elements in the array.

subtype wordChr is character;
type string_chr is array (N downto 0) of wordChr;

end array_pkg;

library ieee;
use ieee.std_logic_1164.all;
use std.textio.all;
use work.array_pkg.all;

--Now start the code for the assembly
entity ASSMBLR is

    port (START : in bit);

end ASSMBLR;

architecture BEHAVIOR_ASSM of ASSMBLR is
begin

process (START)
file infile : text;
file outfile : text;

variable fstatus, fstatus1 : file_open_status;
variable temp : string_chr := (' ', ' ', ' ', ' ', ' ');
variable code, addr : integer;
variable regstr, regstw : line;
variable ctr : integer := -1;

    begin
file_open (fstatus, infile, "asm.txt", read_mode);
```

```
file_open (fstatus1, outfile, "outf.txt", write_mode);

-- Prepare the outfile where the results of the assembler
-- are stored.

write (regstw, "Location Code Address");
writeline (outfile, regstw);

for i in 0 to 11 loop
--while-loop could have been used instead of for-loop.

readline (infile, regstr);

read (regstr, temp);
if (temp = ('O', 'R', 'I', 'G', ' ')) then
read (regstr, ctr);

elsif (ctr = -1)then
-- If the code of the first line in the file is not ORIG
-- report an error

write (regstw, " ERROR: FIRST OPCODE SHOULD BE ORIG");
writeline (outfile, regstw);
exit;

else
read (regstr, addr);
write (regstw, ctr);
write (regstw, " ");
ctr := ctr + 1;

case temp is

when ('H', 'A', 'L', 'T', ' ') =>
code := 0;
write (regstw, code);
write (regstw, " ");
write (regstw, addr);
writeline (outfile, regstw);

when ('A', 'D', 'D', ' ', ' ') =>
code := 1;
```

```
write (regstw, code);
write (regstw, " ");
write (regstw, addr);
writeline (outfile, regstw);
when ('M', 'U', 'L', 'T', ' ') =>
code := 2;
write (regstw, code);
write (regstw, " ");
write (regstw, addr);
writeline (outfile, regstw);
when ('D', 'I', 'V', 'I', 'D') =>
code := 3;
write (regstw, code);
write (regstw, " ");
write (regstw, addr);
writeline (outfile, regstw);
when ('X', 'O', 'R', ' ', ' ') =>
code := 4;
write (regstw, code);
write (regstw, " ");
write (regstw, addr);
writeline (outfile, regstw);

when ('P', 'R', 'I', 'T', 'Y') =>
code := 5;
write (regstw, code);
write (regstw, " ");
write (regstw, addr);
writeline (outfile, regstw);
when ('N', 'A', 'N', 'D', ' ') =>
code := 6;
write (regstw, code);
write (regstw, " ");
write (regstw, addr);
writeline (outfile, regstw);
when ('C', 'L', 'A', ' ', ' ') =>
code := 7;
write (regstw, code);
write (regstw, " ");
write (regstw, addr);
writeline (outfile, regstw);
when ('E', 'N', 'D', ' ', ' ') =>
```

```
write (regstw, "END OF FILE ");
writeline (outfile, regstw);
exit;

when others =>
code := -20;
write (regstw, "ERROR ");
write (regstw, code);
writeline (outfile, regstw);
end case;

end if;

end loop;
file_close(infile);
file_close (outfile);

end process;
end BEHAVIOR_ASSM;
```

Figure 8.10 shows the outfile "outf.txt" after translating Figure 8.8. Notice that in Figure 8.8, the code "HALT" was intentionally miswritten as "HLT." Listing 8.9 spotted this error and reported it in outf.txt (see Figure 8.10).

Location	Code	Address
200	7	0
201	1	9
202	4	10
203	2	11
204	3	12
205	4	13
206	6	14
207	5	0
208	0	0
209	ERROR	-20
210	END OF FILE	

FIGURE 8.10 Contents of the file outf.txt.

Figure 8.11 shows the rewritten assembly program (Figure 8.8) and intentionally omits "ORIG" from the first line of code. According to Listing 8.9, this is an error.

CLA	0
ADD	9
XOR	10
MULT	11
DIVID	12
XOR	13
NAND	14
PRITY	0
HALT	0
HLT	5
END	

FIGURE 8.11 Variation of the infile asm.txt. ORIG is omitted.

Figure 8.12 shows the contents of the outfile according to the infile of Figure 8.11.

Location	Code Address
ERROR: FIRST OPCODE SHOULD BE ORIG	

FIGURE 8.12 Outfile outf.text for translating Figure 8.11.

8.2.2 Examples of Verilog File Processing

Verilog file processing is based on several built-in tasks such as $fopen, $fdisplay, $fmonitor, and $fclose. The following example discusses file processing in Verilog.

EXAMPLE 8.9 **MANIPULATING AND DISPLAYING DATA IN A VERILOG FILE**

In this example, consider a system with one two-bit input, a, and one three-bit output, b. Output b is related to input a as shown in Equation 8.1:

$$b = 2a \tag{8.1}$$

It is desired to record the value of the outputs as the inputs from a file named file4.txt change. This file is located in the same path as the Verilog module that accesses it. Listing 8.10 shows the Verilog code. The file file4.txt is opened using the task $fopen:

```
ch1 = $fopen("file4.txt");
```

where file4.txt is the name of the file, and ch1 is the indicator of the channel that keeps track of the opened file. To write headings to the file, the task $fdisplay is used. For example, the following statement leaves two spaces, one blank line, one tab(t), and writes the heading "This is file4.txt," and then leaves a blank line:

```
$fdisplay (ch1, " \n\tThis is file4.txt\n");
```

To monitor any signals, the task $fmonitor is used. This task monitors the value of the signal and prints this value into the file. For example, the statement:

```
$fmonitor (ch1," %d\t\t%d%b\n",a,b,
b);
```

monitors the value of signals a and b. These values are printed in file4.txt as follows: leave two spaces, print a in decimal, insert two tabs, print the value of b in decimal, leave thirty spaces, print the same value of b in binary.

LISTING 8.10 Verilog Code for Storing b = 2a in file4.txt

```
module file_test (a, b);
input [1:0] a;
output [2:0] b;
reg [2:0] b;
integer ch1;

initial
    begin
        ch1 = $fopen ("file4.txt");
        $fdisplay (ch1, "\n\t\t\t This is file4.txt \n");
        $fdisplay (ch1, " Input a in Decimal\t
            \t Output b in Decimal\t\t Output b in Binary\n ");
/*The above statement when entered in the Verilog module
should be entered in one line without carriage return */

    end
always @ (a)
    begin
        b = 2 * a;
        $fmonitor (ch1,"\t%d\t\t\t\t%d\t\t\t\t%b \n", a,b, b);
    end
endmodule
```

Figure 8.13 shows file4.txt after execution of Listing 8.10.

This is file4.txt		
Input a in decimal	Output b in decimal	Output b in binary
0	0	000
1	2	010
2	4	100
3	6	110

FIGURE 8.13 `File4.txt` of Listing 8.10.

8.3   VHDL Record Type

Record type is a collection of elements; the elements can be of the same type or of different types. An example of `record` is shown in Listing 8.11. The `record` in Listing 8.11 includes elements of type `integer`, `weekdays`, and `weather`.

LISTING 8.11 Example of Record Type

```
Type weather is (rain, sunny, snow, cloudy);
Type weekdays is (Monday, Tuesday, Wednesday,
Thursday, Friday, Saturday, Sunday);

Type forecast is
Record
Tempr : integer range -100 to 100;
Day : weekdays;
Cond : weather;
end record;
```

Another example of implementing `record` is shown in Listing 8.12. The user provides a certain day and a desired unit of temperature (Centigrade or Fahrenheit), and the VHDL program outputs the current temperature and the forecast condition (e.g., rain, cloudy, snowy, or sunny). Let's examine the following code from Listing 8.12:

```
process (Day_in)
variable temp : forecast;

begin

case Day_in is
```

```
when Monday =>
temp.cond := sunny;
if (unit_in = "CEN") then
temp.tempr := 35.6;
else
temp.tempr := 1.2 * 35.6 + 32.0;
end if;
```

The signal Day_in is declared as type weekdays, so possible values for this signal are Monday, Tuesday, Wednesday, Thursday, Friday, Saturday, or Sunday. The variable temp is declared as type forecast. This type is a record, so possible types for this variable are real, string, weekdays, or cast. To select one type out of these four types, we write, for example, temp.cond. Now temp is of type cast and can assume one of the values of this type (i.e., rain, sunny, snow, or cloudy).

LISTING 8.12 VHDL Code for an Example of Record

The following is the code of the package weather_fcst

```
package weather_fcst is
Type cast is (rain, sunny, snow, cloudy);
Type weekdays is (Monday, Tuesday, Wednesday,
    Thursday, Friday, Saturday, Sunday);
Type forecast is
Record
Tempr : real range -100.0 to 100.0;
unit : string (1 to 3);
Day : weekdays;
Cond : cast;

end record;
end package weather_fcst;

-- Now write the program
library ieee;
use ieee.std_logic_1164.all;
use std.textio.all;
use work.weather_fcst.all;
entity WEATHER_FRCST is
    port (Day_in : in weekdays;
            unit_in : in string (1 to 3);
          out_temperature : out real;
```

```
            out_unit : out string (1 to 3);

            out_day : out weekdays; out_cond : out cast);
-- Type string is a predefined

end WEATHER_FRCST;

--Now we write the code
architecture behavoir_record of WEATHER_FRCST is
begin
process (Day_in, unit_in)
variable temp : forecast ;

begin

case Day_in is

when Monday =>
temp.cond := sunny;
if (unit_in = "CEN") then
temp.tempr := 35.6;
elsif (unit_in = "FEH") then
temp.tempr := 1.2 * 35.6 + 32.0;
else
report ("invalid units");
end if;

when Tuesday =>
temp.cond := rain;

if (unit_in = "CEN") then
temp.tempr := 30.2;
elsif (unit_in = "FEH") then
temp.tempr := 1.2 * 30.2 + 32.0;
else
report ("invalid units");
end if;

when Wednesday =>
temp.cond := sunny;
if (unit_in = "CEN") then
temp.tempr := 37.2;
```

```
elsif (unit_in = "FEH") then
temp.tempr := 1.2 * 37.2 + 32.0;
else
report ("invalid units");
end if;

when Thursday =>
temp.cond := cloudy;
if (unit_in = "CEN") then
temp.tempr := 30.2;
elsif (unit_in = "FEH") then
temp.tempr := 1.2 * 30.2 + 32.0;
else
report ("invalid units"); end if;

when Friday =>
temp.cond := cloudy;
if (unit_in = "FEH") then
temp.tempr := 33.9;
elsif  (unit_in = "FEH") then

temp.tempr := 1.2 * 33.9 + 32.0;
else
report ("invalid units");
end if;

when Saturday =>
temp.cond := rain;
if (unit_in = "CEN") then
temp.tempr := 25.1;
elsif (unit_in = "FEH") then
temp.tempr := 1.2 * 25.1 + 32.0;
else
report ("invalid units");
end if;

when Sunday =>
temp.cond := rain;
if (unit_in = "FEH") then
temp.tempr := 27.1;
elsif (unit_in = "FEH") then
temp.tempr := 1.2 * 27.1 + 32.0;
```

```
else
report ("invalid units");
end if;

when others =>
temp.tempr := 99.99;
report ("ERROR-NOT VALID DAY");
end case;

out_temperature <= temp.tempr;
out_unit <= unit_in;
out_day <= Day_in;
out_cond <= temp.cond;
end process;
end behavoir_record;
```

The simulation output is shown in Figure 8.14.

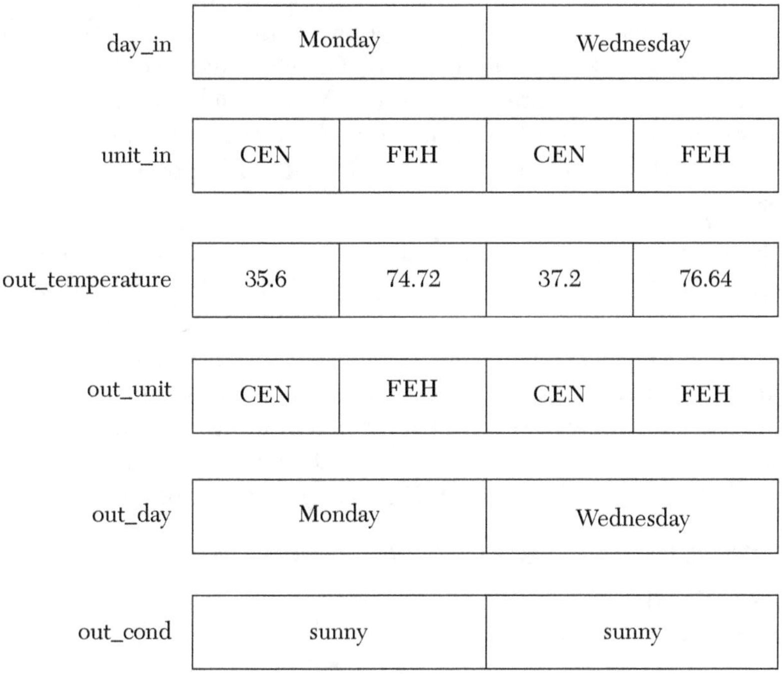

day_in	Monday		Wednesday	
unit_in	CEN	FEH	CEN	FEH
out_temperature	35.6	74.72	37.2	76.64
out_unit	CEN	FEH	CEN	FEH
out_day	Monday		Wednesday	
out_cond	sunny		sunny	

FIGURE 8.14 Simulation output of Listing 8.12.

EXAMPLE 8.10 MEMORY STACK USING ASSERT AND REPORT STATEMENTS

In Chapter 1, the statement `assert` was briefly discussed. The format of this statement is:

```
assert (Boolean condition)
report " optional message display"
severity failure;
```

The severity level can be `note`, `warning`, `error`, or `failure`. The severity level `failure` is the highest priority; it causes the simulation to halt. Here in this example, the `assert` statement is implemented to design a memory stack.

The memory stack consists of a group of memory locations. A special register called the *stack pointer* operates as an address pointer for the stack. The contents of the stack pointer are pointed at the top of the stack. The top of the stack does not necessarily coincide with the physical top of the stack. The lowest address the stack pointer can assume is referred to as the bottom of the stack (see Figure 8.15). The stack has two major operations: push and pop. Push stores data on top of the stack, and the stack pointer is incremented to point to the new top. Pop retrieves data from the top of the stack, and the stack pointer is decremented to point at the new top of the stack.

Usually the stack has two one-bit flags to indicate whether the stack is full or empty. If the stack is full (i.e., the stack pointer is pointing at the highest possible address of the stack), a push operation cannot be executed. If the stack is empty (i.e., the stack pointer is pointing at the lowest possible address of the stack), a pop operation cannot be executed. If the stack is full and the CPU tries to execute a push operation, the full flag is set. If the stack is empty and the CPU tries to execute a pop operation, the empty flag is set. Listing 8.13 shows the VHDL code for stack operation.

LISTING 8.13 VHDL Code for Stack Operation

```
library IEEE;
use IEEE.STD_LOGIC_1164.all;

library IEEE;
use IEEE.STD_LOGIC_1164.all;
package stack_pkg is
constant N : integer := 15;
```

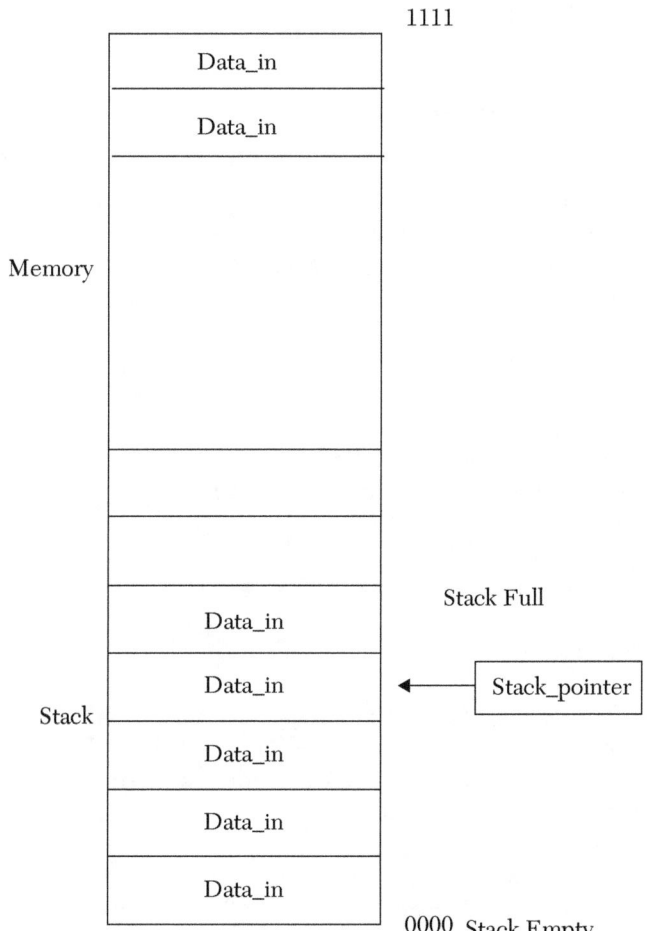

FIGURE 8.15 A block diagram of memory and the memory stack.

```
constant M : integer := 3;
--N+1 is the number of elements in the array.
subtype Memoryword is std_logic_vector (M downto 0);

type Memory is array (N downto 0) of Memoryword;
--The above array represents a 16x4 bits memory
type stack is (push, pop, none);
--The above statement defines three members (push, pop, and
-- none) of the user-defined type stack.
end stack_pkg;
```

```vhdl
library IEEE;
use IEEE.STD_LOGIC_1164.ALL;
use ieee.numeric_std.all;
use work.stack_pkg.all;

entity stck_asrt is
generic (N : integer := 15; M : integer := 3);
    Port (action : in stack; Data_in : in std_logic_vector
        (M downto 0); clk : in std_logic);
end stck_asrt;

architecture Behavioral of stck_asrt is

begin
stk : process (action, data_in, clk)
variable stack_pointer : integer := 0;
variable Mem_comp : Memory;
begin
if (rising_edge (clk)) then

case action is
when push =>
Mem_comp (stack_pointer) := data_in;
stack_pointer := stack_pointer + 1;
--if the operation is push, the stack pointer is
-- incremented as shown above.
--
assert (stack_pointer < 5)
report " stack is full-program halts"
severity Failure;

--The above three statements state that if the stack
-- pointer is not less than 5, the program halts
--and the message "stack is full-program halts" is
-- displayed.
"
when pop =>
stack_pointer := stack_pointer - 1;
--If pop, the stack pointer is decremented

when others => null;
end case;
```

```
end if;
end process;
end Behavioral;
```

Figure 8.16 shows the simulation waveform for Listing 8.13. The figure shows the operation push where the stack pointer is incremented every time a data is pushed. In the figure, the data has a single value = 1011.

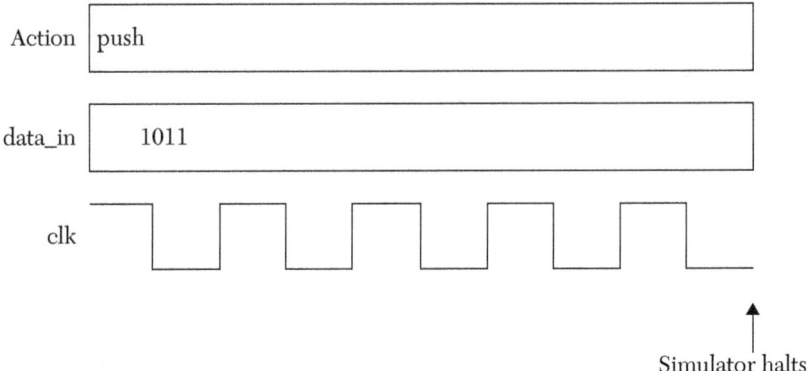

FIGURE 8.16 Simulation waveform of the stack in Listing 8.13.

EXAMPLE 8.11   D-LATCH VHDL DESCRIPTION USING BLOCK STATEMENT

D-latch has been described before using data flow, behavioral, and structural descriptions. Here, the `Block` statement is implemented. The `Block` statement refers to a block of concurrent statements within the architecture. All local declared signals and variables inside the block are visible only inside the block. The `Block` statement has to be labeled. The block can be guarded (accessed on a condition), and signals inside the block can be guarded. A simplified format for the `Block` statement is:

```
label: block (guard_condition)
-- guard_condition can be ommitted
        declarations
begin
        concurrent statements
--above statement can be guarded
end block label;
```

Listing 8.14 shows the VHDL code for describing a four-bit D latch. The block is guarded by the condition E = 1 where E is the enable of the latch; the four output signals Q are guarded by following the input signal D. Be sure that your simulator can handle the block and the block-guarded statements.

Listing 8.14   VHDL Code for a Four-Bit D-Latch Using Block Statement

```
library IEEE;
use IEEE.STD_LOGIC_1164.ALL;

entity BLCKstatement is
port(E: in std_logic; D: in std_logic_vector (3 downto 0);
    Q: out      std_logic_vector ( 3 downto 0));
end BLCKstatement;

architecture blck of BLCKstatement is
begin
Dlatch: block ( E = '1')
begin
Q(0) <= guarded D(0);
Q(1) <= guarded D(1);
Q(2) <= guarded D(2);
Q(3) <= guarded D(3);
end block;
```

8.4   Verilog User-Defined Primitives

Verilog has several built-in primitives such as and, or, and xor gates that have been implemented in Chapter 4. In addition to the built-in, the user can build his or her own primitives to describe combinational and sequential logic. A very simplified format for the user-defined primitive by the name of "sample" is:

```
module identifier(inputs, outputs)
input......
output.....
/*this is the main module where the user defined primitive
is called*/

Sample S1(out, in1, in2, in3)
.....
endmodule

Primitive Sample(outp, inp1, inp2, inp3)
//This is the body of the primitive
Output outp;
Input inp1,inp2, inp3;
table
```

```
//table is a predefined word
Valu1 valu2 valu3 : value4

...........................

endtable
endprimitive
```

The body of the primitive is entered after the end of the module; the primitive is called from within the module. All entries in the `table` of the primitive can have only a single bit. The primitive is allowed to have one output only; primitive `Sample` has an output `outp`. The primitive `Sample` has three inputs `inp1`, `inp2`, and `inp3`. The predefined word `table` allows the user to enter a table that consists of values of the inputs and the output. `Value4` is the value of the output giving the input as `value1`, `value2`, `value3` corresponding to `inp1`, `inp2`, and `inp3`. `Valu1`, `valu2`, `valu3`, and `valu4` can include 0, 1, or x (don't care). Two types of primitives are discussed here: combinational and sequential. In combinational, the output in the table depends on the inputs only; the table resembles the truth table of combinational circuits. In sequential, the output depends on the current state and the input; the table resembles the transition table of sequential circuits. The following examples will clarify the implementation of user-defined primitives.

EXAMPLE 8.12   **DESCRIPTION OF A 2x1 MULTIPLEXER WITH ACTIVE LOW ENABLE USING USER-DEFINED PRIMITIVES**

A description of a 2x1 multiplexer has been written using data-flow description (see Example 2.3a and Figure 2.9), behavioral description (see Example 3.6), and structural description (see Example 4.2). Here, the multiplexer is described using combinational user-defined primitive (UDP). The function of the multiplexer has been shown in Table 2.4 and copied here in Table 8.3.

TABLE 8.3 Truth Table for a 2x1 Multiplexer

Input		Output
SEL	**Gbar**	**Y**
X	H	L
L	L	A
H	L	B

The Verilog code for the description of the muliplexer using UDP is shown in Listing 8.15. Notice that the entered values can be 0, 1, x, or ?. The operator x is the "don't care;" the operator ? can assume the values 0, 1, or x.

LISTING 8.15 Verilog Code 2x1 Multiplexer with Active Low Enable Using Combinational User-Defined Primitive

```
module Mux2x1Prmtv(A, B, SEL, Gbar,Y);
    input A,B,SEL,Gbar;

    output Y;

multiplexer MUX1 (Y, Gbar, SEL,A,B) ;

endmodule
primitive multiplexer (mux, enable, control, dataA, dataB) ;
output mux;
input enable, control, dataA, dataB;
table
// enable control dataA dataB mux
    1   ?   ?      ?      : 0;
    0   0   1      ?      : 1;
    0   0   0      ?      : 0;
    0   1   ?      1      : 1;
    0   1   ?      0      : 0;
    0   x   0      0      : 0;
    0   x   1      1      : 1;
endtable
endprimitive
```

The simulation waveform is the same as in Figure 2.10 but without any delay.

EXAMPLE 8.13   DESCRIPTION OF A ONE-BIT D-LATCH WITH ACTIVE HIGH CLEAR USING SEQUENTIAL USER-DEFINED PRIMITIVE

Example 2.4 covered a one-bit latch; data-flow style was implemented to describe the latch. Table 2.7 showed the transition table for the latch. Table 8.4 shows a transition table for D-latch with active high clear. In the table, Q is the current state, and Q+ is the next state.

TABLE 8.4 Excitation Table of D-Latch with Active High Enable

Inputs				Next State
E	Clr	D	Q	Q⁺
x	1	x	x	0
0	0	x	0	0
0	0	x	1	1
1	0	0	x	0
1	0	1	x	1

Listing 8.16 shows the Verilog code for the latch using sequential user-defined primitive D_latch. The primitive D_latch shows the current state O1 and the next state O1+. Notice that O1 is declared as an output and a register because the primitive needs to know the stored value of the current state.

LISTING 8.16 Verilog Code for a D-Latch with Active High Clear Using Sequential User-Defined Primitive

```
module latchprimitive(E, clr,D, Q, Qbar);
    input E, clr,D;

    output Q,Qbar;

D_latch D1 (Q, E,clr, D) ;
assign Qbar = ~ Q;

endmodule

primitive D_latch(O1, inp1, inp2, inp3) ;
output O1;
reg O1 ;
input inp1,inp2,inp3;
table
// inp1 inp2 inp3 O1    O1+
1   0   1   : ? : 1 ;
1   0   0   : ? : 0 ;
0   0   ?   : ? : - ; // no change
?   1   ?   : ? : 0; //if clear signal is=1, Q=0
endtable
endprimitive
```

Figure 8.17 shows the simulation waveform of the latch.

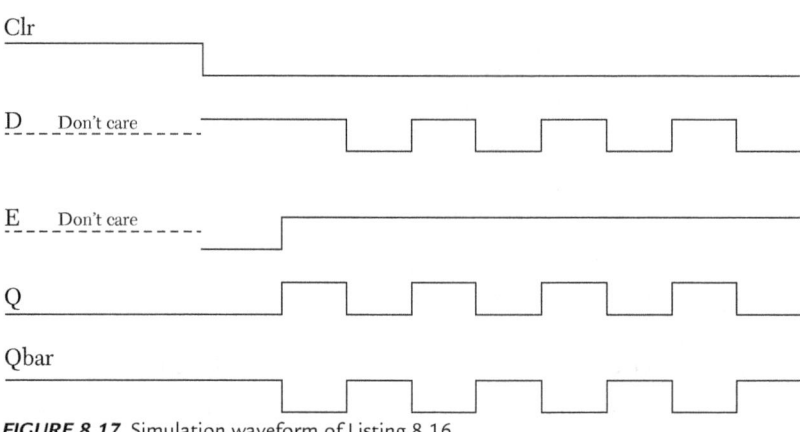

FIGURE 8.17 Simulation waveform of Listing 8.16.

8.5 Cache Memory

Cache memory is a very fast random-access memory (RAM). A typical storage media in a microcomputer consists of a hard disk (the longest access time and the largest size, usually > 100 gigabytes), the main memory (the intermediate access time between the hard disk and the cache with an average size of 1/100 of the hard-disk size), and the cache memory (the shortest access time and the smallest size, usually 1/100 of the size of the main memory). The hard desk is usually the least expensive in terms of the cost per storage byte, followed by the main memory and followed by the cache, so the cache is the most expensive in terms of cost per byte of storage. The cache can be built inside the CPU, usually called L1 cache (level 1), or outside the cache, usually called L2 cache. A computer system may have more than one level of cache. Because the cache is the fastest access memory in the computer, data written into or read from the cache will take a short time, and accordingly, the computer system would be faster than the same system if it did not have cache memory. However, because the size of the cache is the smallest among the storage media, cache memories cannot store all the data available in the memory. The communication protocol between the CPU, main memory, and cache memory assumes that the data requested by the CPU is located in the cache, and the CPU communicates directly with the cache first; if the requested data is not found in the cache, this data has to be moved from the main memory to the cache. Moving the data from the main memory to the cache takes a relatively long time because the main memory is relatively slow compared to the cache. So, to speed up the microcomputer, the requested data by the CPU should be made available in the cache. If the requested data is found in the cache, it is called a hit; otherwise, it is a miss. A parameter called the *hit ratio* calculates the ratio of the number of hits divided by the total number of requests or references. Obviously, to improve the performance of the computer, the hit ratio has to be as high as possible. For the CPU to identify the data in the cache to determine whether a hit or a miss has occurred, a mapping between the data in the main memory and the cache should be established. This mapping is to assure that the data that have moved from the main to the cache can be identified. Several mapping schemes such associative, random, direct, and set-associative mapping are implemented in the cache system. Direct mapping is also known as one-way set-associative mapping. Here, direct and set-associative mapping are discussed; for more information on the cache system, refer to Hayes, 1998 [1] and Patterson, 2011 [2]. To illustrate the mapping schemes, consider a main memory of 16x4 bits and a four-word cache. The width of the cache will be determined according to the mapping scheme. Let's assume that the data in the main memory are as shown in Table 8.5.

TABLE 8.5 Contents of Main Memory in Decimal

Location$_d$	Data$_d$	Location$_d$	Data$_d$
0	3	8	2
1	4	9	1
2	9	10	14
3	10	11	8
4	7	12	6
5	0	13	5
6	13	14	12
7	15	15	11

8.5.1 Direct Mapping

Let's start with an empty cache and assume that the first four words in the main memory are to be moved into the cache. The four words will be moved one word at a time; other applications may move the four words as a block, but here, only a single word movement is considered. To access any word in the main memory, a four-bit address is needed; for the first word, for example, this address is 0000_2. To move the data of this location to the cache using direct mapping, the address is partitioned into two fields, tag and cache address, which is called an index. Because the cache is four words, the index is two bits. Because the index is two bits, the tag is whatever is left from the four-bit memory address, which will be two bits; the memory address is divided into the index of two bits (the least significant two bits) and the tag of two bits (the most significant two bits). For the first memory word, the index is 00 and the tag is 00. For the memory address 0001, the index is 01 and the tag is 00. The data of value 3 will be stored in the cache address of 00; for the same data word, the tag (00) is inserted at the left of the data. In binary, the data is stored in location 00 of the cache, 000011, a total of six bits. The contents of the cache memory after filling the cache with the first four main memory words are shown in Table 8.6. The information in the index and the tag can retrace the data to its memory location. For example, the index 11 (cache memory location 11) and the corresponding tag 00 are pointing at main memory location (0011), and the data in this location is 1010.

TABLE 8.6 Contents of Cache Memory in Binary

Location (index)$_2$	Data$_2$	Location (index)$_2$	Data$_2$
00	000011	10	001001
01	000100	11	001010

Figure 8.18a illustrates the direct-mapping scheme. Listing 8.17 illustrates the direct-mapping Verilog description. The main memory is represented by an array M:

```
reg [3:0] M [0:15];
```

The array M consists of sixteen four-bit words (elements). The data in the main memory are entered for each element of the array; for example, M[12] = 4'd6 means that location 12_d of the main memory is assigned the data 6. The cache memory is represented by the array cache:

```
reg [N:0] cache [0:3];
```

The array cache consists of four words, and each word is four bits.

The CPU requests data from the cache by issuing a memory address where the data is stored. The Listing allows the user to select a one-way or two-way by entering 0 or 1, respectively, for the case-control expression cachemapping. The statement

```
cache[cpuaddress1 [1:0]] =
{cpuaddress1[3:2],M[cpuaddress1]};
```

concatenates ({) bits 2-3 of the issued CPU address with the main-memory data in the address issued by the CPU. For example, if the CPU issues a main-memory address of 0100 (4), then location 00 (index) of the cache will have 01 (tag) concatenated with M (4), which is 7, so location 00 of the cache will have 010111.

8.5.2 Two-Way Set-Associative Mapping

Figure 8.18 illustrates one- and two-way associative mapping. Figure 8.18a illustrates the filling (writing) of the first three main-memory locations (from 0 to 2) in the cache. For this one-way mapping, if another data other than the first four data that have been stored needed to be stored in the cache, one of the cache data has to be deleted. For example, if the data of the main memory location 1111_2 needed to be stored (written) in the cache, the cache data in index 11 has to be deleted because the new tag is 11 while the old tag is 00. In two-way set-associative mapping, each word of the cache is double the width of that of one-way mapping; this allows for storing two memory data with the same index but with different tags in a single word of the cache. Figure 8.18b illustrates two-way mapping. The two main memory words in locations 0010 and 1110 cannot be stored in a single word for one-way mapping because they have the same index (10) but two different tags (00 and 11). However, in two-way mapping, these two data can be stored in location (index) 10 of the cache. The content of this location is 001001111100, which is the first tag (00), the first data (1001), the second tag (11), and the second data (1100). The following statement from Listing 8.17 illustrates the filling of a selected cache location with two-way set-associative mapping:

```
cache2[cpuaddress1 [1:0]] = {({cpuaddress1[3:2],
  M[cpuaddress1]}),
        ({cpuaddress2[3:2],M[cpuaddress2]})};
```

where `cpuaddress1[1:0]` is the index, `cpuaddress1[3:2]` is the first tag,

`M[cpuaddress1]` is the first data, `cpuaddress2[3:2]` is the second tag, and `M[cpuaddress2]` is the second data. Concatenation is used to concatenate all twelve bits into a single cache word.

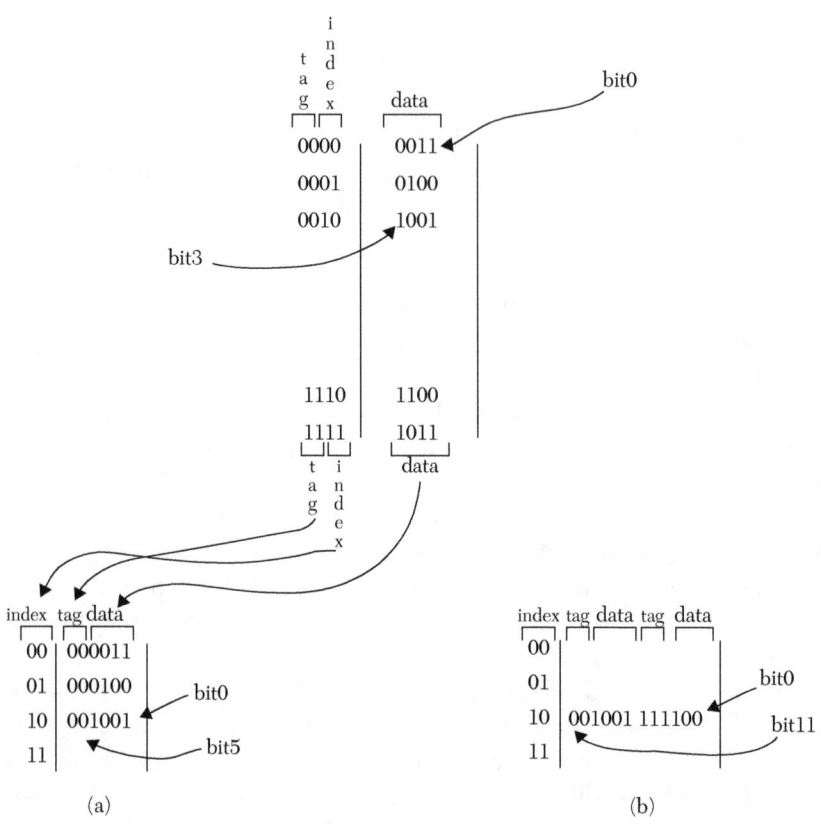

FIGURE 8.18 Mapping schemes. a) One-way set-associative mapping. b) Two-way set-associative mapping.

LISTING 8.17 Verilog Code for One- and Two-Way Set-Associative Mapping

```
//single word mapping, one and two-way
module casheMem(start, cachemapping, cpuaddress1,
                cpuaddress2 );

parameter N= 5;
parameter N1=11;
input start;
input cachemapping;
input [3:0] cpuaddress1, cpuaddress2;

reg [3:0] M [0:15];
reg [N:0] cache [0:3];
reg [N1:0] cache2 [0:3];
```

```
always @(start, cachemapping,cpuaddress1, cpuaddress2)
begin
M[0] = 4'd3; M[1] = 4'd4; M[2] = 4'd9;
M[3] = 4'd10; M[4] = 4'd7; M[5] = 4'd0;
M[6] = 4'd13; M[7] = 4'd15; M[8] = 4'd2;

M[9] = 4'd1; M[10] = 4'd14; M[11] = 4'd8;
M[12] = 4'd6; M[13] = 4'd5; M[14] = 4'd12;
M[15] = 4'd11;

  case (cachemapping)

  1'b0: begin //one-way set-associative (direct mapping)//
cache[cpuaddress1 [1:0]] =
  {cpuaddress1[3:2],M[cpuaddress1]};
  end
1'b1 : begin //two-way associative//
if (cpuaddress1 [1:0] == cpuaddress2 [1:0])
cache2[cpuaddress1 [1:0]] = {({cpuaddress1[3:2],
  M[cpuaddress1]}),
      ({cpuaddress2[3:2],M[cpuaddress2]})};

begin

cache2[cpuaddress1 [1:0]]
={cpuaddress1[3:2],M[cpuaddress1]};
cache2[cpuaddress2 [1:0]]
={cpuaddress2[3:2],M[cpuaddress2]};
end
end
endcase;

end
endmodule
```

Listing 8.18 shows a Verilog code for determining whether a hit or a miss has occurred after the CPU issued a request of data. The request is done by issuing a main-memory address where the required data is located. This address is presented to the cache to see if the data is in its contents. If it is in the cache, a hit has occurred. Otherwise, a miss has occurred. The cache is filled using two-way set-associative mapping. The statement

```
if ((cpuaddress [3:2] == data [11:10])
| (cpuaddress [3:2] == data [5:4]))
```

checks to see if a hit or a miss has occurred for a given index. If the tag cpuaddress[3:2] matches either one of the two tags stored in the cache at the given index, a hit has occurred. The two tags are data [11:10] and data [5:4].

LISTING 8.18 Verilog Code for Determining a Hit or Miss

```
//Determining Hit or Miss without replacement
module hitmiss(cpuaddress, hitORmiss);
parameter N1=11;
input [3:0] cpuaddress;
output [4*8:1] hitORmiss ;
reg [4*8:1] hitORmiss;
reg [N1:0] cache [0:3];
reg [3:0] M [0:15];
reg [N1:0] data;

always @(cpuaddress)
begin
/*fill two-way the cache in order starting from
memory location 0; s0  locations 0000 and 0100
will occupy the first cache location*/

M[0] = 4'd3; M[1] = 4'd4; M[2] = 4'd9;
M[3] = 4'd10; M[4] = 4'd7; M[5] = 4'd0;
M[6] = 4'd13; M[7] = 4'd15; M[8] = 4'd2;

M[9] = 4'd1; M[10] = 4'd14; M[11] = 4'd8;
M[12] = 4'd6; M[13] = 4'd5; M[14] = 4'd12;
M[15] = 4'd11;

cache[0] = 12'b000011010111; cache[1] =12'b000100010000;
cache[2] = 12'b001001011101; cache[3] = 12'b001010011111;
hitORmiss = "miss";
data = cache[cpuaddress[1:0]];

if ((cpuaddress [3:2] == data [11:10])
| (cpuaddress [3:2] == data [5:4]))
hitORmiss = "hit ";
```

```
end
endmodule
```

When a miss occurs in two-way set-associative mapping, new data from the main memory is moved into the cache according to its index and tag. However, if the location in the cache where the new data should be stored is occupied, then one of the cache memory words has to be replaced by the new data. If the cache allows a selection of where to store the new data (such as in two-way mapping, where there are two locations to choose from at each index), then a replacement strategy should be in place. Replacement of old data with new data is not an easy task. If old data is replaced and the computer requests this data again, the data has to be moved again from the main memory to the cache, and this slows the computer. If the replacement strategy is not efficient, the computer with a cache would be slower than the same computer without cache. In Listing 8.18, a simple replacement algorithm is implemented. The replacement is based on the first-in first-out (FIFO) strategy. The new data will replace the oldest data. Here, a bit is added as the least significant bit of the cache word. This bit (if 0) indicates that the least significant four-bit data in the first set of the cache word is the older and should be replaced if needed. If the added bit is 1, it indicates that the most significant four-bit data in the second set of the cache word is older and should be replaced if needed. Figure 8.19 shows the output of Listing 8.19.

LISTING 8.19 Verilog Code for Two-Way Set-Associative Cache System with Replacement

```
module FIFOreplace(cpuaddress, hitORmiss);
parameter N1=11;
input [3:0] cpuaddress;
output [4*8:1] hitORmiss ;
reg [4*8:1] hitORmiss;
reg [N1+1:0] cache [0:3];
reg [3:0] M [0:15];
reg [N1+2:0] data;
initial
begin
/*fill the cache in two-way set associative mapping
in order starting from memory
location 0; so locations 0000 and 0100
will occupy the first cache word*/
```

```
M[0] = 4'd3; M[1] = 4'd4; M[2] = 4'd9;
M[3] = 4'd10; M[4] = 4'd7; M[5] = 4'd0;
M[6] = 4'd13; M[7] = 4'd15; M[8] = 4'd2;

M[9] = 4'd1; M[10] = 4'd14; M[11] = 4'd8;
M[12] = 4'd6; M[13] = 4'd5; M[14] = 4'd12;
M[15] = 4'd11;

/*bit0 (the least significant bit) of the word in the
cache indicates the age of the data in the first and the
second set. If bit0=0 then the least significant
data (set) is the older and should be replaced; otherwise
if bit0 =1; the most significant data(set) is the older
and should be replaced. */

cache[0] = 13'b0000110101110; cache[1] =13'b0001000100000;
cache[2] = 13'b0010010111010; cache[3] =13'b0010100111110;
end

always @(cpuaddress)
begin
hitORmiss = "miss";
data = cache[cpuaddress[1:0]];

if ((cpuaddress [3:2] == data [12:11])
| (cpuaddress [3:2] == data [6:5]))
hitORmiss = "hit ";
  else
  begin
  if (data[0] == 1'b0)
  begin
  data[0] = 1'b1;
  data[6:5] = cpuaddress [3:2];
data[4:1] = M[cpuaddress];
cache[cpuaddress[1:0]] = data;
end
else
data[0] = 1'b0;
data[12:11] = cpuaddress [3:2];
data[10:7] = M[cpuaddress];
cache[cpuaddress[1:0]] = data;
end
```

```
end
endmodule
```

Initial cache contents	01AE	0220	04ba	053E	
cpuaddress[1:0] or Index		0000			hitORmiss = hit
New cache contents	01AE	0220	04ba	053E	
cpuaddress[1:0] or Index		1111			hitORmiss = miss
New cache contents	01AE	0220	04ba	1DF7	
cpuaddress[1:0] or Index		1111			hitORmiss = hit
New cache contents	01AE	0220	04ba	1DF7	
cpuaddress[1:0] or Index		0011			hitORmiss = miss
New cache contents	01AE	0220	04ba	0576	

FIGURE 8.19 Replacement algorithm based on the FIFO strategy.

CASE STUDY 8.1   SIMULATION OF ARTIFICIAL NEURAL NETWORKS

Artificial neural networks (ANNs) are simulated networks that mimic a simplified biological nervous system. To understand how ANNs operate, let us review the operation of an extremely simplified nervous system. The main cells in the nervous system are neurons. A neuron is composed of three major parts: a soma (or body), an axon, and a dendrite (see Figure 8.20). The neuron receives signals from other neurons through its dendrites, so dendrites are the inputs.

The neuron sends signals to other neurons through its axons, so axons are the outputs of the neuron. The connection between the axons of one

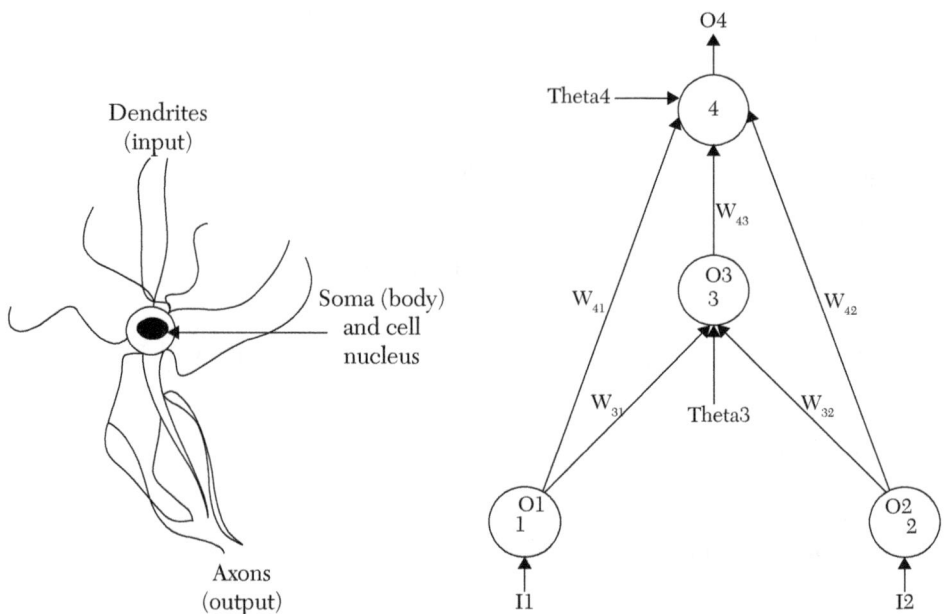

FIGURE 8.20 A biological neuron.

FIGURE 8.21 A simple XOR artificial neural network.

neuron to the dendrites or soma of another neuron is called a *synapse*. The signals that a neuron sends to its neighbors can activate or excite the receiving neurons, or they can deactivate or inhibit them. The signal that the neuron sends can be viewed as a spike. When the neuron sends this signal, the neuron is said to be firing. The neuron sends this signal if it receives enough excitation signals from other neurons. A threshold electric level determines whether or not the excitation signals are high enough for firing. The neuron fires only when the weighted sum of these excitation signals is higher than the threshold. Each neuron asserts different weights on its neighbors.

In the artificial neural network, a node simulates the neuron. Each node has inputs and outputs, and the node is connected to a group of other nodes. The assertion of each node on other nodes is measured by the weight of the connection. The networks are implemented in many applications such as pattern recognition and complex-function generation. In this case study, the network is implemented to generate a simple XOR function. Figure 8.21 shows a simple artificial neural network. The network consists of three layers: input, hidden, and output.

The input layer consists of two nodes, node1 and node2. The hidden layer consists of one node, node3, and the output layer consists of one node,

node4. W_{ij} represents the weight between node j and node i. Because the network, shown in Figure 8.21, functions as a XOR gate, the inputs and the output of the network should satisfy Table 8.7.

TABLE 8.7 Values of the Inputs and the Desired Outputs for an XOR Artificial Neural Network

I_1	I_2	Desired Output (O_4)
0	0	0
1	0	1
0	1	1
1	1	0

By adjusting the weights, the network can be programmed to behave as a XOR gate. Weight adjustment is called *training* the network. Network training is done in the following steps:

Step 1: Initialize the weights and assign random small values to the weights.

Step 2: Select an input with the desired output from Table 8.7.

Step 3: Calculate the output of each node including the output node.

Step 4: Calculate the error of node 4.

Step 5: Select another input and repeat Steps 3–4 and average the four δ_4 errors and the four δ_3 obtained from the four input sets.

Step 6: Update the weights with the new errors calculated in Step 5.

Step 7: Repeat Steps 2–6 until the error δ_4 is lower than the user-defined threshold.

In Step 3, for the input layer (nodes 1 and 2), the output is equal to the input ($O1 = I1$ and $O2 = I2$). For other nodes, the output is calculated as:

$$O_i = f \text{ (weighted sum)} \qquad (8.2)$$

where the weighted sum is the sum of each output of all nodes connected to the node, i, multiplied by the weight. For example, for node 3, the weighted sum is determined as:

$$\text{weighted sum of node 3} = O_1 W_{31} + O_2 W_{32} + \text{Theta3} \times 1$$

Theta is called the *bias* or the *offset*. The weight of all biases theta is equal to 1. The function f (see Equation 8.2) is called the *firing* function. In our example, f is assumed to be a straight line with saturation values in both positive and negative directions (see Figure 8.22). Many other firing functions are implemented in training of artificial neural networks. Some examples of these functions are sigmoid, linear, and relay (zero level or saturation level). More details on artificial neural networks can be found in Haykin, 1999 [3].

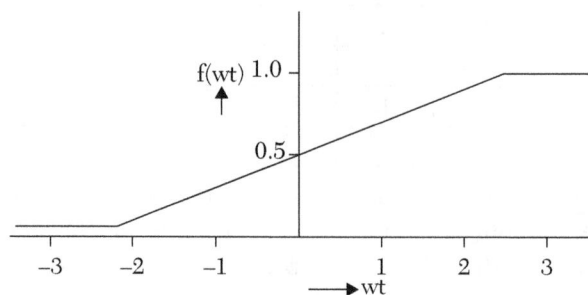

FIGURE 8.22 The firing function.

For Step 4, the outputs of the network (node 4) calculated in Step 2 most likely are not equal to the desired output (see Table 8.3). This error, which resulted from the weights selected in Step 1, is calculated as:

$$\text{Error of node 4} = \delta_4 = (d - O_4)\, O_4\, (1 - O_4) \tag{8.3}$$

Because node 3 is not an output node, its error is calculated with a different formula than Equation 8.3 [3]:

$$\delta_3 = O_3\, (1 - O_3)\, (\delta 4\, W_{43}) \tag{8.4}$$

In Step 5, select another input and repeat Steps 3–4. Average the four 4 errors and the four 3 obtained from the four input sets. You can take the root mean square of the errors instead of the simple average.

In Step 6, use the following equations to update the weights with the new errors calculated in Step 5:

$$W_{4i}\,(\text{new}) = W_{4i}\,(\text{old}) + 0.5\delta_4 O_i \quad i = 1, 3 \tag{8.5}$$

$$W_{3i}\,(\text{new}) = W_{2i}\,(\text{old}) + 0.5\delta_3 O_i \quad i = 1, 2 \tag{8.6}$$

$$\text{Theta}_i\,(\text{new}) = \text{Theta}_i\,(\text{old}) + 0.5\delta_i \tag{8.7}$$

Step 7 repeats Steps 2–6 until the error δ_4 is lower than the user-defined threshold.

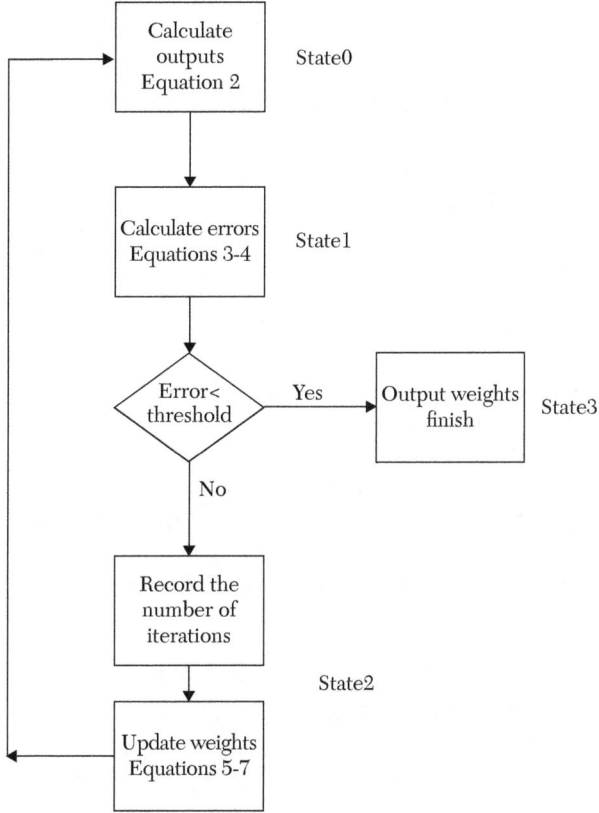

FIGURE 8.23 Flow chart of the state machine.

Listing 8.20 shows the HDL code for the artificial neural network of Figure 8.21. For simplicity, the network is trained only for input $I_1 = 0$ and $I_2 = 1$. The desired output for this set of inputs, as shown in Table 8.7, is 1. The training is done with the help of a finite-state machine. A flow chart of this machine is shown in Figure 8.23. As shown, the machine has four states: state0, state1, state2, and state3. State0 corresponds to Step 3, state1 corresponds to Step 4, state2 corresponds to Step 6, and state3 corresponds to Step 7.

LISTING 8.20 HDL Description of a Simple Artificial Neural Network

```
library IEEE;
use IEEE.STD_LOGIC_1164.all;
-- Write a package to include user-selected type
package types is
```

```vhdl
type state_machine is (state0, state1, state2, state3);
end;

-- Write the code for the state machine
library IEEE;
use IEEE.STD_LOGIC_1164.ALL;
use IEEE.STD_LOGIC_ARITH.ALL;
use ieee.numeric_std.all;
use work.types.all;

entity neural is

    port (clk : in std_logic; I1, I2, Target4,
      threshld : in real;
        W31_O, W32_O, W41_O, W42_O, W43_O, Theta3_O,
        Theta4_O : out real;
        output4 : out real; count_O : out natural);

--The weights could have been entered as an array

end neural;

architecture Behavioral of neural is
-- write the firing function
function firing (wt : in real) return real is
variable wt_rl : real;

begin
--The firing function here is a straight line with
--saturation levels at both the positive and negative ends
if (wt <= -2.2) then
wt_rl := 0.06;
elsif (wt > 2.5) then
wt_rl := 1.0;
else
wt_rl := 0.20 * wt + 0.5;
end if;
return wt_rl;
end firing;

begin
```

```
--Write the code for the state machine
train : process (I1, I2, Target4, threshld, clk)
variable O1, O2, O3, O4, wtsum, delta3, delta4 : real;
variable eita : real := 0.5;
variable pres_st : state_machine := state0;

--Assign initial values for the weights and theta
variable W31 : real := -1.5;
variable W32 : real := -1.5;
variable W41 : real := -1.0;
variable W42 : real := -1.0;
variable W43 : real := -2.0;
variable Theta3 : real := 1.0;
variable Theta4 : real := 1.0;
variable count : natural := 0;
begin

if (clk = '1' and clk'event) then

case pres_st is
    when state0 =>

--Calculate outputs from Equation 2
    O1 := I1;
    O2 := I2;

--Calculate the weighted sum
    wtsum := W31 * O1 + W32*O2 + Theta3;
--Apply the firing function
    O3 := firing (wtsum);

wtsum := W41 * O1 + W42 * O2 + W43 * O3 + Theta4;

O4 := firing (wtsum);

pres_st := state1;

when state1 =>
-- Calculate errors
    delta4 := (Target4 - O4)* O4 * (1.0 - O4);

    delta3 := O3 * (1.0 - O3) * (delta4 * W43);
```

```
if (delta4 < threshld) then
--The threshold is a user-selected value
    pres_st := state3;
    else
    pres_st := state2;
--Record the number of iteration
    count := count + 1;
    count_O <= count;
    end if;

when state2 =>
--Update weights
    W41 := W41 + eita * delta4 * O1;
    W42 := W42 + eita * delta4 * O2;
    W43 := W43 + eita * delta4 * O3;
    Theta4 := Theta4 + eita * O4;
    W31 := W31 + eita * delta3 * O1;
    W32 := W32 + eita * delta3 * O2;
    Theta3 := Theta3 + eita * O3;

    pres_st := state0;
when state3 =>
--Finish; report results
    W41_O <= W41;
    W42_O <= W42;
    W43_O <= W43;
    W32_O <= W32;
    W31_O <= W31;
    Theta3_O <= Theta3;
    Theta4_O <= Theta4;
    output4 <= O4;
end case;
end if;
end process train;

end Behavioral;
```

8.6   Summary

In this chapter, some advanced descriptions were covered. File pro-cessing in both VHDL and Verilog were discussed. To access any file, the file has to be opened before it is accessed. VHDL has several file-processing procedures such as `file_open` to open files, `readline` to read a line from the file, `writeline` to write a line into the file, and `file_close` to close the file.

Verilog has file-processing functions such as `$fopen` to open a file, `$fdisplay` to write data into the file, and `$fmonitor` to monitor an object in the file. The VHDL `record` type was covered; `record` is a collection of different types. Also, Verilog user-defined primitives were covered. Cache memories were briefly discussed and described using Verilog arrays. Finally, artificial neural networks were discussed, as was the complete VHDL code for their training.

8.7   Exercises

1. Write the following data in the VHDL text file `exercise_ch8`. In the file, keep the format and type of the data as it is shown below:

THIS IS THE FILE OF THE EXERCISE OF CHAPTER 8

Training data is 5 3.1 -1.5
Nodes A, B, C, D
Test data 23 12 -5
END

2. Write the VHDL code to store the following words in a file called `greatest.txt`. The words in the file should appear as follows:

ADD
STORE
COMPARE
ZEROS
SUB
STOP

Write the code (in the same module or a new one) to find the word in the above file that has the greatest ASCII value. Also, find its order (e.g., the order of the word STORE is 1).

3. Rewrite Listing 8.9 using a `while-loop` instead of a `for-loop`. Verify your code by simulation.

4. Modify the assembler code of Listing 8.9 to accept labels instead of explicit addresses. Verify your assembler with the program shown below.

Notice that for the statement ADD Data1, Data1 is an address, and the value of this address is 208. Your code should find this address; do not manually substitute 208 for the address.

Label	Code	Address
	ORG	200
	CLA	0
	ADD	Data1
	XOR	Data2
	MULT	Data3
	XOR	Data2
	NAND	Data4
	PRITY	0
	HALT	0
Data1:	7	
Data2:	5	
Data3:	4	
Data4:	2	
	END	

5. Build a package with procedures to find the integer code given the mnemonic code.

6. Use Verilog file processing to compute and display the values of Y when X changes incrementally from 0 to 9. The relationship between X and Y is:

$$Y = X^2 - 2X + 1$$

7. In Listing 8.12, it is desired to output the results to a file. Adjust your code, especially the user-defined types, to conform to the acceptable types that a VHDL file can handle. Rewrite the program and output your results to a text file named `Wthr_forcst`. Each entry of the file should be preceded by a short explanation, such as "The Day is" or "The Temperature is."

8. For Listing 8.12, the following segment of the code has been modified as shown below. The simulation output of the code after modification is not the same as in Figure 8.14. Can you spot the modification and explain why we are not getting the same output as in Listing 8.12?

Listing 8.12 Modified

```
architecture behavoir_record of WEATHER_FRCST is
begin
process (Day_in)
variable temp : forecast;

begin

case Day_in is

when Monday =>
temp.cond := sunny;
if (unit_in = "CEN") then
temp.tempr := 35.6;
elsif    (unit_in = "FEH") then
temp.tempr := 1.2 * 35.6 + 32.0;
else
report ("invalid units");
end if;
```

9. In Listing 8.13, the code for description of a stack operation has been written. An assertion was made on the condition when the stack is full. Do the following:

(a) Adjust the segment of the case for when to push to allow for multiple values of the `data_in`, so your simulation will show different values with the clock instead of just using one value as was done in Figure 8.16.

(b) Use the `assert` statement with `report` to ensure that the stack cannot be popped up if it is empty. Simulate your code and verify.

10. Use Verilog user-defined primitives to describe a D flip-flop with clear and preset inputs. The preset if high Q^+ goes high.

11. Write a Verilog description to find the hit ratio for a cache system. The hit ratio is the ratio between the number of hits divided by the total number of times the cache was referenced.

12. In a FIFO scheme for cache-memory replacement algorithm (Listing 8.19), an additional bit was added to the cache-memory word to indicate the age of the two data stored in the index. Rewrite the code without

using this additional bit or any other extra bits. Your replacement algorithm should still be based on **FIFO**.

13. Simulate the code shown in Listing 8.20 using a threshold of 10^{-8}. What are the final values of the weights? How many cycles does it take the program to reach these final values?

14. In Case Study 8.1, a network was trained for the inputs $I_1 = 0$ and $I_2 = 1$. Here, we want to train the network for all possible inputs. This can be done in the following steps:

Step 1: Initialize the weights (as was done in the case study).

Step 2: Calculate the actual outputs for each input using the same set of weights.

Step 3: Calculate the errors separately for each of the four actual outputs; each input set has its desired output. For example, the input set $I_1 = 1$, $I_2 = 1$ has a desired output of 0.

Step 4: Take the average of the four errors and consider this average the ERROR.

Step 5: Update the weights using the ERROR as was done in the case study.

Step 6: Repeat Steps 2–5 until the ERROR is lower than the threshold.

8.8 References

Hayes, J. P., *Computer Architecture and Organization*, 3rd ed. McGraw Hill, 1998.

Patterson, A., and Hennessy, J. *Computer Organization and Design,* 4th ed. Morgan Kaufmann, 2011

Haykin, S., *Neural Networks*, 2nd ed. Prentice Hall, 1999.

CHAPTER 9

MIXED-LANGUAGE DESCRIPTION

Chapter Objectives

- Understand the concept of mixed-language description
- Learn the advantages of mixing VHDL and Verilog modules
- Learn how to invoke a Verilog module from a VHDL module
- Learn how to invoke a VHDL module from a Verilog module
- Learn the current limitations of mixed-language description

9.1 Highlights of Mixed-Language Description

Mixed-Language Description is a powerful tool in writing HDL code. The mixing here is referring to an HDL code with VHDL and Veilog extracts in the same module. Highlights of the mixed-language description can be summarized in the following facts.

Facts

- To write HDL code in mixed language, the simulator used with the HDL package should be able to handle a mixed-language environment.

- In the mixed-language environment, both VHDL and Verilog module files are made visible to the simulator.

- In the mixed-language environment, both VHDL and Verilog libraries are made visible to the simulator.

- At the present time, the mixed-language environment has some limitations, but the development of simulators that can handle mixed-language environments with minimal constraints is underway. One of these major constraints is that a VHDL module can only invoke the entire Verilog module, and a Verilog module can only invoke a VHDL entity. For example, we cannot invoke a VHDL procedure from a Verilog module. Check your simulator to see if it has recent updates that may not have such restrictions.

- Mixed-language description can combine the advantages of both VHDL and Verilog in one module. For example, VHDL has more extensive file operations than Verilog including `write` and `read`. By writing mixed language, the VHDL file operations can be incorporated in Verilog modules.

9.2 How to Invoke One Language From the Other

As mentioned, when writing VHDL code you can invoke (import) a Verilog module; if you are writing Verilog code, you can invoke (import) a VHDL entity. The process is similar in concept to invoking procedures, functions, tasks, and packages. For example, by instantiating a VHDL package in a Verilog module, the contents of this package are made visible to the module (see Section 9.2.1). Similarly, by invoking a Verilog module in a VHDL module, all information in the Verilog module is made visible to the VHDL module (see Section 9.2.2).

9.2.1 How to Invoke a VHDL Entity From a Verilog Module

In Verilog, invoke a VHDL entity by entering its name (identifier) and its ports in the Verilog module. The parameters of the module should match the type and port directions of the entity. VHDL ports that can be mapped to Verilog modules are: `in`, `out`, and `inout`; `buffer`, in some simulators, is not allowed. Only the entire VHDL entity can be made visible to the Verilog module. Listing 9.1 shows an example of how to invoke a VHDL entity from a Verilog module.

LISTING 9.1 Invoking a VHDL Entity From a Verilog Module

```
//This is the Verilog module
module mixed (a, b, c, d);
input a, b;
output c, d;
...........
VHD_enty V1 (a, b, c, d);
/*The above module VHD_enty is the VHDL entity to be
invoked in this module*/
...........
endmodule
--This is the VHDL entity
library IEEE;
use IEEE.STD_LOGIC_1164.ALL;

entity VHD_enty is
    port (x, y : in std_logic; O1, O2 : out std_logic);
end VHD_enty;

architecture VHD_enty of VHD_enty is
begin
...........


end VHD_enty;
```

Consider the following statement in Listing 9.1:

```
VHD_enty V1 (a,b,c,d)
```

The simulator looks first in the Verilog module to see if there are any Verilog modules by the name of VHD_enty. If it could not find one, the simulator looks in the VHDL entities. When the simulator finds an entity with the name VHD_enty, it binds this entity to the Verilog module. In Listing 9.1, input a is passed to input port x; input b is passed to input y. The VHDL entity calculates the outputs O1 and O2; these two outputs are passed to the Verilog outputs c and d, respectively. Invoking a VHDL module is very similar to invoking a function or a task.

9.2.2 How to Invoke a Verilog Module From a VHDL Module

In the VHDL module, declare a component with the same name as the Verilog module to be invoked (see Chapter 4); the name and port modes of the component should be identical to the name and input/output modes of the Verilog module. Remember that Verilog is case sensitive, so be sure to match the case. Listing 9.2 shows an example of how to invoke a Verilog module from a VHDL module.

LISTING 9.2 Invoking a Verilog Module From a VHDL Module

```
-- This is the VHDL Project

library IEEE;
use IEEE.STD_LOGIC_1164.ALL;
entity Ver_VHD is
    port (a, b : in std_logic; c : out std_logic);
end Ver_VHD;

architecture Ver_VHD of Ver_VHD is
component V_modl
    port (x, y : in std_logic; z : out std_logic);

-- The name of the Component V_modl should be
-- identical to the name of the
-- Verilog module; also, the ports should be
-- identical in name and mode
-- with the inputs and outputs of the Verilog module

end component;


.......
end Ver_VHD;
//This is the Verilog module
module V_modl (x, y, z);

    input x, y;
        output z;

endmodule
```

Referring to Listing 9.2, the component statement in the VHDL module

```
component V_modl
port (x, y : in std_logic; z : out std_logic);
end component;
```

declares a component by the name of `v_mod1` with two input ports, `x` and `y`, and an output port `z`. The Verilog module `v_mod1` has the same name (including the case) as the component and identical inputs and outputs. Accordingly, the Verilog module `v_mod1` is bound to the VHDL component `v_mod1`. If the Verilog module describes, for example, a two-input XOR gate, then in the VHDL module, component `v_mod1` is a two-input XOR gate. In the following sections, complete examples of mixed-language descriptions are covered.

9.3 Mixed-Language Description Examples

This section presents mixed-language examples. Section 9.3.1 covers examples of invoking VHDL entities from Verilog modules, and Section 9.3.2 covers examples of invoking Verilog modules from VHDL modules.

9.3.1 Invoking a VHDL Entity From a Verilog Module

As previously mentioned, a VHDL entity is invoked in a Verilog module by instantiating the Verilog module with a name that is identical to the entity's name. No other construct should have the same name as the entity. A discussion of complete examples follows.

EXAMPLE 9.1 **MIXED-LANGUAGE DESCRIPTION OF A FULL ADDER**

Here, a full adder is constructed from two half adders, as was done in Chapter 4. The logic diagram shown in Figure 4.6 is copied here into Figure 9.1 for convenience. The code of the half adder is written in VHDL. A Verilog module is written to describe a full adder using the VHDL code of the half adder. Listing 9.3 shows a mixed-language code for the full adder.

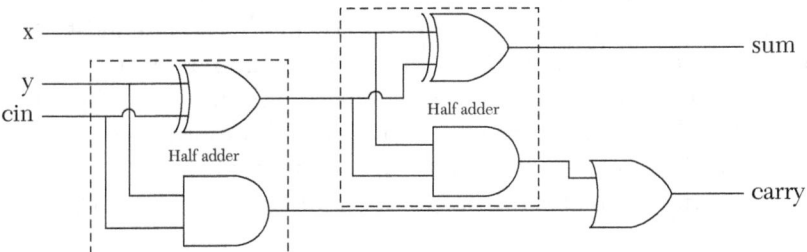

FIGURE 9.1 Full adder as two half adders.

LISTING 9.3 Mixed-Language Description of a Full Adder

```
--This is the Verilog module
module Full_Adder1 (x, y, cin, sum, carry);
    input x, y, cin;
    output sum, carry;
    wire c0, c1, s0;

HA H1 (y, cin, s0, c0);
HA H2 (x, s0, sum, c1);

// Description of HA is written in VHDL in the entity HA
    or (carry, c0, c1);
endmodule

library IEEE;
use ieee.std_logic_1164.all;
entity HA is

--For correct binding between this VHDL code and the above
--Verilog code, the entity has to be named HA.
    port (a, b : in std_logic; s, c : out std_logic);
end HA;
architecture HA_Dtflw of HA is
begin
    s <= a xor b;
    c <= a and b;
end HA_Dtflw;
```

Referring to Listing 9.3, the Verilog statement

```
HA H1 (y, cin, s0, c0);
```

invokes a module by the name of HA. Because there is no Verilog module by this name, the simulator looks at the VHDL modules attached to the Verilog modules. The simulator finds an entity by the name of HA; accordingly, this entity and its bound architecture(s) are made visible to the Verilog module. The architecture here is a data-flow description of a half adder. The inputs y and cin are passed to the input ports of HA, a and b. The VHDL entity calculates the outputs s and c as:

```
s <= a xor b; c <= a and b;
```

The outputs of the entity s and c are passed to the outputs of the module HA, s0 and c0.

EXAMPLE 9.2   MIXED-LANGUAGE DESCRIPTION OF A NINE-BIT ADDER

In this example, a nine-bit adder consisting of three adder slices is described. Each adder slice is a three-bit carry-lookahead adder. Figure 9.2 shows a block diagram of the adder.

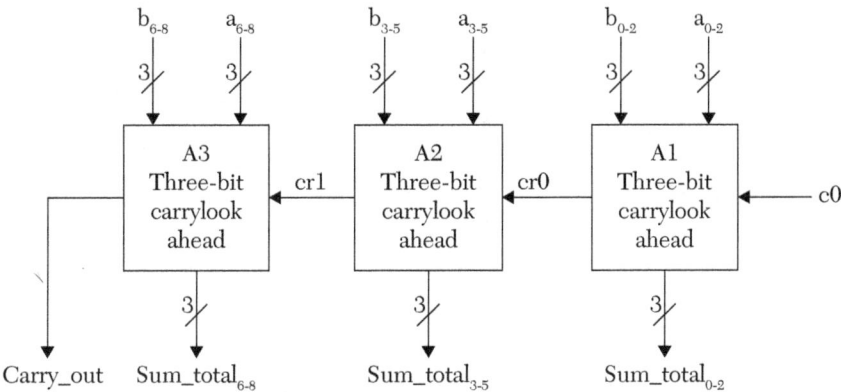

FIGURE 9.2 Block diagram of nine-bit adder.

Listing 9.4 shows the mixed-language description of the nine-bit adder. The three-bit carry-lookahead is described by a VHDL module, and the Verilog module invokes the VHDL entity three times. The VHDL entity `adders_RL` is a data-flow description of a three-bit lookahead adder (see Chapter 2). The delay-propagation time in Listing 9.2 is taken as 0. In the Verilog module, the VHDL entity is invoked by the statement:

```
adders_RL A1 (a [2:0], b [2:0], c0, sum_total [2:0], cr0);
```

The statement above passes the inputs $(a_2\ a_1\ a_0)$, $(b_2\ b_1\ b_0)$, and c0 to the input ports of the entity `adders_RL`, $(x_2\ x_1\ x_0)$, (y2 y1 y0), and `cin`. The entity calculates the three-bit output $(sum_2\ sum_1\ sum_0)$ and the one-bit output cout. The outputs $(sum_2\ sum_1\ sum_0)$ and cout are passed to the outputs of the Verilog module $(sum\_total_2\ sum\_total_1\ sum\_total_0)$, and cr0, respectively. Each time the VHDL entity is invoked, three bits are added, and the output is passed to the Verilog module. Invoking the VHDL entity generates a nine-bit adder.

LISTING 9.4 Mixed-Language Description of a Nine-Bit Adder

```
module Nine_bitAdder (a, b, c0, sum_total, carry_out);
    input [8:0] a, b;
    input c0;
    output [8:0] sum_total;
```

```
    output carry_out;
    wire cr0, cr1;

    //Invoke the VHDL entity
    adders_RL A1 (a [2:0], b [2:0], c0,
                  sum_total [2:0], cr0);
    adders_RL A2 (a [5:3], b [5:3], cr0,
                  sum_total [5:3], cr1);
    adders_RL A3 (a [8:6], b [8:6], cr1,
                  sum_total [8:6], carry_out);

//adders_RL is the name of the VHDL entity

endmodule
library IEEE;
use IEEE.STD_LOGIC_1164.ALL;
-- This is a VHDL data-flow code for a 3-bit
--carry-lookahead adder.

entity adders_RL is
port (x, y : in std_logic_vector (2 downto 0);
cin : in std_logic;
sum : out std_logic_vector (2 downto 0);
cout : out std_logic);

--The entity name is identical to that of the Verilog
-- module. The input and output ports have
-- the same mode as the inputs. and outputs of the Verilog
-- module.

end adders_RL;

architecture lkh_DtFl of adders_RL is

signal c0, c1 : std_logic;
signal p, g : std_logic_vector (2 downto 0);
constant delay_gt : time := 0 ns;
--The gate propagation delay here is equal to 0.
begin

g(0) <= x(0) and y(0) after delay_gt;
g(1) <= x(1) and y(1) after delay_gt;
g(2) <= x(2) and y(2) after delay_gt;
```

```
p(0) <= x(0) or y(0) after delay_gt;
p(1) <= x(1) or y(1) after delay_gt;
p(2) <= x(2) or y(2) after delay_gt;
c0 <= g(0) or (p(0) and cin) after 2 * delay_gt;

c1 <= g(1) or (p(1) and g(0)) or (p(1) and
    p(0) and cin) after 2 * delay_gt;
cout <= g(2) or (p(2) and g(1)) or (p(2) and p(1)
and g(0)) or(p(2) and p(1) and p(0) and cin)
    after 2 * delay_gt;

sum(0) <= (p(0) xor g(0)) xor cin after delay_gt;
sum(1) <= (p(1) xor g(1)) xor c0 after delay_gt;
sum(2) <= (p(2) xor g(2)) xor c1 after delay_gt;
end lkh_DtFl;
```

EXAMPLE 9.3 **MIXED-LANGUAGE DESCRIPTION OF A THREE-BIT ADDER WITH ZERO FLAG**

In this example, a mixed-language description of a three-bit adder is written. The adder has a one-bit flag. If the output of the adder is 0, the flag is set to 1; otherwise, it is set to 0. Figure 9.3 shows the logic diagram of the adder. A VHDL entity is written to describe the one-bit adder using structural description (see Chapter 4). The VHDL entity is invoked in the Verilog module three times.

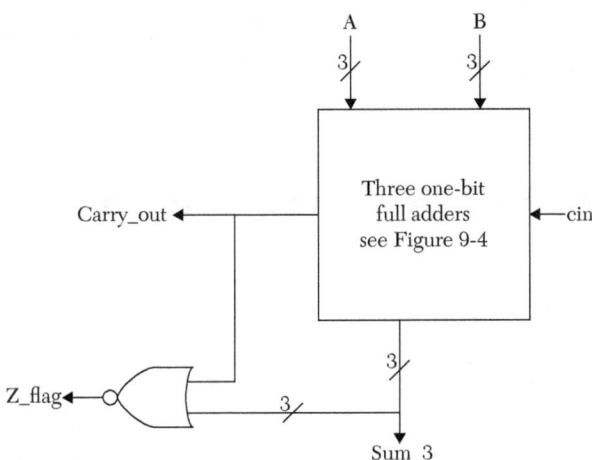

FIGURE 9.3 Block diagram of a three-bit adder with a zero flag.

Listing 9.5 shows the mixed-language description of the adder. The VHDL one-bit adder is built from AND_OR_NOT gates (see Figure 9.4).

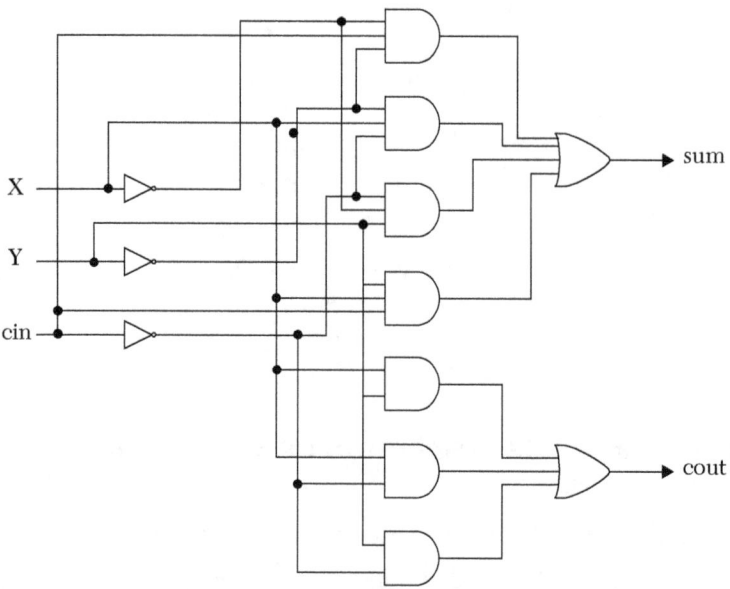

FIGURE 9.4 Logic diagram of a one-bit adder.

The Verilog module

```
full_add FA0 (A[0], B[0], cin, Sum_3[0], cr0);
```

invokes the VHDL entity `full_add`. This entity describes, in structural description, a one-bit full adder. Invoking this entity three times from the Verilog module generates a three-bit adder. The VHDL module looks very long due to the fact that standard VHDL, in contrast to Verilog, does not have built-in primitive gates.

LISTING 9.5 Mixed-Language Description of a Three-Bit Adder with a Zero Flag

```
module three_bitAdd (A, B, cin, Sum_3, Carry_out, Z_flag);
    input [2:0] A, B;
    input cin;
    output [2:0] Sum_3;
    output Carry_out;
    output Z_flag;
    wire cr0, cr1;
```

```
    full_add FA0 (A[0], B[0], cin, Sum_3[0], cr0);
    full_add FA1 (A[1], B[1], cr0, Sum_3[1], cr1);
    full_add FA2 (A[2], B[2], cr1, Sum_3[2], Carry_out);
```

--The above modules invoke the VHDL entity full_add

```
    assign Z_flag = ~(Sum_3[0] | Sum_3[1] | Sum_3[2] |
                      Carry_out);
endmodule

library IEEE;
use IEEE.STD_LOGIC_1164.ALL;
entity full_add is
    Port (X, Y, cin : in std_logic;
          sum, cout : out std_logic);
```
--This is a 1-bit full adder component built from
--AND-OR-NOT gates; see Figure 9.4.

```
end full_add;

architecture beh_vhdl of full_add is
```
--Instantiate the components of a 1-bit adder;
--see Figure 9.4.
```
component inv
    port(I1 : in std_logic; O1 : out std_logic);
end component;
component and2
    port(I1, I2 : in std_logic; O1 : out std_logic);
end component;
component and3
    port(I1, I2, I3 : in std_logic; O1 : out std_logic);
end component;
component or3
    port(I1, I2, I3 : in std_logic; O1 : out std_logic);
end component;
component or4
    port(I1, I2, I3, I4 : in std_logic;
O1 : out std_logic);
end component;
for all : inv use entity work.bind1 (inv_0);
for all : and2 use entity work.bind2 (and2_0);
for all : and3 use entity work.bind3 (and3_0);
```

```
for all : or3 use entity work.bind3 (or3_0);
for all : or4 use entity work.bind4 (or4_0);

--The above five "for" statements are to bind the inv,
-- and3, and2, or3, and or4 with the architecture
-- beh_vhdl. See Chapter 4, "Structural Descriptions."
    signal Xbar, Ybar, cinbar, s0, s1, s2,
               s3, s4, s5, s6 : std_logic;
begin
Iv1 : inv port map (X, Xbar);
Iv2 : inv port map (Y, Ybar);
Iv3 : inv port map (cin, cinbar);
A1 : and3 port map (X, Y, cin, s0);
A2 : and3 port map (Xbar, Y, cinbar, s1);
A3 : and3 port map (Xbar, Ybar, cin, s2);
A4 : and3 port map (X, Ybar, cinbar, s3);
A5 : and2 port map (X, cin, s4);
A6 : and2 port map (X, Y, s5);
A7 : and2 port map (Y, cin, s6);
O1 : or4 port map (s0, s1, s2, s3, sum);
O2 : or3 port map (s4, s5, s6, cout);
end beh_vhdl;

--The following is the behavioral description of the
--components instantiated in the entity full_add.

library IEEE;
use IEEE.STD_LOGIC_1164.ALL;
entity bind1 is
   port (I1 : in std_logic; O1 : out std_logic);
end bind1;
architecture inv_0 of bind1 is
begin
   O1 <= not I1;
end inv_0;
library IEEE;
use IEEE.STD_LOGIC_1164.ALL;

entity bind2 is
   port (I1, I2 : in std_logic; O1 : out std_logic);
end bind2;
architecture and2_0 of bind2 is
```

```
begin
O1 <= I1 and I2;
end and2_0;
architecture or2_0 of bind2 is
begin
O1 <= I1 or I2;
end or2_0;
library IEEE;
use IEEE.STD_LOGIC_1164.ALL;
entity bind3 is
    port (I1, I2, I3 : in std_logic; O1 : out std_logic);
end bind3;

architecture and3_0 of bind3 is
begin
    O1 <= I1 and I2 and I3;
end and3_0;

architecture or3_0 of bind3 is
begin
    O1 <= I1 or I2 or I3;
end or3_0;
library IEEE;
use IEEE.STD_LOGIC_1164.ALL;

entity bind4 is
    Port (I1, I2, I3, I4 : in std_logic;
    O1 : out std_logic);
end bind4;
architecture or4_0 of bind4 is
begin
    O1 <= I1 or I2 or I3 or I4;
end or4_0;
```

EXAMPLE 9.4   MIXED-LANGUAGE DESCRIPTION OF A MASTER-SLAVE D FLIP-FLOP

In Chapter 4, a structural description of a master-slave flip-flop was written. The flip-flop was built from two D-latches (see Figure 9.5).

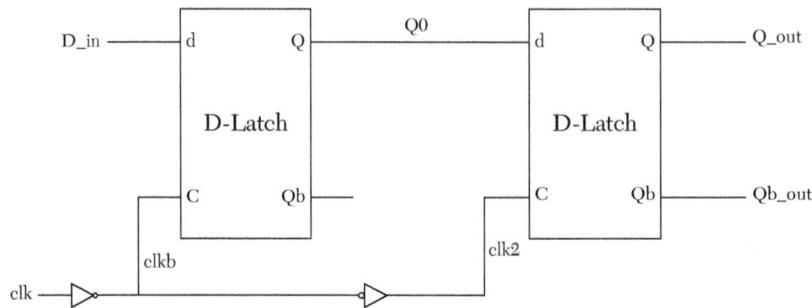

FIGURE 9.5 Logic diagram of a master-slave D flip-flop.

In this example, a mixed-language description of the flip-flop is written. Instead of structural description, VHDL data-flow description is used to simulate the D-latch (see Chapter 2). The master-slave flip-flop is described in a Verilog module. The VHDL entity is invoked to import the description of a D-latch. Listing 9.6 shows the mixed-language description of a master-slave flip-flop.

LISTING 9.6 Mixed-Language Description of a Master-Slave D Flip-Flop

```
//This is the Verilog module
module D_Master (D_in, clk, Q_out, Qb_out);
    input D_in, clk;
    output Q_out, Qb_out;
    wire Q0, Qb, clkb; /* wire statement here can be omitted.*/
    assign clkb = ~ clk;
    assign clk2 = ~ clkb;
    D_Latch D0 (D_in, clkb, Q0, Qb);

//D_Latch is the name of a VHDL entity describing a D-Latch

D_Latch D1 (Q0, clk2, Q_out, Qb_out);

endmodule
library IEEE;
use IEEE.STD_LOGIC_1164.ALL;

entity D_Latch is
--The entity has the same name as
--the calling Verilog module

port (D, E : in std_logic;
      Q, Qbar : buffer std_logic);
```

```
end D_Latch;

architecture DL_DtFl of D_Latch is
--This architecture describes a D-latch using
--data-flow description
constant Delay_EorD : Time := 9 ns;
constant Delay_inv : Time := 1 ns;

begin

Qbar <= (D and E) nor (not E and Q) after Delay_EorD;
Q <= not Qbar after Delay_inv;
end DL_DtFl;
```

The simulation waveform is shown in Figure 9.6.

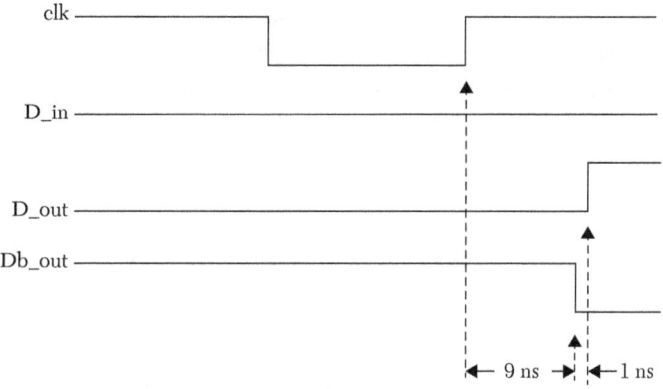

FIGURE 9.6 Simulation waveform of a master-slave D flip-flop.

EXAMPLE 9.5 MIXED-LANGUAGE DESCRIPTION OF A 4x4 COMPARATOR

In Chapter 4, an HDL structural description of a 3x3 comparator was introduced. The comparator was built from three one-bit adders (see Figure 4.8). Here, mixed-language description is used. A VHDL behavioral module (see Chapter 3) to describe a one-bit full adder is written. A Verilog module invokes this VHDL module four times. Listing 9.7 shows the mixed-language description of a 4x4 comparator. Consider the Verilog code:

```
generate

genvar i;
for (i = 0; i <= N; i = i + 1)
    begin : u
```

```
        and (eq[i+1], sum[i], eq[i]);
end
endgenerate
```

If N = 3, the above Verilog code constitutes four two-input AND gates (see Figure 9.7). The input to each gate is `sum(i)` and `eq(i)`; the output is `eq(i+1)`. The output of the fourth AND gate, `eq(4)`, is equal to 1 if and only if all sum(i), i = 1, 3 are equal to 1. Otherwise, it is equal to 0. If `eq(4)` = 1, this means that X = Y.

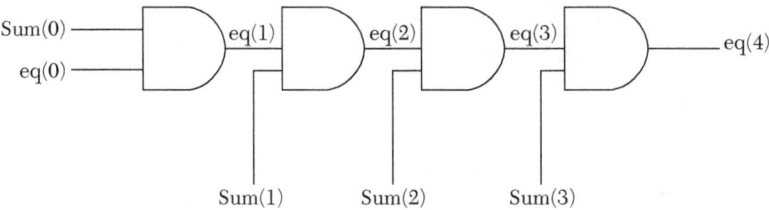

FIGURE 9.7 Logic diagram of the Verilog-generated statements in Listing 9.7.

LISTING 9.7 Mixed-Language Description of a 4x4 Comparator

```
module compr_genr (X, Y, xgty, xlty, xeqy);
parameter N = 3;
input [N:0] X, Y;
output xgty, xlty, xeqy;
wire [N:0] sum, Yb;
wire [N+1:0] carry, eq;
assign carry[0] = 1'b0;
assign eq[0] = 1'b1;
assign Yb = ~Y;

FULL_ADDER FA (X[0], Yb[0], carry[0],
               sum[0], carry[1]);

-- The module FULL_ADDER has the same name
-- as the VHDL entity FULL_ADDER

FULL_ADDER FA1 (X[1], Yb[1], carry[1],
                sum[1], carry[2]);
FULL_ADDER FA2 (X[2], Yb[2], carry[2],
                sum[2], carry[3]);
FULL_ADDER FA3 (X[3], Yb[3], carry[3],
                sum[3], carry[4]);
generate
```

```
genvar i;
for (i = 0; i <= N; i = i + 1)
    begin : u
and (eq[i+1], sum[i], eq[i]);

end
endgenerate

assign xgty = carry [N+1];
assign xeqy = eq [N+1];
nor (xlty, xeqy, xgty);
endmodule

library IEEE;
use IEEE.STD_LOGIC_1164.ALL;
entity FULL_ADDER is
    Port (A, B, cin : in std_logic;
          sum_1, cout : out std_logic);
end FULL_ADDER;

architecture beh_vhdl of FULL_ADDER is

--This architecture is a behavioral
--description of a full adder.

begin

oneBit : process (A, B, cin)
    variable y : std_logic_vector (2 downto 0);
    begin
        Y := (A & B & Cin);

--The above statement is a concatenation of
--three bits A, B, and Cin

case y is
    when "000" => sum_1 <= '0'; cout <= '0';
    when "110" => sum_1 <= '0'; cout <= '1';
    when "101" => sum_1 <= '0'; cout <= '1';
    when "011" => sum_1 <= '0'; cout <= '1';
    when "111" => sum_1 <= '1'; cout <= '1';
    when others => sum_1 <= '1'; cout <= '0';
--Others here refer to 100, 001, 010
```

```
end case;
   end process;
   end beh_vhdl;
```

9.3.2 Invoking a Verilog Module From a VHDL Module

As mentioned, a Verilog module can be invoked from a VHDL module by instantiating a component in the VHDL module that has the same name and ports as the Verilog module. The Verilog module should be the only construct that has the same name as the component. Presently, this is the only way Verilog modules can be invoked from VHDL. Several examples are discussed below.

EXAMPLE 9.6   INSTANTIATING AN AND GATE FROM A VHDL MODULE

A basic VHDL does not have built-in gates such as AND, OR, and XOR, unless the user attaches a vendor's package that contains a description of the gates. Standard Verilog, on the other hand, has built-in descriptions of primitive gates of which we can take advantage. Using mixed-language description, a Verilog module is invoked in the VHDL module, and the gates that we want to use are instantiated. Listing 9.8 shows a mixed-language description of instantiating an AND gate in a VHDL module. The description of the AND gate is provided by the Verilog module. Referring to Listing 9.8, the VHDL statements

```
component and2
    port (x, y : in std_logic; z : out std_logic);
end component;
```

declare a component by the name of `and2`. The component has two input ports, `x` and `y`, and one output port, `z`. To link this component to a Verilog module, the module has to have the same name and ports as the component. The Verilog module is written as:

```
module and2 (x, y, z);
```

It has the same name and the same ports, so all Verilog statements pertaining to `x`, `y`, and `z` are visible to the VHDL module. In the Verilog module, write:

```
and(z,x,y);
```

The statement above describes an AND relationship between `x`, `y`, and `z`.

LISTING 9.8 Mixed-Language Description of an AND Gate

```
library IEEE;
use IEEE.STD_LOGIC_1164.ALL;
use IEEE.STD_LOGIC_ARITH.ALL;

--This is the VHDL module

entity andgate is
    port (a, b : in std_logic; c : out std_logic);
end andgate;

architecture andgate of andgate is
component and2

--For correct binding with the Verilog module,
--the name of the component should be identical
--to that of the Verilog module.

    port (x, y : in std_logic; z : out std_logic);

--The name of the ports should be identical to the name
--of the inputs/outputs of the Verilog module.

end component;

begin
    g1 : and2 port map (a, b, c);
end andgate;
//This is the Verilog module
module and2 (x, y, z);

    input x, y;
    output z;
    and( z, x , y);
endmodule
```

EXAMPLE 9.7   **MIXED-LANGUAGE DESCRIPTION OF A JK FLIP-FLOP WITH A CLEAR SIGNAL**

In this example, a mixed-language description of a JK flip-flop is written. JK flip-flops were covered in Chapters 3 and 4. The excitation table of a JK flip-flop with a clear signal is shown in Table 9.1.

TABLE 9.1 Excitation Table for a JK Flip-Flop

Clear	J	K	clk	q (next state)
1	x	x	↑	q = 0
0	0	0	↑	No change (hold), next = current
0	1	0	↑	1
0	0	1	↑	0
0	1	1	↑	Toggle (next state) = invert of (current state)

The flip-flop is declared as a VHDL component, and a Verilog behavioral description of the flip-flop based on Table 9.1 is written. The Verilog is linked to the VHDL component by giving the Verilog module the same name as the VHDL component. The ports of the component should also be the same as those of the Verilog module. Listing 9.9 shows the mixed-language description of the flip-flop. The JK flip-flop is declared as a component with the statement:

```
component jk_verilog
    port(j, k, ck, clear : in std_logic;
q, qb : out std_logic);
end component;
```

The above component is linked to a Verilog module by the statement:

```
module jk_verilog (j, k, ck, clear, q, qb);
```

The above module has the same name and ports as the VHDL component `jk_verilog`. Accordingly, the relationship between the input and output ports described in the Verilog module is visible to the VHDL component. The Verilog module describes, in behavioral style, a JK flip-flop with an active high clear. Hence, the VHDL component `jk_verilog` is also a JK flip-flop with an active high clear.

LISTING 9.9 Mixed-Language Description of a JK Flip-Flop

```
library IEEE;
use IEEE.STD_LOGIC_1164.ALL;

entity JK_FF is
    Port (Jx, Kx, clk, clx : in std_logic;
          Qx, Qxbar : out std_logic);
end JK_FF;
```

```
architecture JK_FF of JK_FF is

--The JK flip flop is declared as a component
component jk_verilog
    port(j, k, ck, clear : in std_logic;
         q, qb : out std_logic);
end component;
begin

jk1 : jk_verilog port map (Jx, Kx, clk, clx, Qx, Qxbar);

end JK_FF;
module jk_verilog (j, k, ck, clear, q, qb);
// The module name jk_verilog matches
// the name of the VHDL components

input j, k, ck, clear;
output q, qb;
--The input and output ports match those of the
--VHDL component, jk_verilog

reg q, qb;
reg [1:0] JK;
always @ (posedge ck, clear)
begin
    if (clear == 1)
        begin
            q = 1'b0;
            qb = 1'b1;
        end
        else
            begin
                JK = {j, k};
                case (JK)
                2'd0 : q = q;
                2'd1 : q = 0;
                2'd2 : q = 1;
                2'd3 : q = ~q;
                endcase
                qb = ~q;
            end
end
endmodule
```

The simulation waveform of the JK flip-flop is shown in Figure 9.8.

FIGURE 9.8 Simulation waveform of a JK flip-flop with an active high clear.

EXAMPLE 9.8   MIXED-LANGUAGE DESCRIPTION OF A THREE-BIT SYNCHRONOUS COUNTER WITH CLEAR

This example was first covered in Chapter 4. Figure 4.20 shows the logic diagram of the counter, and Listing 4.23 shows the VHDL and the Verilog descriptions. Here, the code of the counter is written using mixed language. For convenience, Figure 4.20 is presented again here as Figure 9.9. As shown, the counter consists of three JK flip-flops, and OR, AND, and INVERT gates.

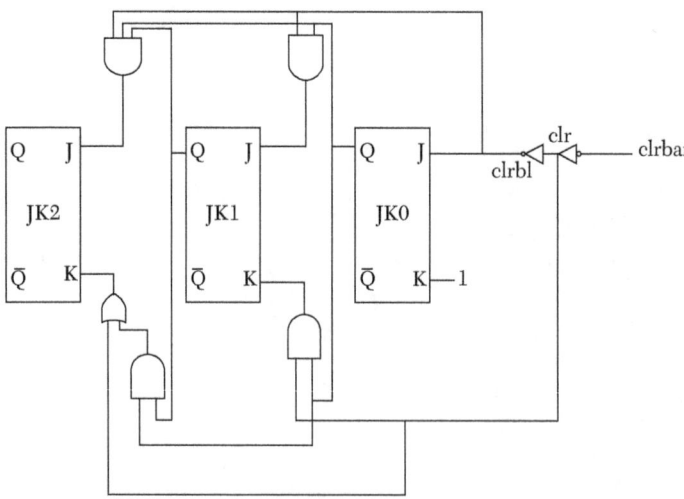

FIGURE 9.9 Three-bit synchronous counter with clear.

Listing 9.10 shows the mixed-language description of the counter. A Verilog module that describes a JK flip-flop and an AND, OR, and INVERT is written. The Verilog module is invoked from a VHDL module three times. In the VHDL module, a component declaration for the flip-flop and the gates is written. The names of the components are the same as the corresponding Verilog modules. For example, the VHDL statement:

```
component JK_FF
port (I1, I2, I3 : in std_logic;
O1, O2 : inout std_logic);
end component;
```

declares a component by the name of JK_FF. The Verilog module by the name of JK_FF describes a JK flip-flop. Accordingly, the VHDL component JK_FF is a JK flip-flop. To facilitate the link between the Verilog and VHDL modules, we slightly modify the VHDL module from Listing 4.23, in which the instantiation statement for flip-flop FF0 was written as:

```
FF0 : JK_FF port map (clrb1, '1', clk, q(0), qb(0));
```

The Verilog module can accept a signal, variable, or constant, but it may not accept the value 1. Therefore, we declare a signal named high and assign it a value of 1 as follows:

```
high <= '1';
FF0 : JK_FF port map (clrb1, High, clk, q(0), qb(0));
```

LISTING 9.10 Mixed-Language Description of Three-Bit Counter with Clear

```
library IEEE;
use IEEE.STD_LOGIC_1164.ALL;

entity countr_3 is
port (clk, clrbar : in std_logic;
    q, qb : inout std_logic_vector (2 downto 0));
end countr_3;

architecture CNTR3 of countr_3 is

component JK_FF
    port (I1, I2, I3 : in std_logic;
        O1, O2 : inout std_logic);
end component;

component inv
```

```
        port (I1 : in std_logic; O1 : out std_logic);
end component;

component and2
        port (I1, I2 : in std_logic; O1 : out std_logic);
end component;

component or2
        port (I1, I2 : in std_logic; O1 : out std_logic);
end component;

signal J1, K1, J2, K2, clr, clrb1, s1, high : std_logic;
begin

        high <= '1';

FF0 : JK_FF port map (clrb1, High, clk, q(0), qb(0));
A1 : and2 port map (clrb1, q(0), J1);
inv1 : inv port map (clr, clrb1);
inv2 : inv port map (clrbar, clr);

r1 : or2 port map (q(0), clr, K1);
FF1 : JK_FF port map (J1, K1, clk, q(1), qb(1));
A2 : and2 port map (q(0), q(1), s1);
A3 : and2 port map (clrb1, s1, J2);
r2 : or2 port map (s1, clr, K2);
FF2 : JK_FF port map (J2, K2, clk, q(2), qb(2));
end CNTR3 ;

module and2 (I1, I2, O1);
//This Verilog module represents an AND function

input I1, I2;
output O1;
assign O1 = I1 & I2;
endmodule

module inv (I1, O1);
//This Verilog module represents an INVERT function

input I1;
output O1;
```

```
assign O1 = ~I1;
endmodule

module or2 (I1, I2, O1);
//This Verilog module represents an OR function

input I1, I2;
output O1;
assign O1 = I1 | I2;
endmodule

module JK_FF (I1, I2, I3, O1, O2);
//This Verilog module represents a JK flip-flop.
input I1, I2, I3;
output O1, O2;

reg O1, O2;
reg [1:0] JK;
initial
    begin
    O1 = 1'b0;
    O2 = 1'b1;
    end
always @ (posedge I3)
begin
    JK = {I1, I2};
    case (JK)
    2'd0 : O1 = O1;
    2'd1 : O1 = 0;
    2'd2 : O1 = 1;
    2'd3 : O1 = ~O1;
    endcase
    O2 = ~O1;
end
endmodule
```

EXAMPLE 9.9   MIXED-LANGUAGE DESCRIPTION OF AN *N*-BIT COUNTER WITH RIPPLE CARRY-OUT

In this example, we discuss an n-bit asynchronous counter with a ripple carry-out (RCO). Figure 9.10 shows the logic diagram of the counter. As shown, the ripple carry-out is 1 when all Qs are 1s. In Chapter 4, asynchronous counters were discussed and described using the generate statement. Here, we use mixed-language description to invoke a Verilog module from a VHDL module. Listing 9.11 shows the mixed-language description of the counter.

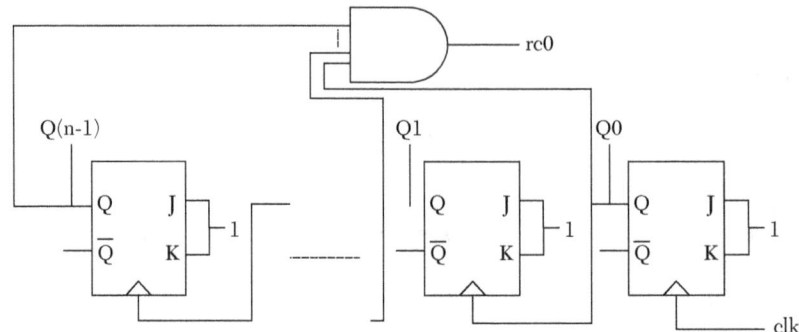

FIGURE 9.10 Logic diagram of an n-bit synchronous counter with ripple carry-out.

As shown in Figure 9.10, to construct the counter, we need n-JK flip-flops and n-input and gates. Two Verilog modules, jkff and andgate, are implemented to describe a JK flip-flop and a three-input AND gate, respectively. The module jkff is written in behavioral description, and the module andgate is written in data-flow description.

LISTING 9.11 Mixed-Language Description of an N-Bit Asynchronous Counter

```
--This is a VHDL module
library IEEE;
use IEEE.STD_LOGIC_1164.ALL;

entity asynch_ctrMx is
Generic (N : integer := 3);

port (clk, clear : in std_logic;
     C, Cbar : out std_logic_vector (N-1 downto 0);
rco : out std_logic);

end asynch_ctrMx;

architecture CT_strgnt of asynch_ctrMx is

component jkff is
--This is a JK flip-flop with a clear bound to Verilog
```

```
-- module jkff

    port (j, k, clk, clear : in std_logic;
          q, qb : out std_logic);
end component;
component andgate is
--This is a three-input AND gate bound to Verilog module
-- andgate

    port (I1, I2, I3 : in std_logic; O1 : out std_logic);
end component;

signal h, l : std_logic;
signal s : std_logic_vector (N downto 0);
signal s1 : std_logic_vector (N downto 0);
signal C_tem : std_logic_vector (N-1 downto 0);

begin
h <= '1';
l <= '0';
s <= (C_tem & clk);

-- s is the concatenation of Q and clk. We need this
-- concatenation to describe the clock of
--each JK flip-flop.

s1(0) <= not clear;

Gnlop : for i in (N - 1) downto 0 generate

G1 : jkff port map (h, h, s(i), clear, C_tem(i), Cbar(i));

end generate GnLop;
C <= C_tem;
rc_gen : for i in (N - 2) downto 0 generate
--This loop to determine the ripple carry-out
rc : andgate port map (C_tem(i), C_tem(i+1),
s1(i), s1(i+1));
end generate rc_gen;
rco <= s1(N-1);
end CT_strgnt;
module jkff (j, k, clk, clear, q, qb);
// This is a behavioral description of a JK flip-flop
```

```
input j, k, clk, clear;
output q, qb;
reg q, qb;
reg [1:0] JK;
always @ (posedge clk, clear)
begin
    if (clear == 1)
        begin
            q = 1'b0;
            qb = 1'b1;
        end
        else
            begin
                JK = {j,k};
                case (JK)
                2'd0 : q = q;
                2'd1 : q = 0;
                2'd2 : q = 1;
                2'd3 : q = ~q;
                endcase
                qb = ~q;
            end
end
endmodule
module andgate (I1, I2,I3, O1);
//This is a three-input AND gate
    input I1, I2, I3;
    output O1;
    assign O1 = (I1 & I2 & I3);
    endmodule
```

The simulation waveform is shown in Figure 9.11.

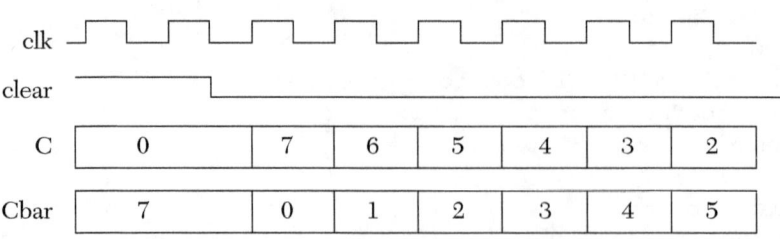

FIGURE 9.11 Simulation waveform for an n-bit asynchronous counter. The simulation pattern might be different than shown due to the presence of transient states (hazards).

EXAMPLE 9.10   MIXED-LANGUAGE DESCRIPTION OF A SWITCH-LEVEL MULTIPLEXER

In Chapter 5, several combinational and sequential logics were described using VHDL or Verilog switch-level description. We also saw that the basic VHDL package, in contrast to Verilog, does not have built-in switch-level primitives. Here, mixed-language description is used to describe a 2x1 multiplexer; a switch-level Verilog description is invoked from a VHDL module. By invoking Verilog modules, the VHDL module behaves as if it possesses built-in switch-level primitives. Listing 9.12 shows the mixed-language description of a 2x1 multiplexer. The statement

```
component pmos_verlg
port (O1 : out std_logic; I1, I2 : in std_logic);
end component;
```

declares a VHDL component by the name `pmos_verlg`. The name of the component is the same as the name of the Verilog module that uses the built-in primitive `pmos` to describe a `pmos` switch. In this way, the switch is made visible to the VHDL module.

LISTING 9.12 Mixed-Language Description of a 2x1 Multiplexer

```
--This is the VHDL module.
library IEEE;
use IEEE.STD_LOGIC_1164.ALL;
entity mux2x1_mxd is
Port (a, b, Sel, E : in std_logic; ybar : out std_logic);
end mux2x1_mxd;

architecture mux2x1switch of mux2x1_mxd is

component nmos_verlg
--This component, after linking to a
--Verilog module, behaves as an nmos switch
port (O1 : out std_logic; I1, I2 : in std_logic);

end component;

component pmos_verlg
--This component, after linking to a Verilog module,
-- behaves as a pmos switch.
port (O1 : out std_logic; I1, I2 : in std_logic);
```

```
end component;
--constant vdd : std_logic := '1';
--constant gnd : std_logic := '0';

-- In Chapter 5 we wrote Vdd and gnd as constants.
-- Some VHDL/Verilog simulators do not transfer constants
-- between VHDL and Verilog. So we wrote them as signals.

signal vdd, gnd, Selbar, s0, s1, s2, s3 : std_logic;
begin
    vdd <= '1';
    gnd <= '0';

--Invert signal Sel. If the complement of Sel is available,
--then no need for the following pair of transistors.

v1 : pmos_verlg port map (Selbar, vdd, Sel);
v2 : nmos_verlg port map (Selbar, gnd, Sel);
--Write the pull-down combination
n1 : nmos_verlg port map (s0, gnd, E);
n2 : nmos_verlg port map (s1, s0, Sel);
n3 : nmos_verlg port map (ybar, s1, a);
n4 : nmos_verlg port map (s2, s0, Selbar);
n5 : nmos_verlg port map (ybar, s2, b);
--Write the pull-up combination
p1 : pmos_verlg port map (ybar, vdd, E);
p2 : pmos_verlg port map (ybar, s3, Sel);
p3 : pmos_verlg port map (ybar, s3, a);
p4 : pmos_verlg port map (s3, vdd, Selbar);
p5 : pmos_verlg port map (s3, vdd, b);

end mux2x1switch;

// This is the Verilog Module

module nmos_verlg (O1, I1, I2);
    input I1, I2;
    output O1;
nmos (O1, I1, I2);
endmodule

module pmos_verlg (O1, I1, I2);
    input I1, I2;
```

```
    output O1;
pmos (O1, I1, I2);
endmodule
```

EXAMPLE 9.11 **INSTANTIATING CASEX IN VHDL**

Chapter 3 covered the `casex` statement for both VHDL and Verilog. We have seen that `casex` ignores the "don't care" (x) in the values of the control expression. Consider the following `casex`:

```
casex (a)
    4'bxxx1 : b = 4'd1;
    4'bxx10 : b = 4'd2;

    ..................
    endcase;
```

All xs are ignored; for example, `b` = 1 if and only if the least significant bit of `a` is 1, regardless of the value of the high-order bits of `a`. Another Verilog variation of `case` is the `casez` (see Chapter 3), where z is the high impedance. VHDL does not have an exact replica of `casex` or `casez`. With mixed-language description, we can instantiate a command similar to `casex` and `casez` in the VHDL module. Listing 9.13 shows a mixed-language description that instantiates a command by the name of `cas_x` in the VHDL module; this command performs the same function as the Verilog `casex`. Listing 9.13 represents a four-bit priority encoder. This encoder was discussed in Chapter 3. The truth table of the encoder is shown in Table 9.2.

TABLE 9.2 Truth Table for a Four-Bit Encoder

Input	Output
a	b
xxx1	1
xx10	2
x100	4
1000	8
Others	0

LISTING 9.13 Instantiating casex in a VHDL Module

```
library IEEE;
use IEEE.STD_LOGIC_1164.all;

entity P_encodr is
```

```
      Port (X : in std_logic_vector (3 downto 0);
          Y : out std_logic_vector (3 downto 0));
end P_encodr;
architecture P_encodr of P_encodr is

component cas_x
--The name of the component is identical to the name of the
--Verilog module

port (a : in std_logic_vector (3 downto 0);
      b : out std_logic_vector (3 downto 0));

end component;

begin

ax : cas_x port map (X, Y);

end P_encodr;

module cas_x (a, b);
    input [3:0] a;
    output [3:0] b;
    reg [3:0] b;
    always @ (a)
        begin
            casex (a)
            4'bxxx1 : b = 4'd1;
            4'bxx10 : b = 4'd2;
            4'bx100 : b = 4'd4;
            4'b1000 : b = 4'd8;
            default : b = 4'd0;

            endcase
        end
endmodule
```

EXAMPLE 9.12   MIXED-LANGUAGE DESCRIPTION OF A LOW-PASS RC FILTER

The function of an electronic filter is to block a certain frequency band in a signal. There are several types of simple filters such as low pass, high pass, and band pass. Low-pass filters allow frequencies below a certain threshold (called the *cutoff frequency*) to pass with or without minimal attenuation. All frequencies above the threshold are attenuated; frequencies close to the cutoff are less attenuated than those frequencies far from the cutoff. High-pass filters pass frequencies higher than the cutoff with or without minimal attenuation. Frequencies lower than the cutoff are attenuated, and frequencies close to the cutoff are less attenuated than those signals far from the cutoff.

Figure 9.12 shows a low-pass filter consisting of a resistance (R) connected in serial with a capacitance (C). The impedance of the capacitance is (1/jwC) where w = 2πf, f is the frequency, and j = $\sqrt{-1}$. The ratio of the output signal (Vo) to the input signal (Vi) is:

$$\frac{Vo}{Vi} = \frac{1/jwC}{R + 1/jwC} = \frac{1}{jwCR + 1} \qquad (9.1)$$

(Vo/Vi) is called the *transfer function* of the filter (H(w)). The square of the amplitude of the transfer function can be written as:

$$[H(w)]^2 = \frac{1}{w^2 CR + 1} \qquad (9.2)$$

$$\text{The cutoff frequency } w_c = (1/RC) \qquad (9.3)$$

Substitute Equation 9.3 into Equation 9.2 to get:

$$[H(w)]^2 = \frac{1}{(w/w_c)^2 + 1} \qquad (9.4)$$

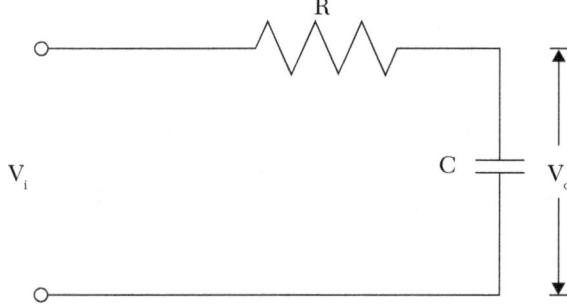

FIGURE 9.12 Simple low-pass RC filter.

We want to simulate Equation 9.4 using mixed-language description and output the value of $[H(w)]^2$ as it changes with w; this value will be stored in a file. Because VHDL has extensive file operations, VHDL is implemented here to handle the file operations. The Verilog module will handle the calculations. Listing 9.14 shows the mixed-language description of a simple RC filter. In the VHDL module, the inputs and outputs are described as:

```
entity Filter_draw is
    Port (w, w_ctoff : in std_logic_vector (3 downto 0);
    Hw_vhd : out std_logic_vector (7 downto 0));
end Filter_draw;
```

As shown in the entity `Filter_draw`, the inputs and outputs are selected to be of type `std_logic_vector`. The output $[H(w)]^2$ in Equation 9.4 is represented by the signal `Hw_vhd`. The inputs w and wc in Equation 9.4 are represented by `w` and `w_ctoff`. To simplify the description, all inputs and outputs are assumed to be integers. We could have selected the type of inputs and outputs in the entity to be integer, but here we want to practice converting from one type to another. Also, we want an easy link between the VHDL and Verilog ports because integer ports are not allowed to be mapped from Verilog to VHDL. If the output `Hw_vhd` is calculated as in Equation 9.4, using integer division, the output would be zero for all values of w because the numerator is always less than the denominator. Instead, we calculate Equation 9.4 as real division and then scale it up by multiplying it by 100. For example, if w = 3 units, and the cutoff = 4 units, then from Equation 9.4:

$$\text{Real (Hw\_vhd)} = \frac{1}{(3/4)^2 + 1} = 0.64$$

After scaling up by 100, then Hw_vhd = 64

Because VHDL files accept only integers, real values, and characters, we write a VHDL function to convert from `std_logic_vector` to integer. In Listing 9.14, the user-defined function `TO_Intgr` converts `std_logic_vector` to integer. To invoke the Verilog module from the VHDL module, we write a component declaration in the VHDL module:

```
component flter_RC
    port (I1, I2 : in std_logic_vector (3 downto 0);
        O1 : out std_logic_vector (7 downto 0));
end component;
```

The name of the component is `flter_RC`; it has two input ports, `I1` and `I2` of type `std_logic_vector`, and one output port, `O1` of type `std_logic_vector`. To invoke the Verilog module, we declare the module as follows:

```
module flter_RC (I1, I2, O1);
input [3:0] I1, I2;
output [7:0] O1;
```

The above Verilog module has the same name and ports as the VHDL component; thus, the module is visible to the VHDL module. In the Verilog module, we perform the real division $O1 = 1/[(I1/I2)^2 + 1]$. Because `I1` and `I2` are not declared as real, the division will be performed as integer, and O1 will be zero for all values of `I1` and `I2`. To avoid this, we multiply `I1` and `I2` by 1.0:

```
s1 = ((1.0 * I1) / (1.0 * I2)) ** 2;
S = 1.0 / (1.0 + s1);
```

`s` and `s1` are declared as real; the value of `s` is the real value of the division $1/[(I1/I2)^2 + 1]$. The output of the Verilog module, O1, is calculated by multiplying S by 100. This output is passed to the VHDL module. As can be seen, Verilog, in contrast to VHDL, is flexible in handling different data types. We would not have been able to easily perform the real division in VHDL. After calculating the output, it is entered into a text file. All of the data in `std_logic_vector` to be entered into the file must be converted to integers because files cannot take the type `std_logic_vector`.

LISTING 9.14 Mixed-Language Description of a Simple RC Filter

```
library IEEE;
use IEEE.STD_LOGIC_1164.ALL;
use std.textio.all;
use ieee.numeric_std.all;

entity Filter_draw is
Port (w, w_ctoff : in std_logic_vector (3 downto 0);
      Hw_vhd : out std_logic_vector (7 downto 0));

end Filter_draw;

architecture Filter_draw of Filter_draw is

Function TO_Intgr (a : in std_logic_vector) return
                      integer is
```

```
--This Function converts std_logic_vector type to integer

variable result : integer;

begin
    result := 0;
    lop1 : for i in a' range loop
    if a(i) = '1' then
    result := result + 2**i;
    end if;
    end loop;
return result;
end TO_Intgr;

component flter_RC
--The name of the component "flter_RC" is the same name as
-- the Verilog module.
    port (I1, I2 : in std_logic_vector (3 downto 0);
          O1 : out std_logic_vector (7 downto 0));

end component;
signal Hw_tmp : std_logic_vector (7 downto 0);
begin
dw : flter_RC port map (w, w_ctoff, Hw_tmp);

//output the data on a file
f1 : process (w, w_ctoff, Hw_tmp)
file outfile : text;
variable fstatus : file_open_status;
variable temp : line;
variable Hw_int, w_int, w_ctoffintg : integer;

begin
--Files can take integer, real, or character;
--they cannot take std-logic-vector; so convert to integer.

Hw_int := TO_Intgr (Hw_tmp);
w_int := TO_Intgr (w);
w_ctoffintg := TO_Intgr (w_ctoff);
file_open (fstatus, outfile, "Wfile_int.txt", write_mode);
--The file name is Wfile_int.txt
```

```
--Write headings. Be sure your simulator supports
-- formatted output. otherwise take out all formatted
-- output statements

write (temp, " This is a Simple R-C Low Pass Filter");
--The above statement when entered in the VHDL module
-- should be entered in one line without carriage return.

writeline (outfile, temp);
write (temp, " ");
writeline (outfile, temp);
write (temp, " FREQUENCY
CUTOFF Amplitude Square");
--The above statement when entered in the VHDL module
--should be entered in one line without carriage return.

writeline (outfile, temp);
write (temp, " ");

--write the values of the filter parameters
write (temp, w_int);
write (temp, " ");
write (temp, w_ctoffintg);
write (temp, " ");
write (temp, Hw_int);
writeline (outfile, temp);

file_close (outfile);
Hw_vhd <= Hw_tmp;
end process fl;
end Filter_draw;

// Next we write the Verilog module;
// the module performs a real division
module flter_RC (I1, I2, O1);

/*The module performs the real division
O1 = 1/[(I1/I2)**2 + 1]*/

input [3:0] I1, I2;
output [7:0] O1;
reg [7:0] O1;
```

```
real S, s1;
always @ (I1, I2)
begin
s1 = ((1.0*I1)/(1.0 * I2))**2 ;
/*we multiply by 1.0 so the division is done in real
format.*/

S = 1.0 / (1.0 + s1);
O1 = 100.00 * S;
end
endmodule
```

The file `Wfile_int.txt`, after simulation, is shown in Figure 9.13.

This is a Simple R-C Low Pass Filter		
FREQUENCY	CUTOFF	Amplitude Square°100
3	4	64

FIGURE 9.13 The file `Wfile_int.txt` after simulation.

EXAMPLE 9.13 MIXED-LANGUAGE DESCRIPTION OF A 2x1 MULTIPLEXOR WITH ACTIVE-LOW ENABLE USING USER-DEFINED PRIMITIVE

Example 8.12 introduced a Verilog code for the description of a 2x1 multiplexor with active-low enable using user-defined primitive (UDP). Here, a VHDL code is written that invokes the Verilog code, so the VHDL code appears as if it can use the Verilog UDP. Listing 9.15 shows the mixed-language description of the multiplexer. The first part is a VHDL code declaring the inputs and the output of the multiplexer as ports of the entity muxVHDL:

```
port(G1,SL1,A1,B1: in std_logic; Y1: out std_logic);
```

The same VHDL code then declares a component by the name Mux2x1Prmtvvlog. The name of the component and its ports have to be the same as the name of the Verilog module and its ports. The Verilog code is identical to Listing 8.15. The simulation of the VHDL code is identical to that of Listing 8.15, except the inputs now are G1, SL1, A1, and B1, and the output is Y1. The VHDL simulation shows that the VHDL code can implement the Verilog UDP.

LISTING 9.15 Mixed-Language Description of a 2x1 Multiplexor with Active-Low Enable Using Verilog User-Defined Primitive

```
--This is the VHDL code

library IEEE;
use IEEE.STD_LOGIC_1164.ALL;

entity muxVHDL is
port(G1,SL1,A1,B1: in std_logic; Y1: out std_logic);
end muxVHDL;

architecture Behavioral of muxVHDL is
component Mux2x1Prmtvvlog
port(Gbar, SEL,A,B: in std_logic; Y: out std_logic);
end component;

begin
pl1: Mux2x1Prmtvvlog port map(G1,SL1,A1,B1,Y1);

end Behavioral;

//This is the Verilog code that should be
//attached in the same project as the VHDL code

module Mux2x1Prmtvvlog(Gbar, SEL,A,B,Y);
input Gbar, SEL,A,B;
output Y;
    multiplexer MUX1 (Y, Gbar, SEL,A,B) ;
endmodule

primitive multiplexer (mux, enable, control, dataA, dataB) ;
output mux;
input enable, control, dataA, dataB;
table
// enable control dataA dataB mux

    1    ?    ?         ?        : 0;
    0    0    1         ?        : 1;
    0    0    0         ?        : 0;
    0    1    ?         1        : 1;
    0    1    ?         0        : 0;
```

```
    0   x   0      0        : 0;
    0   x   1      1        : 1;
endtable
endprimitive
```

9.4   Limitations of Mixed-Language Description

As previously mentioned, mixed-language description is somehow limited at present time. These limitations can be summarized as follows:

- Not all VHDL data types are supported in mixed-language description. Only bit, bit_vector, std_logic, std_ulogic, std_logic_vector, and std_ulogic_vector are supported.

- The VHDL port type buffer is not supported.

- Only a VHDL component construct can invoke a Verilog module. We cannot invoke a Verilog module from any other construct in the VHDL module.

- A Verilog module can only invoke a VHDL entity. It cannot invoke any other construct in the VHDL module such as a procedure or function.

9.5   Summary

This chapter discussed mixed-language descriptions: HDL code that includes constructs from both VHDL and Verilog. To be able to write in mixed-language style, the simulator should be able to handle mixed-language description. Presently, mixed-language description has some limitations. The main limitation is that in the VHDL module, only the entire Verilog module can be invoked; conversely, in the Verilog module, only the entire VHDL entity can be invoked. We have seen how to invoke/instantiate a VHDL entity from a Verilog module and how to invoke/instantiate a Verilog module from a VHDL component. To invoke a VHDL entity from a Verilog module, the module statement is written in Verilog. The name of the module should be identical to the name of the entity, and the parameter types of the module should match the types of the ports of the entity. For example, the module statement:

```
HA H1 (y, cin, s0, c0);
```

written in a Verilog module invokes a VHDL entity named HA. In the Verilog module, no other module should have the name HA. On the other hand, invoking a Verilog module from VHDL is done by declaring a component in the VHDL module with the same name as the Verilog module. The component ports should have the same names and types as the ports of the Verilog module. For example, the VHDL component:

```
component V_mod1
port (x, y : in std_logic; z : out std_logic);
end component;
```

invokes a Verilog module named V_mod1.

9.6 Exercises

1. Consider the code shown in Listing 9.16.

LISTING 9.16 Code for Exercise 9.1

```
module mixed (a, b, c, d);
input a, b;
output c, d;
lgic L1 (c, d, a, b)
endmodule
entity lgic is
    port (x, y : in std_logic; O1, O2 : buffer std_logic);
end lgic;

architecture lgic of lgic is
begin
O1 <= x and y;
O2 <= not x;

end lgic;
```

Without using a computer, find any error(s) in Listing 9.16. Correct the errors (if any), and write the values of c and d if a = 1 and b = 0. Verify your answer by simulating the program.

2. In Listing 9.4, set the gate delay to 8 ns. Simulate the adder with the new gate delay and measure the total delay. Analytically justify the delay that you measured.

3. In Listing 9.7, we wrote the Verilog module as behavioral description. Repeat Example 9.5, but use Verilog gate-level description instead of behavioral description. Verify your description by simulation.

4. In Listing 9.13, HDL code was written to instantiate the Verilog command `casex` in a VHDL module. Repeat the same steps to instantiate `casez` in a VHDL module. The truth table for `casez` is as shown in Table 9.3.

TABLE 9.3 Truth Table for `casez`

Input	Output
a	b
zzz1	1
zz10	2
z100	4
1000	8
Others	0

5. In Example 9.12, a low-pass RC filter was simulated. Repeat the same steps for a high-pass RC filter.

6. Add another output Y1bar in the VHDL code in Example 9.13. Y1bar is the invert of Y1. Write the mixed code and simulate.

9.7   Reference

Reed, M., and R. Rohrer, *Applied Introductory Circuit Analysis for Electrical and Computer Engineers*, Prentice Hall, Upper Saddle River, New Jersey, 1999.

10

SYNTHESIS BASICS

Chapter Objectives

- Understand the concept of synthesis
- Learn how to map behavioral statements into logical gates and components
- Learn how to verify your synthesis
- Review and understand the fundamentals of digital-logic design for digital systems, such as adders, multiplexers, decoders, comparators, encoders, latches, flip-flops, counters, and memory cells
- Understand the concept of sequential finite-state machines

10.1 Highlights of Synthesis

This chapter covers the fundamentals of synthesis. Synthesis here converts HDL behavioral code into logical gates or components. These logical gates and components can be downloaded into electronic chips such as field programmable gate arrays (FPGAs).

Facts

- Synthesis maps the simulation (software) domain into the hardware domain.

- In this chapter, synthesis can be viewed as reverse engineering. The user is provided with the behavioral code and is asked to develop the logic diagram.

- Not all HDL statements can be mapped into the hardware domain. The hardware domain is limited to signals that can take zeroes, ones, or are left open. The hardware domain cannot differentiate, for example, between signals and variables, as does the simulation (software) domain.

- To successfully synthesize behavior code into certain electronic chips, the mapping has to conform to the requirements and constraints imposed by the electronic-chip vendor.

- Several synthesis packages are available on the market. These packages can take behavior code, map it, and produce a net list that can be downloaded into the chip. This chapter focuses on learning how to synthesize the code manually, rather than on how to use the available synthesizers.

- Two synthesizers may synthesize the same code using a different number of the same gates. This is due to the different approaches taken by the two synthesizers to map the code. Consider, for example, the VHDL statement $y := 2x$. One synthesizer might approach this statement as a shift to the left of x; another might approach it as a multiplication and might use a multiplier, which usually results in more gates than the mere shift.

General synthesis steps can be summarized (see Figure 10.1), as follows:

Step 1: If the behavioral description of the system is available, go to Step 3. Otherwise, formulate a flowchart for the behavior of the system.

Step 2: Use the flowchart to write a behavioral description of the system. Be sure to review the instructions of your synthesis tools to see if there are constraints on any of the behavioral statements you plan to use.

Step 3: Simulate the behavioral code and verify that the simulation correctly describes the system.

Step 4: Map the behavioral statements into components or logic gates (this chapter shows you how to do that). Be sure that the components used are downloadable into the selected chip.

Step 5: Write a structural- or gate-level description of the components and logic gates of Step 4. Simulate the structural description and verify that this simulation is similar to that of Step 3.

Step 6: Use CAD tools to download the gates and components of Step 4 into the electronic chip, usually a FPGA chip.

Step 7: Test the chip by inputting signals to the input pins of the chip and observe the output from the output pins. This step is similar to the verification done in Step 5, except the test here is on real, physical signals.

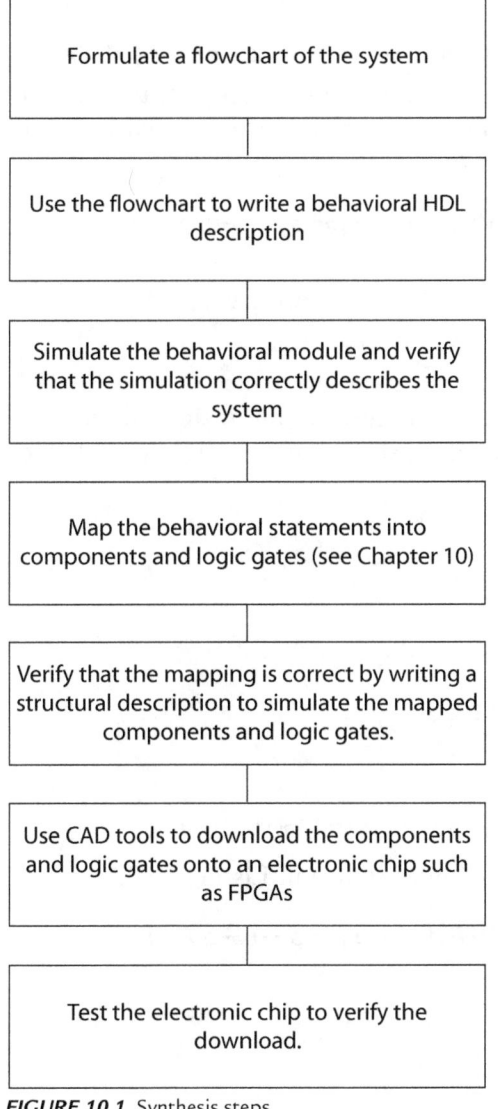

FIGURE 10.1 Synthesis steps.

10.2   Synthesis Information From `Entity` and `Module`

`Entity` (VHDL) or `Module` in (Verilog) provide information on the inputs and outputs and their types for the system to be synthesized. For all the following examples, unless otherwise explicitly stated, the digital hardware domain in which the HDL code is synthesized consists of binary signals; their values can be 0, 1, or tristate (open). The domain does not include analog or multilevel signals.

10.2.1   Synthesis Information From `Entity` (VHDL)

In all of the examples shown here, libraries are not shown in the code since they provide no information to the hardware domain. Consider the VHDL code shown in Listing 10.1.

LISTING 10.1 VHDL Code for `System1` Entity

```
entity system1 is
port (a, b : in bit; d : out bit);
end system1;
```

The synthesis information extracted from Listing 10.1 is summarized in Figure 10.2; `system1` has two input signals, each one bit, and one output signal of one bit. Each signal can take 0 (low) or 1 (high).

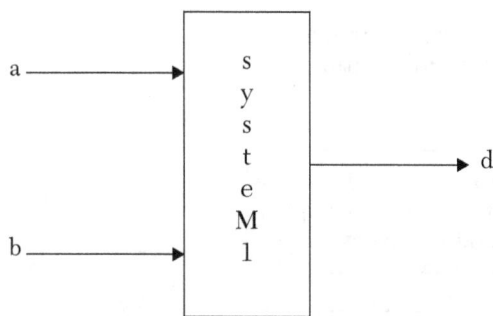

FIGURE 10.2 Synthesis information extracted from Listing 10.1.

Consider the entity shown in Listing 10.2.

LISTING 10.2 VHDL Code for `System2` Entity

```
entity system2 is
port (a, b : in std_logic; d : out std_logic);
end system2;
```

System2 also has two one-bit input signals and one one-bit output signal. However, because the type is std_logic, each signal can take 0 (low), 1 (high), or high impedance (open).

Consider the entity shown in Listing 10.3.

LISTING 10.3 VHDL Code for System3 Entity

```
entity system3 is
port (a, b : in std_logic_vector (3 downto 0);
d : out std_logic_vector (7 downto 0));
end entity system3;
```

System3 has two four-bit input signals and one eight-bit output signal. The input signals can be binary or left open. Figure 10.3 illustrates the information extracted from Listing 10.3.

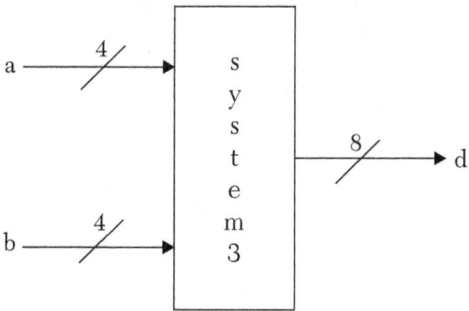

FIGURE 10.3 Synthesis information extracted from Listing 10.3.

Consider the entity shown in Listing 10.4.

LISTING 10.4 VHDL Code for System4 Entity

```
entity system4 is
port (a, b : in signed (3 downto 0);
    d : out std_logic_vector (7 downto 0));
end entity system4;
```

System4 has two four-bit signals and one eight-bit signal. The input signals are binary; the output signal can be binary or high impedance.

Consider the entity shown in Listing 10.5.

LISTING 10.5 VHDL Code for System5 Entity

```
entity system5 is
port (a, b : in unsigned (3 downto 0);
```

```
    d : out std_logic_vector (7 downto 0));
end entity system5;
```

Synthesis information extracted from Listing 10.5 is identical to that extracted from Listing 10.4. Now consider the entity shown in Listing 10.6.

LISTING 10.6 VHDL Code for *System6* Entity

```
entity system6 is
port (a, b : in unsigned (3 downto 0);
    d : out integer range -10 to 10);
end entity system6;
```

System6 has two four-bit input signals and one five-bit output signal. In the hardware domain, the integer is represented by binary, so five bits is adequate for representing d. Figure 10.4 illustrates the information extracted from Listing 10.6.

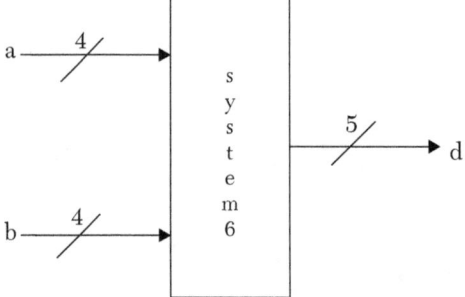

FIGURE 10.4 Synthesis information extracted from Listing 10.6.

Consider the entity in Listing 10.7.

LISTING 10.7 VHDL Code for *System7* Entity

```
entity system7 is
    generic (N : integer := 4; M : integer := 3);
        Port (a, b : in std_logic_vector (N downto 0);
            d : out std_logic_vector (M downto 0));
end system7;
```

Because N = 4 and M = 3, system7 has two five-bit input signals and one four-bit output signal. All signals are binary. N and M have no explicit hardware mapping. Figure 10.5 illustrates the synthesis information extracted from the code of Listing 10.7.

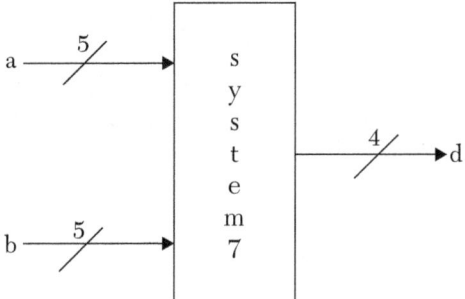

FIGURE 10.5 Synthesis information extracted from Listing 10.7.

Consider the entity in Listing 10.8.

LISTING 10.8 VHDL Code for `ALUS2` Entity

```
package codes is
type op is (add, mul, divide, none);
end;
use work. codes;

entity ALUS2 is
    port (a, b : in std_logic_vector (3 downto 0);
        cin : in std_logic; opc : in op;
        z : out std_logic_vector (7 downto 0);
        cout : out std_logic; err : out Boolean);
end ALUS2;
```

The package `codes` defines type `op`. Signal `opc` is of type `op`. In our digital hardware domain, there are only zeros and ones. Packages and libraries have no explicit mapping into the hardware domain; they are simulation tools. To map the signal `opc` into the hardware domain, the signal is decoded. Because the signal can take one of four values (`add`, `mul`, `divide`, or `none`), it is decoded into two bits. A possible decoding is shown in Table 10.1. Better decoding could be used; choose the one that yields the minimum number of components after minimization.

TABLE 10.1 Decoding of Signal `opc`

Code	Binary Code
add	00
mul	01
divide	10
none	11

Figure 10.6 illustrates the information extracted from the Listing 10.8. As shown, entity ALUS2 has two input signals, a and b, each of four bits, one input signal cin of one bit, one input signal opc of two bits, one output signal z of eight bits, one output signal cout of one bit, and one output signal err of one bit. The Boolean type is mapped to binary 0 or 1.

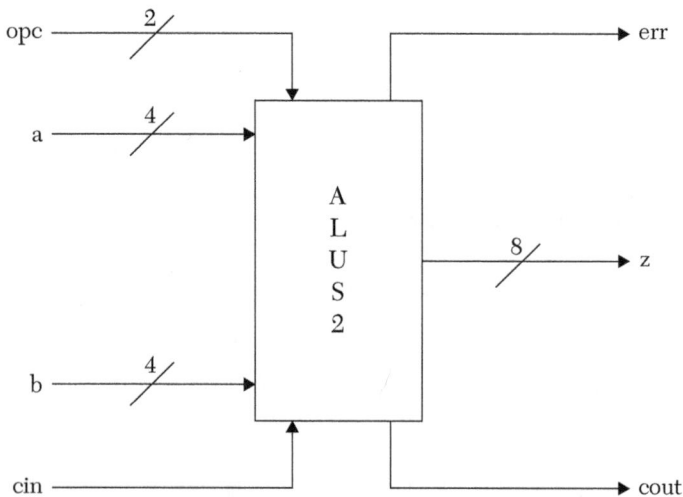

FIGURE 10.6 Synthesis information extracted from Listing 10.8.

Consider the entity and package shown in Listing 10.9.

LISTING 10.9 VHDL Code for Array1 Entity

```
package array_pkg is
constant N : integer := 4;
constant M : integer := 3;
subtype wordN is std_logic_vector (M downto 0);
type strng is array (N downto 0) of wordN;
end array_pkg;
library IEEE;
use IEEE.STD_LOGIC_1164.ALL;
use IEEE.STD_LOGIC_ARITH.ALL;
use IEEE.STD_LOGIC_UNSIGNED.ALL;
use work.array_pkg.all;

entity array1 is
    generic (N : integer := 4; M : integer := 3);
    Port (a : in strng; z : out std_logic_vector (M downto 0));
end array1;
```

From the package, type `strng` is an array of five elements, and each element is four bits wide, so entity `array1` has five input signals, each of four bits. The output of `array1` is a four-bit signal. Figure 10.7 illustrates the synthesis information extracted from the code of Listing 10.9.

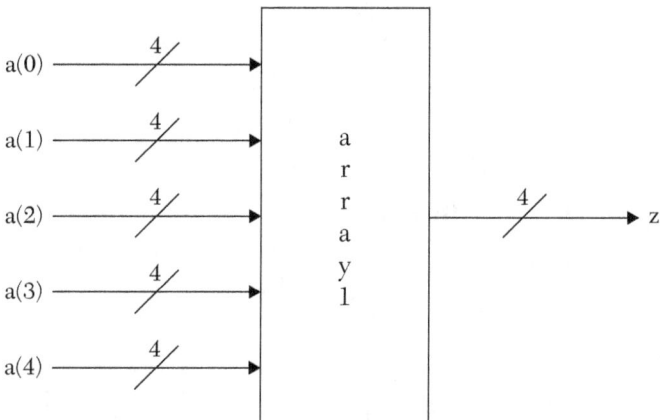

FIGURE 10.7 Synthesis information extracted from Listing 10.9.

Now consider the entity shown in Listing 10.10.

LISTING 10.10 VHDL Code for `Weather_frcst` Entity

```
package weather_fcst is
Type cast is (rain, sunny, snow, cloudy);
Type weekdays is (Monday, Tuesday, Wednesday,
    Thursday, Friday, Saturday, Sunday);
end package weather_fcst;
library ieee;
use ieee.std_logic_1164.all;
use std.textio.all;
use work.weather_fcst.all;
entity WEATHER_FRCST is
port (Day_in : in weekdays; out_temperature : out integer
    range -100 to 100; out_day : out weekdays;
    out_cond : out cast);
end WEATHER_FRCST;
```

Elements of type `cast` in package `weather_fcst` can be decoded by two bits, as shown in Table 10.2.

TABLE 10.2 Decoding Elements of Type cast

Code	Binary Code
rain	00
sunny	01
snow	10
cloudy	11

The elements of type weekdays need three bits to be decoded. Table 10.3 shows a possible decoding of these elements.

TABLE 10.3 Decoding Elements of Type weekdays

Code	Binary Code
Monday	000
Tuesday	001
Wednesday	010
Thursday	011
Friday	100
Saturday	101
Sunday	110

Accordingly, entity WEATHER_FRCST has one input signal, Day_in, which is three bits, an output signal, out_temperature, of seven bits, an output signal, out_day, of three bits, and an output signal, out_cond, of two bits. Figure 10.8 illustrates the synthesis information extracted from the code of Listing 10.10.

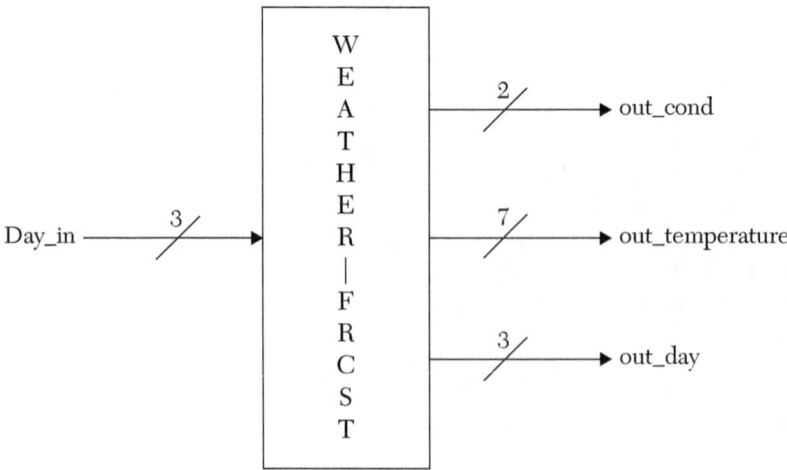

FIGURE 10.8 Synthesis information extracted from Listing 10.10.

Consider the code shown in Listing 10.11

LISTING 10.11 VHDL Code for Entity `Procs_Mchn`

```
library ieee;
use ieee.std_logic_1164.all;

package state_machine is
Type machine is (state0, state1, state2, state3);
Type st_machine is
record
state : machine;
weight : natural range 3 to 16;
Measr : std_logic_vector (5 downto 0);
end record;
end package state_machine;

library ieee;
use ieee.std_logic_1164.all;
use std.textio.all;
use work.state_machine.all ;

entity Procs_Mchn is
port (S : in machine; Y : in st_machine;
      Z : out integer range -5 to 5);
end Procs_Mchn;
```

The entity `Procs_Mchn` has two inputs, `S` and `Y`, and one output, `Z`. Input `S` is of type `machine`; this type has four elements, so input `S` is mapped to two bits. Input `Y` is of type `st_machine`; this type is `record` (a collection of different types). The `record` includes type `state`, which is mapped to a two-bit signal, type `weight`, which is mapped to a five-bit signal, and type `Measr`, which is mapped to a six-bit signal. So, signal `Y` is mapped to six bits (the largest out of two, five, and six). Output `Z` is mapped to a four-bit signal. Figure 10.9 shows the synthesis information extracted from the code of Listing 10.11.

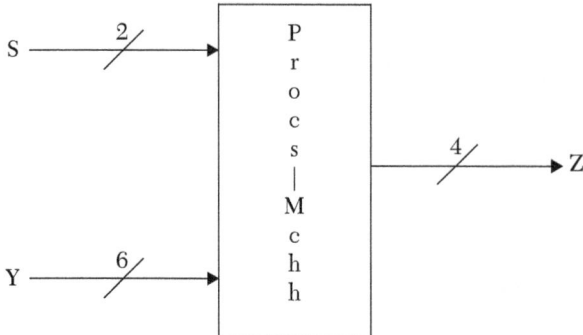

FIGURE 10.9 Synthesis information extracted from Listing 10.11.

10.2.2 Verilog Synthesis Information From Module Inputs/Outputs

Verilog, in contrast to VHDL, does not have a large variety of types. In the following, we discuss synthesis information that can be extracted from the inputs and outputs of a module. Consider the code shown in Listing 10.12.

LISTING 10.12 Verilog Code for Module `System1v`

```
module system1v (a, b, d);
input a, b;
output d;
endmodule
```

From Listing 10.12, `system1v` has two input signals, a and b, each of one bit, and one output signal d of one bit. All signals can take 0, 1, or high impedance. Figure 10.10 shows the synthesis information extracted from Listing 10.12.

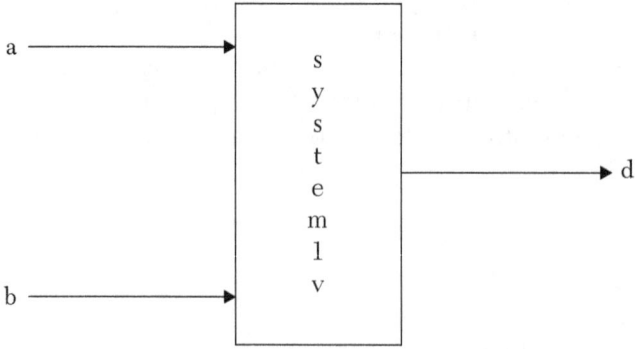

FIGURE 10.10 Synthesis information extracted from Listing 10.12.

Consider the Verilog code shown in Listing 10.13.

LISTING 10.13 Verilog Code for Module System2v

```
module system2v (X, Y, Z);
input [3:0] X, Y;
output [7:0] Z;
reg [7:0] Z
. . . . . . . .
endmodule
```

Listing 10.13 describes system2v with two input signals, X and Y, each of four bits, and one output signal, Z, of eight bits. The statement reg [7:0] Z; does not convey any additional information to the hardware domain; its use is solely for simulation. Figure 10.11 illustrates the information extracted from Listing 10.13.

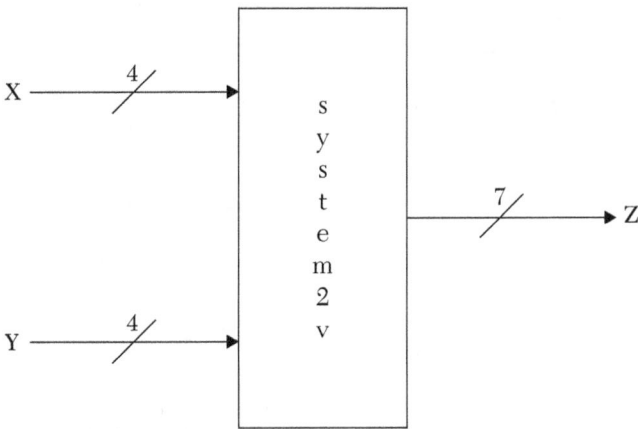

FIGURE 10.11 Synthesis information extracted from Listing 10.13.

Consider the code shown in Listing 10.14.

LISTING 10.14 Verilog Code for Module System3v

```
module system3v (a, b, c);
parameter N = 4;
parameter M = 3;
input [N:0] a;
output [M:0] c;
input b;
. . . . . . . . .
endmodule
```

Module `system3v` has two input signals, `a` and `b`, and one output signal `c`. Input `a` is a five-bit signal, input `b` is one bit, and output `c` is a four-bit signal. `Parameter` has no explicit mapping in the hardware domain; it is just a simulation tool to instantiate `N` and `M`. Figure 10.12 illustrates the synthesis information extracted from Listing 10.14.

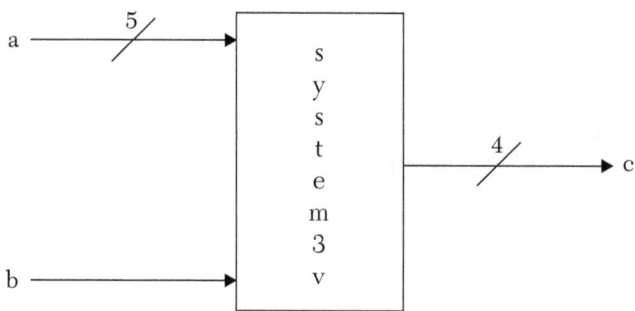

FIGURE 10.12 Synthesis information extracted from Listing 10.14.

Consider the code shown in Listing 10.15.

LISTING 10.15 Verilog Code for Module `Array1v`

```
module array1v (start, grtst);
parameter N = 4;
parameter M = 3;
input start;
output [3:0] grtst;
reg[M:0] a[0:N];
. . . . . . . . . . . . .
endmodule
```

Module `array1v` has one one-bit input signal (`start`) and one four-bit output signal (`grtst`). The register `a` is an array of five elements, each of four bits. This register is mapped to five signals, each of four bits. Figure 10.13 illustrates the synthesis information extracted from Listing 10.15.

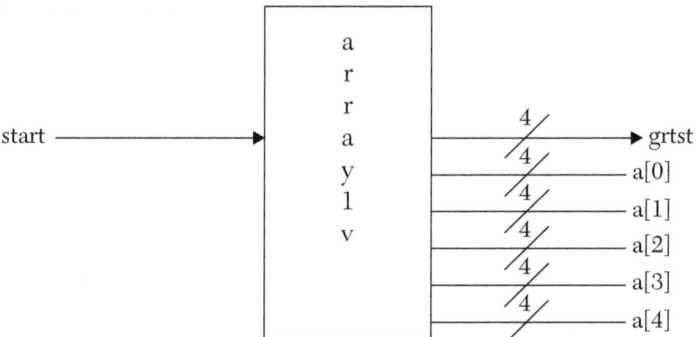

FIGURE 10.13 Synthesis information extracted from Listing 10.15.

10.3   Mapping Process and Always in the Hardware Domain

Process (VHDL) and Always (Verilog) are the major behavioral statements. These statements are frequently used to model systems with data storage such as counters, registers, and CPUs. The first line in both statements declares, among other factors, the sensitivity list. This list determines the signals that activate process or always. The following examples illustrate the mapping of process and always.

10.3.1   Mapping the Signal-Assignment Statement to Gate Level

Consider the entity (module) shown in Listing 10.16.

LISTING 10.16a Mapping VHDL Code for Signal-Assignment Statement Y <= X

```
library ieee;
use ieee.std_logic_1164.all;

entity SIGNA_ASSN is
port (X : in bit; Y : out bit);
end SIGNA_ASSN;

architecture BEHAVIOR of SIGNA_ASSN is
begin
    P1 : process (X)
    begin

        Y <= X;
    end process P1;
    end BEHAVIOR;
```

LISTING 10.16b Mapping Verilog Code for Signal-Assignment Statement Y = X

```
module SIGNA_ASSN (X, Y);
input X;
output Y;
reg y;
always @ (X)
    begin
    Y = X;
    end
endmodule
```

The code in Listing 10.16 describes a one-bit input signal x and a one-bit output signal y (see Figure 10.14a). In VHDL Listing 10.16a, the entity is bound to architecture BEHAVIOR. The process has x as the sensitivity list. The signal-assignment statement states that Y = X. In the hardware domain, this statement is mapped to a buffer. Other statements such as begin, end, and architecture have no hardware mapping. The same applies for Listing 10.16b; the hardware is a buffer. Figure 10.14b shows this mapping: if x changes, y is updated. This mimics the process activation in Listing 10.16 when an event occurs on x.

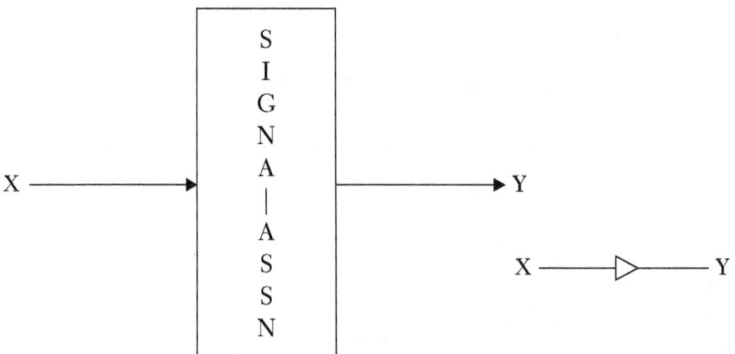

FIGURE 10.14 Gate-level synthesis of Listing 10.16. a) Logic symbol. b) Gate-level logic diagram.

Consider the entity (module) shown in Listing 10.17.

LISTING 10.17 VHDL Code for Signal-Assignment Statement Y = 2 * X + 3: VHDL and Verilog

```
VHDL Signal-Assignment Statement Y = 2 * X + 3
library ieee;
use ieee.std_logic_1164.all;
use ieee.numeric_std.all;
```

```
entity sign_assn2 is
    port (X : in unsigned (1 downto 0);
          Y : out unsigned (3 downto 0));
end ASSN2;
architecture BEHAVIOR of sign_assn2 is
begin

P1 : process (X)
    begin
        Y <= 2 * X + 3;
end process P1;
end BEHAVIOR;
```

Verilog Signal-Assignment Statement Y = 2 * X + 3
```
module sign_assn2 (X, Y);
input [1:0] X;
output [3:0] Y;
reg [3:0] Y;
always @ (X)
    begin
        Y = 2 * X + 3;
    end
endmodule
```

Listing 10.17 shows an entity (sign_assn2) with one input, x, of two bits and one output, Y, of four bits (see Figure 10.15a). The architecture that is bound to the entity and the Verilog module includes one process (P1) and one always, respectively. The process (always) contains one signal-assignment statement: Y <= 2 * X + 3; (VHDL) or Y = 2 * X + 3 (Verilog). To synthesize the code, construct a truth table to find the logic diagram of sign_assn2 and use gate-level synthesis. Table 10.4 shows the truth table of sign_assn2.

TABLE 10.4 Truth Table for Listing 10.17

Input X		Output Y			
X_1	X_0	Y_3	Y_2	Y_1	Y_0
0	0	0	0	1	1
0	1	0	1	0	1
1	0	0	1	1	1
1	1	1	0	0	1

From Table 10.4:

$$Y(0) = 1$$

$$Y(1) = \overline{X(0)}$$

$$Y(2) = \overline{X(1)}\ X(0) + X(1)\ \overline{X(0)}$$

$$Y(3) = X(1)\ X(0)$$

Figure 10.15b shows the gate-level logic diagram of Listing 10.17.

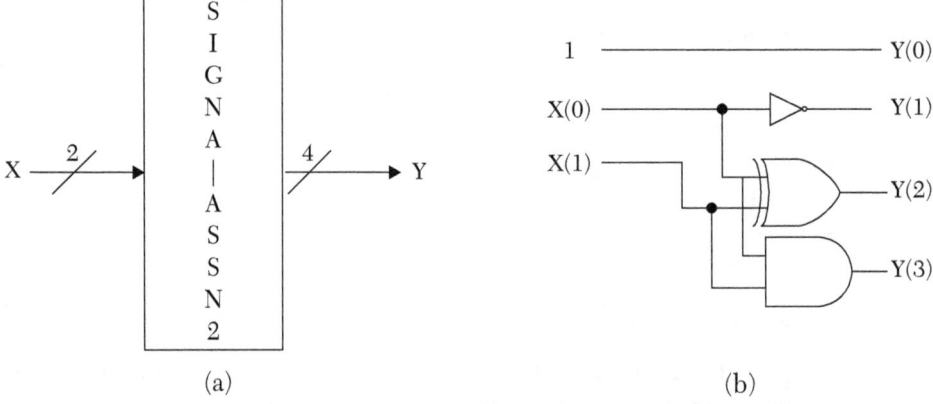

(a)   (b)

FIGURE 10.15 Gate-level synthesis of Listing 10.17. a) Logic symbol. b) Gate-level logic diagram.

To verify the synthesis, write the structural code for the logic diagram shown in Figure 10.15b and then simulate it. If the simulation waveform is the same as the simulation waveform in Listing 10.17, then the synthesis is correct. The simulation waveform for Listing 10.17 is shown in Figure 10.16. The Verilog structural code is shown in Listing 10.18.

LISTING 10.18 Structural Verilog Code for the Logic Diagram in Figure 10.15b.

```
module sign_struc(X, Y);
input [1:0] X;
output [3:0] Y;
reg [3:0] Y;
always @ (X)
    begin
        Y[0] = 1'b1;
        Y[1] = ~ X[0];
        Y[2] = X[0] ^ X[1];
```

```
        Y[3] = X[1] & X[0];
end
endmodule
```

After simulating the code in Listing 10.18, the simulation is identical to Figure 10.16. We conclude that the synthesis is correct.

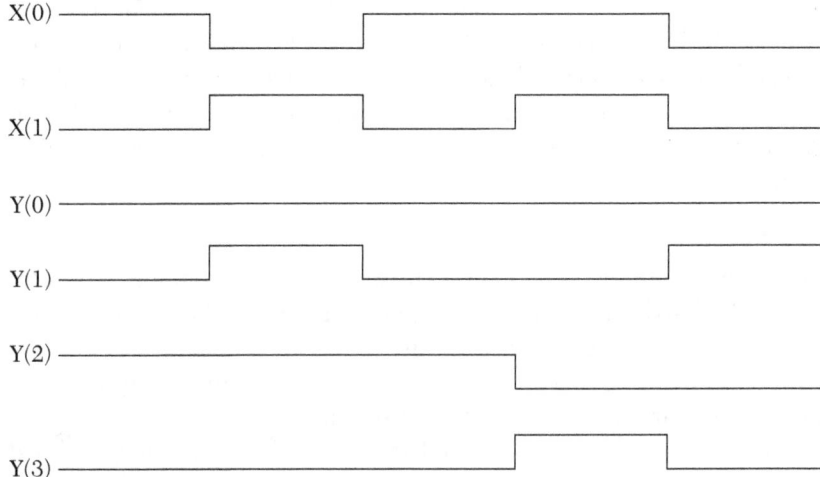

FIGURE 10.16 Simulation waveform for Listing 10.17.

10.3.2   Mapping the VHDL Variable-Assignment Statement to Gate Level

The variable-assignment statement is a VHDL statement. Verilog does not distinguish between signal- and variable-assignment statements. Consider the VHDL code shown in Listing 10.19.

Listing 10.19 *VHDL Variable-Assignment Statement*

```
library ieee;
use ieee.std_logic_1164.all;

entity parity_even is
    port (x : in std_logic_vector (3 downto 0);
          C : out std_logic);

end parity_even;

architecture behav_prti of parity_even is
begin

P1 : process (x)
```

```
variable c1 : std_logic;
    begin
        c1 := (x(0) xor x(1)) xor (x(2) xor x(3));
        C <= c1;
        end process P1;
end behav_prti;
```

Listing 10.19 shows an entity with one four-bit input and one one-bit output (see Figure 10.17a). The architecture `behav_prti` is bound to the entity and consists of one process (`P1`). The process contains one variable declaration, `variable c1 : std_logic;` and two assignment statements. One of the assignment statements is a signal, `C <= c1;`, and the other is a variable assignment:

```
c1 := (x(0) xor x(1)) xor (x(2) xor x(3));
```

The hardware domain cannot distinguish between signal and variable; all we have in the hardware domain are signals. To synthesize the code, notice that signal `c` takes the value of variable `c1`, so in the hardware domain, `c1` and `c` are one signal. The variable-assignment statement includes three XOR functions that are mapped to three XOR gates. More details on logical operators are covered in Section 10.3.3. Figure 10.17b shows the gate-level synthesis of Listing 10.19.

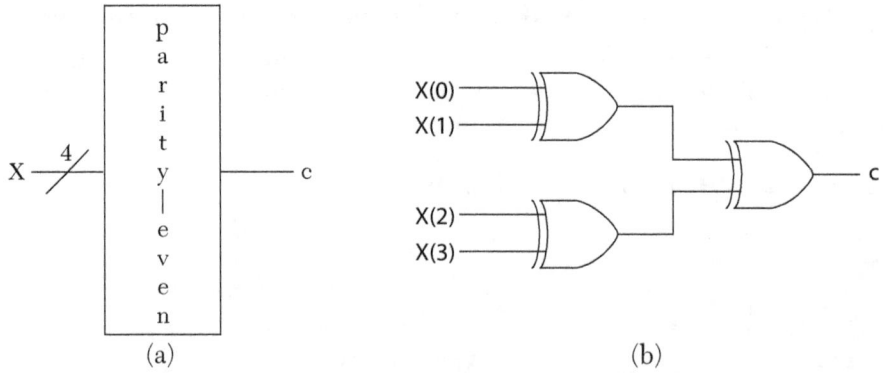

FIGURE 10.17 Gate-level synthesis of Listing 10.19. a) Logic symbol. b) Gate-level logic diagram.

10.3.3 Mapping Logical Operators

Mapping logical operators is relatively straightforward because finding the gate counterpart of a logical operator is very easy. For example, the mapping of logical operator `and` (VHDL) or `&` (Verilog) is an AND gate. Table 10.5 shows the logical operators in VHDL and Verilog and their gate-level mappings.

TABLE 10.5 Logical Operators and Their Gate-Level Mappings

Logical Operator		Gate-Level Mapping
VHDL	**Verilog**	
and	&	AND
or			OR
not	~	INVERTER
xor	^	XOR
xnor	^~	XNOR

To illustrate the mapping of logical operators, consider the code in Listing 10.20.

LISTING 10.20 Mapping Logical Operators: VHDL and Verilog

```
VHDL: Mapping Logical Operators
library IEEE;
use IEEE.STD_LOGIC_1164.ALL;

entity decod_var is
    port (a : in std_logic_vector (1 downto 0);
          D : out std_logic_vector (3 downto 0));
end decod_var;

architecture Behavioral of decod_var is

begin
dec : process (a)
variable a0bar, a1bar : std_logic;
    begin
        a0bar := not a(0);
        a1bar := not a(1);
        D(0) <= not (a0bar and a1bar);
        D(1) <= not (a0bar and a(1));
        D(2) <= not (a(0) and a1bar);
        D(3) <= not (a(0) and a(1));
    end process dec;

end Behavioral;

Verilog: Mapping Logical Operators
module decod_var (a, D);
```

```
input [1:0] a;
output [3:0] D;
reg a0bar, a1bar;
reg [3:0] D;
always @ (a)

    begin
        a0bar = ~ a[0];
        a1bar = ~ a[1];
        D[0] = ~ (a0bar & a1bar);
        D[1] = ~ (a0bar & a[1]);
        D[2] = ~ (a[0] & a1bar);
        D[3] = ~ (a[0] & a[1]);
    end
endmodule
```

The statements

```
a0bar := not a(0); -- VHDL
a0bar = ~ a[0]; // Verilog
```

represent an inverter. The input to the inverter is the least significant bit of the input a. The statements

```
D[3] = ~ (a[0] & a[1]); -- VHDL
D(3) <= not (a(0) and a(1)); // Verilog
```

represent a two-input NAND gate. The input is a, and the output is the most significant bit of D.

Figure 10.18 shows the synthesis of the code in Listing 10.20.

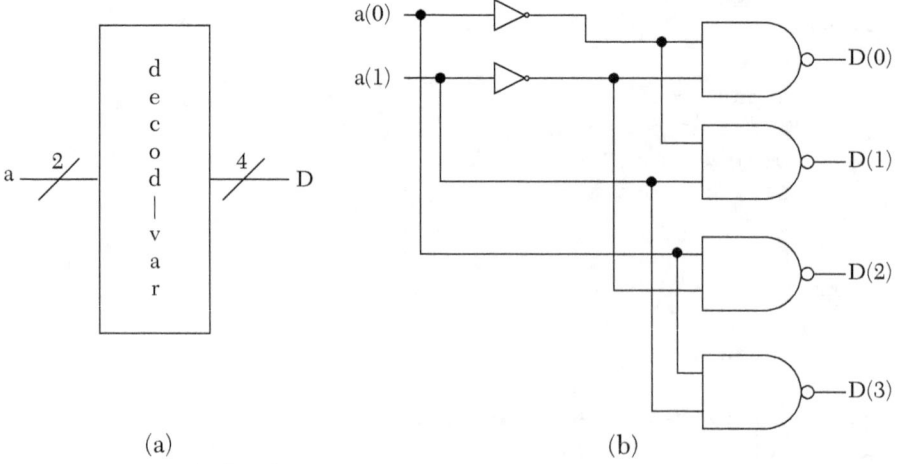

(a)   (b)

FIGURE 10.18 Gate-level synthesis of Listing 10.20. a) Logic symbol. b) Gate-level logic diagram.

10.3.4 Mapping the IF Statement

Consider the HDL IF-else statement shown in Listing 10.21.

LISTING 10.21 Example of IF-else Statement: VHDL and Verilog

VHDL IF-else Description
```
process (a, x)
begin
    if (a = '1') then
    Y <= X;
    else
    Y <= '0';
    end if;
end process;
```

Verilog IF-else Description
```
always @ (a, X)
begin
    if (a == 1'b1)
    Y = X;
    else
    Y = 1'b0;
end
```

The IF statement in Listing 10.21 is synthesized by just an AND gate, as shown in Figure 10.19.

FIGURE 10.19 Gate-level synthesis of Listing 10.21.

Now, consider the IF statement shown in Listing 10.22.

LISTING 10.22 Example of Multiplexer IF-else Statement: VHDL and Verilog

VHDL Multiplexer IF-else Description

```
process (a, X, X1)
begin
    if (a = '1') then
    Y <= X;
```

```
        else
        Y <= X1;
        end if;
end process;
```

Verilog Multiplexer IF-else Description
```
always @ (a, X, X1)
begin
    if (a == 1'b1)
    Y = X;
    else
Y = X1;
        end
```

The IF statement in Listing 10.22 represents a 2x1 multiplexer. Figure 10.20 shows the synthesis of Listing 10.22.

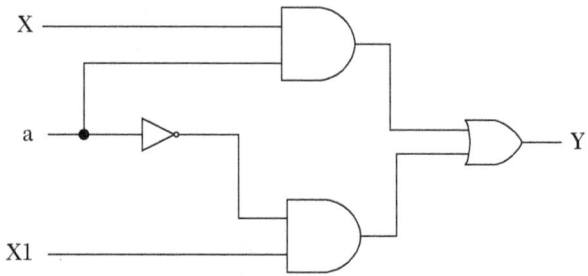

FIGURE 10.20 Gate-level synthesis of Listing 10.22.

Consider the IF statement shown in Listing 10.23.

LISTING 10.23 Example of Comparison Using IF-else Statement: VHDL and Verilog

VHDL IF-else Statement
```
library IEEE;
use IEEE.STD_LOGIC_1164.ALL;

entity IF_st is
    port (a : in std_logic_vector (2 downto 0);
        Y : out Boolean);
end IF_st;

architecture IF_st of IF_st is
begin

IfB : process (a)
```

```
variable tem : Boolean;
begin
    if (a < "101") then
    tem := true;
    else
    tem := false;
    end if;
Y <= tem;
end process;
end IF_st;
```

Verilog IF-else Statement
```
module IF_st (a, Y);
input [2:0] a;
output Y;
reg Y;
always @ (a)
begin
if (a < 3'b101)
Y = 1'b1;
else
Y = 1'b0;
end
endmodule
```

To find the gate-level mapping of Listing 10.23, construct a truth table (see Table 10.6).

TABLE 10.6 Truth Table for Listing 10.23

Input a			Output Y
a_2	a_1	a_0	Y
0	0	0	1
0	0	1	1
0	1	0	1
0	1	1	1
1	0	0	1
1	0	1	0
1	1	0	0
1	1	1	0

Figure 10.21 shows the K-map of Listing 10.23. From the figure, the Boolean function of Y is:

$$Y = \overline{a(2)} + \overline{a(1)}\,\overline{a(0)}$$

a1a0 a2	00	01	11	10
0	1	1	1	1
1	1	0	0	0

Y

FIGURE 10.21 K-map for Listing 10.23.

From the Boolean function, draw the gate-level synthesis for Listing 10.22 as shown in Figure 10.22.

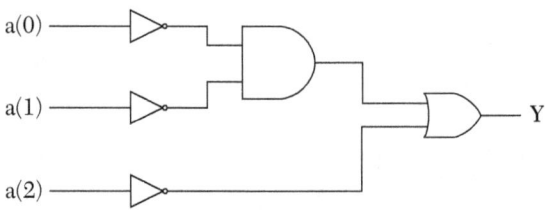

FIGURE 10.22 Gate-level synthesis of Listing 10.23.

Now consider the `elseif` (VHDL) and `Else-If` (Verilog) statements in Listing 10.24.

LISTING 10.24 Example of `elseif` and `Else-If`: VHDL and Verilog

VHDL elseif Description

```
library IEEE;
use IEEE.STD_LOGIC_1164.ALL;
use IEEE.STD_LOGIC_ARITH.ALL;

entity elseif is
port (BP : in natural range 0 to 7;
     ADH : out natural range 0 to 15);
   end;
```

```
architecture elseif of elseif is

    begin
    ADHP : process(BP)
        variable resADH : natural := 0;
        begin
            if BP <= 2 then resADH := 15;
            elsif BP >= 5 then resADH := 0;
            else
            resADH := BP * (-5) + 25;
            end if;

ADH <= resADH;
end process ADHP;
end elseif;
```

Verilog Else-If Description
```
module elseif (BP, ADH);
input [2:0] BP;
output [3:0] ADH;
reg [3:0] ADH;
always @ (BP)
begin
    if (BP <= 2) ADH = 15;
        else if (BP >= 5) ADH = 0;
        else
        ADH = BP * (-5) + 25;
    end
endmodule
```

Notice that the variable resADH in Listing 10.24 (VHDL) is identical in value to the output ADH. Accordingly, resADH is not mapped into the hardware domain. To synthesize the code, construct the truth table (see Table 10.7).

TABLE 10.7 Truth Table for Listing 10.24

BP bit210	ADH bit3210
000	1111
001	1111
010	1111
011	1010

BP bit210	ADH bit3210
100	0101
101	0000
110	0000
111	0000

From Table 10.7, construct K-maps to find ADH (see Figure 10.23).

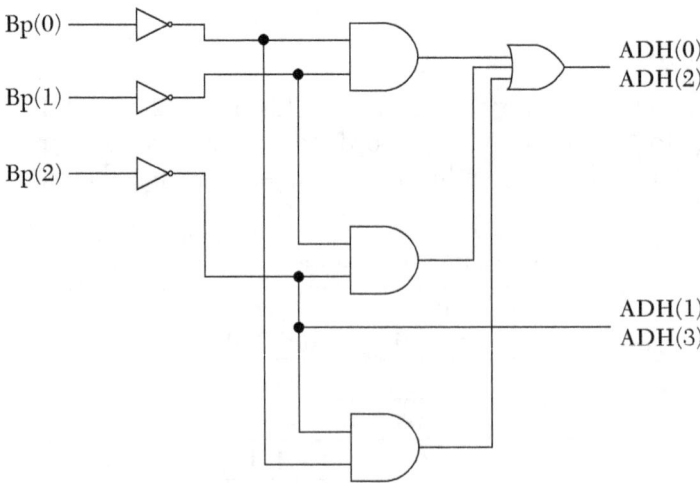

Bp1Bp0 \ Bp2	00	01	11	10
0	1	1	0	1
1	1	0	0	0

ADH(0)

Bp1Bp0 \ Bp2	00	01	11	10
0	1	1	1	1
1	0	0	0	0

ADH(1)

FIGURE 10.23 K-maps of Table 10.7.

From the K-maps, we find:

$$ADH(0) = ADH(2) = \overline{Bp(1)}\,\overline{Bp(0)} + \overline{Bp(2)}\,\overline{Bp(1)} + \overline{Bp(2)}\,\overline{Bp(0)}$$

$$ADH(1) = ADH(3) = \overline{Bp(2)}$$

Figure 10.24 shows the gate-level synthesis of Listing 10.24.

FIGURE 10.24 Gate-level synthesis of Listing 10.24.

Now consider the code in Listing 10.25.

LISTING 10.25 Example of IF Statement with Storage: VHDL and Verilog

VHDL IF Statement with Storage

```
library IEEE;
use IEEE.STD_LOGIC_1164.ALL;

entity If_store is
port (a, X : in std_logic; Y : out std_logic);
end If_store;

architecture If_store of If_store is

begin
    process (a, X)
    begin
        if (a = '1') then
        Y <= X;

        end if;
    end process;
end If_store;
```

Verilog IF Statement with Storage

```
module If_store (a, X, Y);
input a, X;
output Y;
reg Y;
always @ (a, X)
    begin
        if (a == 1'b1)
        Y = X;
    end
endmodule
```

The IF statement in Listing 10.25 is similar to that of Listing 10.22, except when a = 0. In Listing 10.22, the value of the output Y is explicitly stated when a = 0. In Listing 10.25, the code states that when a = 0, there should be no change in the values of any signal. This means that the value of all signals should be stored during the execution of the IF statement. To store signals in the hardware domain, latches or flip-flops are used. In Listing 10.25, signal a is implemented as a clock to a D-latch; the input to the latch is the signal x. If a = 0, then the output of the latch stays the same. If a = 1, then the output follows the input x. Figure 10.25 shows the mapping of Listing 10.25 to the hardware domain.

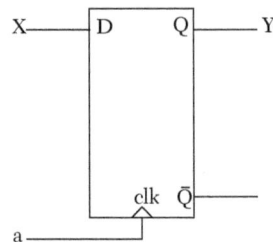

FIGURE 10.25 Synthesis of Listing 10.25.

Consider the code in Listing 10.26.

Listing 10.26 *Else-If Statement with Gate-Level Logic*

```
package weather_fcst is
Type unit is (cent, half, offset);

end package weather_fcst;

library ieee;
use ieee.std_logic_1164.all;
use work.weather_fcst.all;

entity weather is
    port (a : in unit; tempr : in integer range 0 to 15;
          z : out integer range 0 to 15);
end weather;

architecture weather of weather is

begin
T : process (a, tempr)
variable z_tem : integer range 0 to 15;
    begin
        if ((tempr <= 7) and (a = cent)) then
        z_tem := tempr;
```

```
        elsif ((tempr <= 7) and (a = offset)) then
        z_tem := tempr + 4;

        elsif ((tempr <= 7) and (a = half)) then
        z_tem := tempr /2;

        else
        z_tem := 15;

        end if;

        z <= z_tem;
    end process T;
end weather;
```

From the entity (module), we can summarize the extracted information as follows:

- Input `a` is a two-bit signal.

- Input `tempr` is a four-bit signal.

- Output `z` is a four-bit signal.

The code can be summarized as shown in Table 10.8.

TABLE 10.8 Summary of the Code in Listing 10.26

a	tempr	z
00 (cent)	0–7	z = tempr
01 (offset)	0–7	z = tempr + 4
10 (half)	0–7	z = tempr/2
11	xx	z = 15
xx	>7	z = 15

If we want to construct a truth table, it will be a $(2 + 4 = 6)$ six-bit input and four-bit output; this table will be huge and cannot be analyzed easily. Accordingly, the code in Listing 10.26 is analyzed logically. Input `a` can be the select lines of a multiplexer. The multiplexer has four inputs; each input is a four-bit signal representing one of the four values `tempr`, `tempr+4`, `tempr/2`, or the constant 15. Figure 10.26 shows this analysis using register transfer level (RTL) logic.

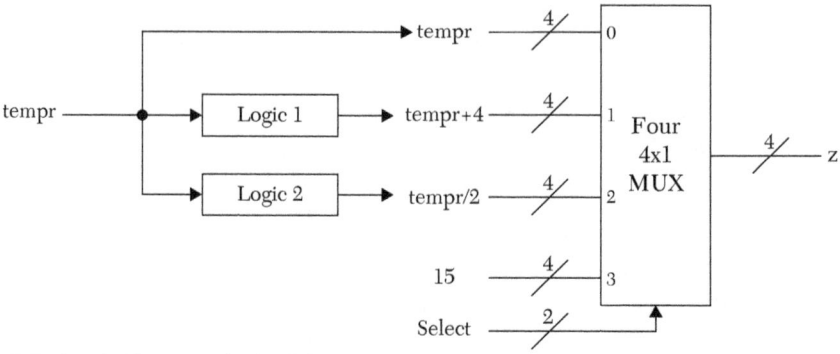

FIGURE 10.26 RTL synthesis of Listing 10.26.

To find the gate-level synthesis of Logic 1 in Figure 10.26, construct a truth table as shown in Table 10.9.

TABLE 10.9 Truth Table for Logic 1

tempr	tempr +4
bit3210	**bit3210**
0000	0100
0001	0101
0010	0110
0011	0111
0100	1000
0101	1001
0110	1010
0111	1011
1000–1111	dddd

Inspecting Table 10.8, `tempr +4` can be written as:

$$\text{tempr} +4(0) = \text{tempr} (0)$$

$$\text{tempr} +4(1) = \text{tempr} (1)$$

$$\text{tempr} +4(2) = \overline{\text{tempr}(2)}$$

$$\text{tempr} +4(3) = \text{tempr} (2)$$

For logic 2, do the same as for Logic 1. Table 10.10 shows the truth table of Logic 2.

TABLE 10.10 Truth Table for Logic 2

tempr	tempr/2
bit3210	**bit3210**
0000	0000
0001	0000
0010	0001
0011	0001
0100	0010
0101	0010
0110	0011
0111	0011
1000-1111	dddd

After inspecting Table 10.10:

$$\text{tempr/2}(0) = \text{tempr} \, (1)$$

$$\text{tempr} \, /2(1) = \text{tempr} \, (2)$$

$$\text{tempr} \, /2(2) = 0$$

$$\text{tempr} \, /2(3) = 0$$

For the select in Figure 10.26 to satisfy the condition temp ≤ 7, tempr (3) must be equal to 0. Accommodating the values of a, construct a truth table as shown in Table 10.11.

TABLE 10.11 Truth Table for Figure 10.26 Select

Tempr(3)	a(1)	a(0)		Select
0	0	0		00
0	0	1		01
0	1	0		10
0	1	1		11
1	0	0		11
1	0	1		11
1	1	0		11
1	1	1		11

Figure 10.27 shows the K-maps of Table 10.11. From the K-maps:

$$Select(0) = temp(3) + a(0)$$
$$Select(1) = temp(3) + a(1)$$

Incorporating the gate-level logic of Logic 1, Logic 2, and Select in Figure 10.26, the synthesis diagram of Listing 10.26 is shown in Figure 10.28.

a(1)a(0) tempr(3)	00	01	11	10
0	0	1	1	0
1	1	1	1	1

Select (0)

a(1)a(0) tempr(3)	00	01	11	10
0	0	0	1	1
1	1	1	1	1

Select (1)

FIGURE 10.27 K-maps for Table 10.11.

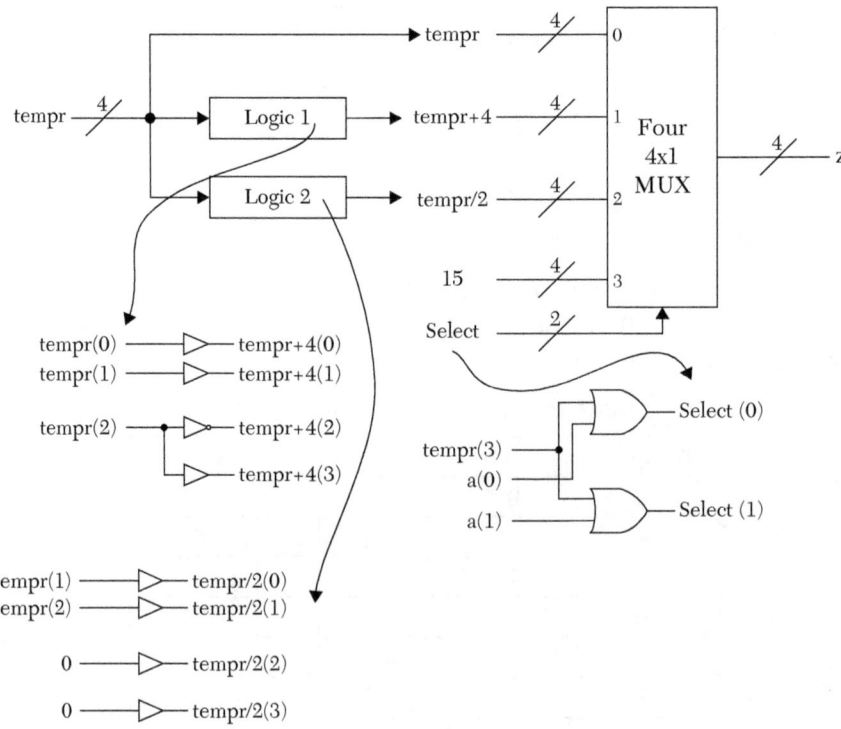

FIGURE 10.28 Synthesis of Listing 10.26.

10.3.5 Mapping the `case` Statement

Mapping the `case` statement is very similar to mapping the `IF` statement. The `case` statement is treated as a group of `IF` statements. Consider the `case` statement in Listing 10.27.

LISTING 10.27 Example of `case` Mapping

```
module case_nostr (a, b, ct, d);
input [3:0] a, b;
input ct;
output [4:0] d;
reg [4:0] d;
always @ (a, b, ct)
begin
case (ct)
1'b0 : d = a + b;

1'b1 : d = a - b;
endcase
end

endmodule
```

To synthesize the above code, construct a truth table. This table would have (4 + 4 +1 = 9) nine bits input for a, b, and ct, and five bits for the output d. This table would yield a minimum number of gates for the code in Listing 10.27; however, the table would be very large and hard to analyze. Another approach is to logically analyze the code using RTL blocks. Listing 10.27 includes two operations: four-bit addition and four-bit subtraction. The result is expressed in a five-bit output, d. Signal ct selects whether to add or subtract. To add, use four one-bit ripple-carry adders. To subtract, use four one-bit subtractors, but the number of components can be reduced by noticing that the full adder can be used as a subtractor, as shown below:

$$d = a - b = a + (-b) = a + \overline{b} + 1$$

Figure 10.29 shows the RTL synthesis of Listing 10.27. The XOR gate is implemented to generate the complement of signal b.

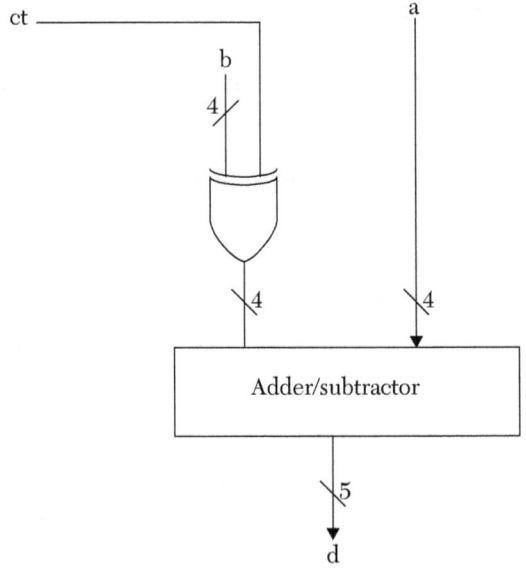

FIGURE 10.29 RTL synthesis of Listing 10.27.

Now, slightly change the code of Listing 10.27 to that shown in Listing 10.28.

LISTING 10.28 *case Statement with Storage*

```
module case_str (a, b, ct, d);
input [3:0] a, b;
input ct;
output [4:0] d;
reg [4:0] d;
always @ (a, b, ct)
begin
    case (ct)
        1'b0: d = a + b;
        1'b1: ; /* This is a blank statement with
                no operation (null in VHDL)*/
    endcase
end

endmodule
```

The `case` in Listing 10.28 does not specify an action when `ct` = 1, so a latch is used to store the value of `d` when `ct` = 1. Figure 10.30 shows the RTL synthesis of Listing 10.28.

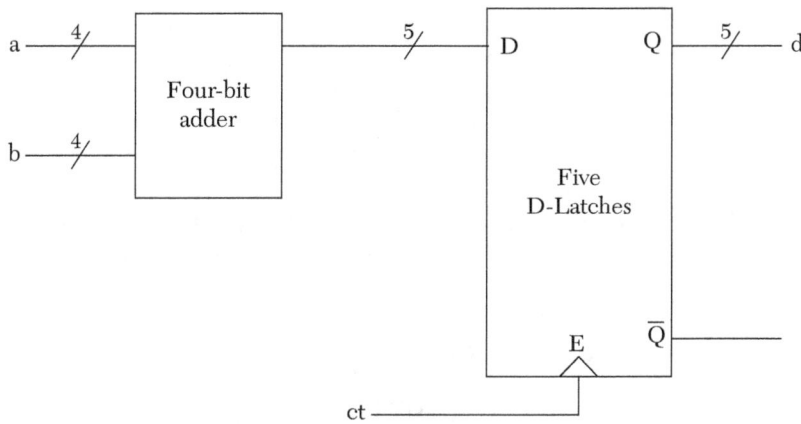

FIGURE 10.30 RTL synthesis of Listing 10.28.

As mentioned in Chapter 3, Verilog has a variation of the command `case`, `casex`. Listing 10.29 shows a Verilog code using `casex`.

LISTING 10.29 Verilog `casex`

```
module Encoder_4 (IR, RA);
input [3:0] IR;
output [3:0] RA;
    reg [3:0] RA;
    always @ (IR)
        begin
            casex (IR)
                4'bxxx1 : RA = 4'd1;
                4'bxx10 : RA = 4'd2;
                4'bx100 : RA = 4'd4;
                4'b1000 : RA = 4'd8;
                default : RA = 4'd0;
            endcase
        end
endmodule
```

To synthesize the code in Listing 10.29, build a truth table as shown in Table 10.12.

TABLE 10.12 Truth Table for the Code in Listing 10.29

Input	Output
IR	RA
xxx1	0001
xx10	0010
x100	0100
1000	1000
Others	0000

Notice the input IR has explicit value for all of its entries, so synthesis does not need storage. By inspecting Table 10.12, the Boolean function of the output can be written as:

$$RA(0) = IR(0)$$
$$RA(1) = \overline{IR(0)} \; IR(1)$$
$$RA(2) = \overline{IR(0)} \; \overline{IR(1)} \; IR(2)$$
$$RA(3) = \overline{IR(0)} \; \overline{IR(1)} \; \overline{IR(2)} \; IR(3)$$

Figure 10.31 shows the logic diagram of Listing 10.29.

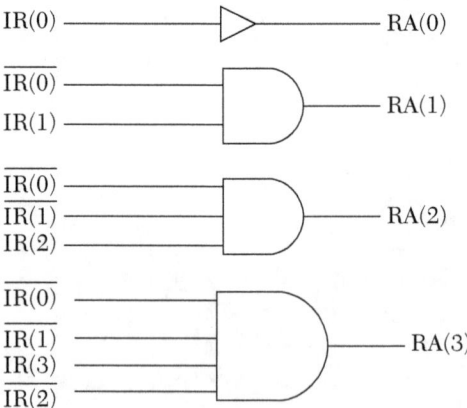

FIGURE 10.31 Logic diagram of Listing 10.29.

Now consider the code shown in Listing 10.30. This code is slightly different from that of Listing 7.9 (see Chapter 7).

LISTING 10.30 Example of `case` *with Storage*

```
library IEEE;
use IEEE.STD_LOGIC_1164.all;
package types is
type states is (state0, state1, state2, state3);
end;
library IEEE;

use IEEE.STD_LOGIC_1164.ALL;
use work.types.all;
entity state_machine is
    port (A, clk : in std_logic; pres_st : buffer states;
          Z : out std_logic);
end state_machine;

architecture st_behavioral of state_machine is

begin

FM : process (clk, pres_st, A)
variable present : states := state0;
begin
if (clk = '1' and clk'event) then
--clock'event is an attribute to the signal clk;
--the above if Boolean expression means the positive
-- edge of clk
--
    case pres_st is
        when state0 =>
        if A ='1' then
        present := state1;
        Z <= '0';
        else
        present := state0;
        Z <= '1';
        end if;

        when state1 => if A ='1' then
        present := state2;
        Z <= '0';
        else
        present := state3;
```

```
        Z <= '0';
        end if;

        when state2 => if A ='1' then
        present := state3;
        Z <= '1';
        else
        present := state0;
        Z <= '0';
        end if;

        when state3 => if A ='1' then
        present := state0;
        Z <= '1';
        else
        present := state2;
        Z <= '1';
        end if;
     end case;

pres_st <= present;
end if;
end process FM;
end st_behavioral;
```

In Listing 10.30, the package `types` declares user-select types `state0`, `state1`, `state2`, and `state3`. To decode these user-selected types into the hardware domain, two bits are needed. So, `state0` is decoded as 00, `state1` as 01, `state2` as 10, and `state3` as 11. The libraries are software constructs that have no mapping into the hardware domain.

Now let us summarize the information collected from the entity. The name of the system or entity is `state-machine`. The system has a one-bit input `A`, a one-bit input `clk`, two-bit input/output states, and a one-bit output `z`. The architecture consists of `case` and `IF` statements. Let us see if we need to use a storage element. Consider the `case` statement:

```
case pres_st is
    when state0 => if A ='1' then
    present := state1;
    Z <= '0';
    else
    present := state0;
    Z <= '1';
    end if;
```

In order to know to which state to go, we need to know the present state. For example, if the present state is state0, then the next state can be state1 or state0. The code implies that the current state must be remembered, so, accordingly, storage elements are needed to synthesize the code. The best approach here is to follow the same steps covered in Chapter 4 for analyzing state machines. Write the excitation table of the machine and use D flip-flops. Table 10.13 shows the excitation table for Listing 10.30.

TABLE 10.13 Excitation Table for Listing 10.30

Present State Input			Next State		Output	D Flip-Flop	
Q1	Q0	A	Q1+	Q0+	Z	D1	D0
0	0	0	0	0	1	0	0
0	0	1	0	1	0	0	1
0	1	0	1	1	0	1	1
0	1	1	1	0	0	1	0
1	0	0	0	0	1	0	0
1	0	1	1	1	1	1	1
1	1	0	1	0	1	1	0
1	1	1	0	0	1	0	0

From Table 10.13, construct K-maps to minimize the outputs. Figure 10.32 shows the K-maps.

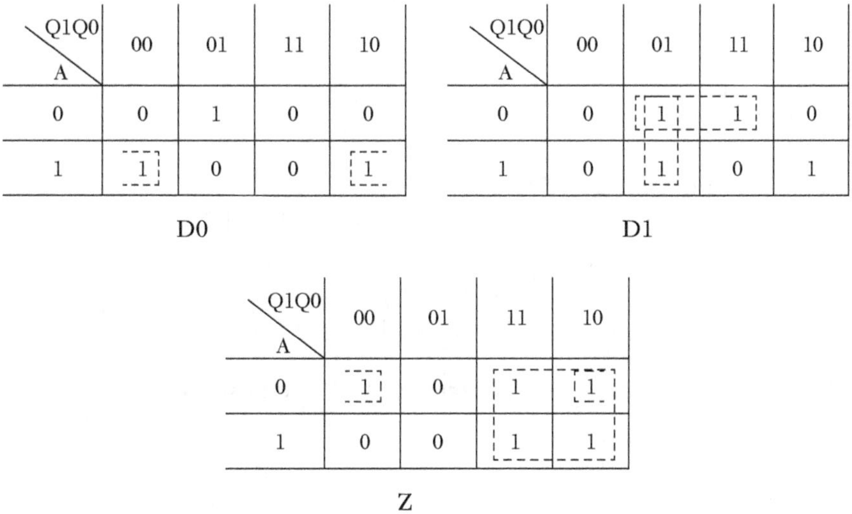

FIGURE 10.32 K-maps for Table 10.13.

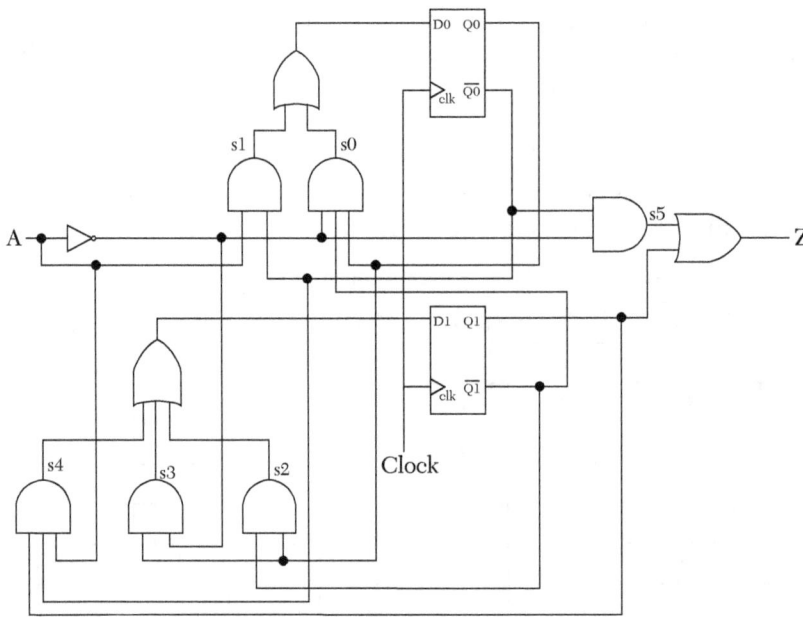

FIGURE 10.33 RTL logic diagram of Listing 10.30.

From the K-maps, find the Boolean function of the system as:

$$D0 = \overline{A}\,\overline{Q1}\,\overline{Q0} + A\,\overline{Q0}$$

$$D1 = Q0\,\overline{Q1} + \overline{A}\,Q0 + AQ1\,\overline{Q0}$$

$$Z = Q1 + \overline{A}\,\overline{Q0}$$

From the Boolean function, the logic diagram of the system is drawn. Figure 10.33 shows the logic diagram of Listing 10.30.

10.3.6 Mapping the `Loop` Statement

`Loop` in HDL description is an essential tool for behavioral modeling. It is, however, easier to code than it is to synthesize in the hardware domain. The problem is the repetition involved in the loop. For example, consider the VHDL `Loop` statement shown in Listing 10.31.

LISTING 10.31 A `For-Loop` Statement: VHDL and Verilog

VHDL For-Loop Statement
```
for i in 0 to 63 loop
temp(i) := temp(i) + b(i);
end loop;
```
Verilog For-Loop Statement

```
for i = 0; i <= 63; i = i + 1
   begin
   temp[i] = temp[i] + b[i];
   end
```

As shown in Listing 10.31, the loop repeats the statement `temp(i) =`
`temp(i) + b(i)` 64 times. This statement can be synthesized using adders.
Each time the statement repeats, the index of the operands to be added is
incremented. So the three lines of code in Listing 10.31 result in 64 adders.
The straightforward approach to synthesizing a loop is to expand the loop
into statements and synthesize each statement individually. For example,
the loop in Listing 10.31 can be logically written as:

$$temp(0) = temp(0) + b(0)$$
$$temp(1) = temp(1) + b(1)$$
$$temp(2) = temp(2) + b(2)$$
$$\dots\dots\dots\dots\dots\dots\dots\dots\dots\dots$$
$$temp(63) = temp(63) + b(63)$$

Each statement is synthesized as a one-bit adder.

EXAMPLE 10.1   **SYNTHESIS OF THE LOOP STATEMENT**

Consider the VHDL behavioral code shown in Listing 10.32.

LISTING 10.32 VHDL Code Includes *For-Loop*

```
library IEEE;
use IEEE.STD_LOGIC_1164.ALL;
entity Listing10_32 is

port (a : in std_logic_vector (3 downto 0);
   c : in integer range 0 to 15;
   b : out std_logic_vector (3 downto 0));
   end Listing10_32;
   architecture Listing10_32 of Listing10_32 is
   begin
   shfl : process (a, c)
   variable result, j : integer;
   variable temp : std_logic_vector (3 downto 0);
   begin

      result := 0;
```

```
lop1 : for i in 0 to 3 loop
    if a(i) = '1' then
    result := result + 2**i;
    end if;
end loop;
    if result > c then
    lop2 : for i in 0 to 3 loop
        j := (i + 2) mod 4;
        temp (j) := a(i);
    end loop;
        else
        lop3 : for i in 0 to 3 loop
            j := (i + 1) mod 4;
            temp (j) := a(i);
        end loop;
        end if;
    b <= temp;
end process shfl;
end Listing10_32;
```

The code in Listing 10.32 describes a system with one four-bit input a, one integer input c, and a four-bit output b. In the hardware domain, there are only bits, so the integer c (because its range is from 0 to 15), is represented by four bits. If you are using a vendor's synthesizer, be sure to specify the integer range; otherwise, the synthesizer, because it does not know the range, will allocate more than 32 bits for the integer. Figure 10.34 summarizes the information retrieved from the entity.

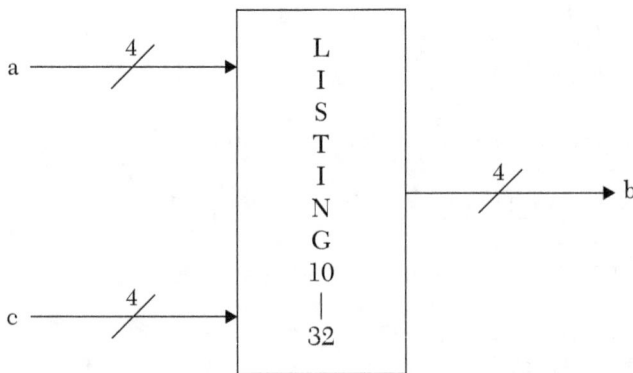

FIGURE 10.34 Information retrieved from entity Listing 10.32.

The simulation output of the system described by Listing 10.32 is shown in Figure 10.35. From the figure, the system shuffles input a with two shuffling patterns, depending on whether or not a is greater than c.

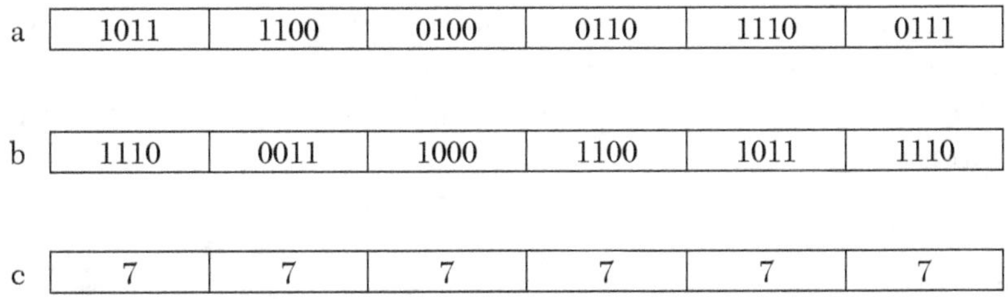

a	1011	1100	0100	0110	1110	0111

b	1110	0011	1000	1100	1011	1110

c	7	7	7	7	7	7

FIGURE 10.35 Simulation output of Listing 10.32.

The code in Listing 10.32 included a process labeled shfl. The process has an IF statement and three For-Loops: lop1, lop2, and lop3. The first For-Loop, lop1, converts the std_logic_vector a to an integer. This conversion is ignored by the hardware domain; the main goal of this conversion is to be able to compare a with the integer c. The hardware views the variable result and a as the same signal. The IF statement that starts with

```
if result > c then
```
is complete; if result > c, then loop lop2 is executed. Otherwise, loop lop3 is executed. Accordingly, latches are not needed to synthesize this IF statement. For loop lop2, expand the loop as shown in Table 10.14.

TABLE 10.14 Expanding the Loop lop2

i	j	temp(j) = a(i)
0	2	temp(2) = a(0)
1	3	temp(3) = a(1)
2	0	temp(0) = a(2)
3	1	temp(1) = a(3)

Notice from Listing 10.32 that the variable temp is identical to signal b; the hardware domain views b and temp as the same signal. For loop lop2, expand the loop as shown in Table 10.15.

TABLE 10.15 Expanding Loop lop3

i	j	temp(j) = a(i)
0	1	temp(1) = a(0)
1	2	temp(2) = a(1)
2	3	temp(3) = a(2)
3	0	temp(0) = a(3)

From Tables 10.14 and 10.15, the logic diagram of the system consists of a four-bit magnitude comparator and four 2x1 multiplexers (see Figure 10.36). The four-bit comparator can be built from four-bit adders (see Chapter 4).

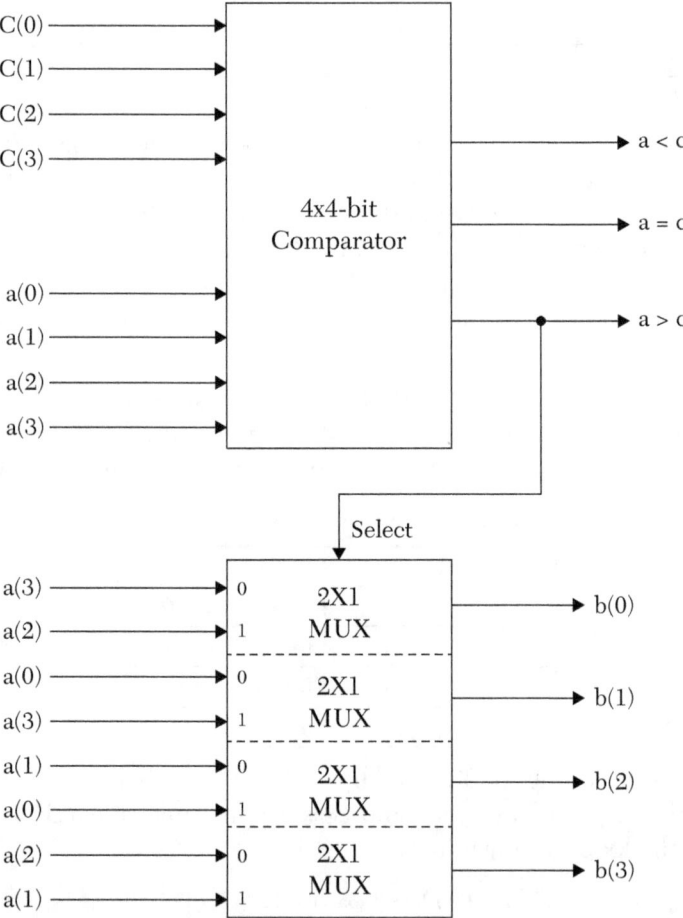

FIGURE 10.36 RTL synthesis of Listing 10.32.

10.3.7 Mapping Procedures or Tasks

As mentioned in Chapter 6, procedures, tasks, and functions are code constructs that optimize HDL module writing. In the hardware domain, there is no logic for procedures or tasks; they are incorporated in the entity or the module that calls them. Consider the Verilog code for a task shown in Listing 10.33.

Listing 10.33 A Verilog Example of a Task

```verilog
module example_task (a1, b1, d1);
input a1, b1;
output d1;
reg d1;
always @ (a1, b1)
begin

xor_synth (d1, a1, b1);
end

task xor_synth;
output d;
input a, b;
begin
d = a ^ b;
end
endtask

endmodule
```

The task is performing a logical XOR operation on two operands, `a` and `b`. By incorporating this information in the module `example_task`, the module can be summarized as a system with two one-bit inputs, `a1` and `b1`, and one one-bit output, `d1`. The relationship between `d1` and `a1` and `b1` is:

$$d1 = a1 \oplus b1$$

The synthesis of this module is shown in Figure 10.37.

FIGURE 10.37 Synthesis of Listing 10.33.

Now consider the code shown in Listing 10.34.

LISTING 10.34 An Example of a Procedure

```vhdl
library IEEE;
use IEEE.STD_LOGIC_1164.ALL;
```

```vhdl
entity Int_Bin is
generic (N : integer := 3);
port (X_bin : out std_logic_vector (N downto 0);
      Y_int : in integer;
      flag_even : out std_logic);
end Int_Bin;

architecture convert of Int_Bin is

procedure itb (bin : out std_logic_vector;
               signal flag : out std_logic;
               N : in integer; int : inout integer) is

begin
if (int MOD 2 = 0) then
    flag <= '1';
    else
    flag <= '0';
end if;
for i in 0 to N loop

    if (int MOD 2 = 1) then
    bin (i) := '1';
    else
    bin (i) := '0';
    end if;

    int := int / 2;

end loop;
end itb;

begin
process (Y_int)
variable tem : std_logic_vector (N downto 0);
variable tem_int : integer;

begin
    tem_int := Y_int;
    itb (tem, flag_even, N, tem_int);
    X_bin <= tem;
end process;
end convert;
```

Let's analyze the procedure `itb`. This procedure has two outputs (`flag` and `bin`), one input (`N`) and one inout (`int`). In the hardware domain, there is no distinction between variables and signals: all are signals. Also, type integer has to be converted to binary. The signal `flag` checks to see if signal `int` is divisible by two (even) or not (odd). This is done by the statements:

```
if (int MOD 2 = 0) then
    flag <= '1';
    else
    flag <= '0';
end if;
```

The procedure also includes a `For-Loop`:

```
for i in 0 to N loop

    if (int MOD 2 = 1) then
        bin (i) := '1';
        else
        bin (i) := '0';
    end if;

    int := int / 2;
end loop;
```

The loop is converting type integer `int` to binary `bin`. This conversion is not mapped to the hardware domain. As mentioned above, all signals in the hardware domain are binary; we cannot have an integer signal in the hardware domain. So, for our synthesis, the procedure is performing a test to see whether the signal is even or odd.

Now let's analyze the entity `Int_Bin`. The entity has two outputs: a four-bit signal `X_bin` (because `N` = 3) and a one-bit signal `flag_even`. The entity has one input of type integer, `Y_int`. The entity has one process:

```
process (Y_int)
variable tem : std_logic_vector (N downto 0);
variable tem_int : integer;

begin
    tem_int := Y_int;
    itb (tem, flag_even, N, tem_int);
    X_bin <= tem;
end process;
```

The process is calling the procedure itb, the integer Y_int is converted to binary X_bin, and flag_even is assigned a value of 1 if Y_int is even or 0 if it is odd. To find the hardware logic of flag_even, notice that if a binary number is even, its least significant bit is 0. Otherwise, the number is odd. So, flag_even = $\overline{\text{X\_bin}(0)}$. That is all there is to the synthesis of Listing 10.34. Figure 10.38 shows the synthesis of Listing 10.34; it is just a single inverter.

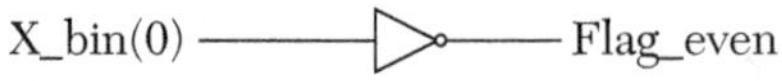

X_bin(0) ————————▷∘———— Flag_even

FIGURE 10.38 Synthesis of Listing 10.34.

10.3.8 Mapping the Function Statement

Functions, like procedures, are simulation constructs; they optimize the HDL module writing style. Consider the Verilog code shown in Listing 10.35.

LISTING 10.35 Verilog Example of a Function

```
module Func_synth (a1, b1, d1);
input a1, b1;
output d1;
reg d1;

always @ (a1, b1)
begin
d1 = andopr (a1, b1);
end

function andopr;
input a, b;
begin

andopr = a ^ b;
end
endfunction

endmodule
```

In the hardware domain, there is no distinction between the main module and a function; we look to see what the function is performing and then incorporate this information in the entity or module where the function is being called. For example, in Listing 10.35, the function `andopr` is performing an AND logical operation on two operands. The result is a single operand. In the module `Func_synth`, this function is called to perform an AND operation on the two inputs of the module, `a1` and `b1`; the result is stored in the output of the module `d1`. Listing 10.35 is synthesized as shown in Figure 10.39; it has an AND gate with two one-bit inputs, `a1` and `b1`, and a one-bit output, `d1`.

x	011	000	100	001	101	0111

y	1011	0101	1101	0111	0111	0111

FIGURE 10.39 Synthesis of Listing 10.35.

Another example of function synthesis is shown in Listing 10.36.

LISTING 10.36 Example of Function Synthesis

```
module Function_Synth2 (x, y);

input [2:0] x;
output [3:0] y;
reg [3:0] y;
always @ (x)
begin
y = fn (x);
end

function [3:0] fn;
input [2:0] a;
begin

if (a <= 4)

fn = 2 * a + 5;
end
endfunction

endmodule
```

The function in Listing 10.36 has one three-bit input `a` and one four-bit output `fn`. If the value of the input is less than or equal to four, the output is calculated as `fn = 2 * a + 5`. If the input is greater than four, the function does not change the previous value of the output. Incorporating the function into the module `Function_Synth2`, we summarize the module as representing a system with one three-bit input x and one four-bit output y. If x is less than or equal to four, y = 2 * a + 5. If x is greater than four, y retains its previous value. This means that latches must be used to retain the previous value.

Figure 10.40 shows the simulation output of the module Function_ Synth2. As is shown, if x is greater than four, y retains its previous value. To synthesize this module, we use four high-level triggered D-latches because output y is four bits. If x is from zero to four, these latches should be transparent; if x is from five to seven, these latches should be inactive. We design a signal clk connected to the clock of the latches; if x is from zero to four, the clk is high; otherwise, it is low. Table 10.16 shows the truth table of signal clk.

x	011	000	100	001	101	0111

y	1011	0101	1101	0111	0111	0111

FIGURE 10.40 Simulation output of Listing 10.36.

TABLE 10.16 Truth Table for Signal clk

x(2)	x(1)	x(0)	clk
0	0	0	1
0	0	1	1
0	1	0	1
0	1	1	1
1	0	0	1
1	0	1	0
1	1	0	0
1	1	1	0

From Table 10.16, the signal clk can be written as:

$$clk = \overline{x(2)} + \overline{x(0)}\,\overline{x(1)}$$

The truth table of output y when clk is high is shown in Table 10.17.

TABLE 10.17 Truth Table for Output y When clk is High

x(2)	x(1)	x(0)	y(3)	y(2)	y(1)	y(0)
0	0	0	0	1	0	1
0	0	1	0	1	1	1

x(2)	x(1)	x(0)	y(3)	y(2)	y(1)	y(0)
0	1	0	1	0	0	1
0	1	1	1	0	1	1
1	0	0	1	1	0	1
1	0	1	d	d	d	d
1	1	0	d	d	d	d
1	1	1	d	d	d	d

By inspecting Table 10.17, we find:

$$y(0) = 1$$
$$y(1) = x(0)$$
$$y(2) = \overline{x(1)}$$
$$y(3) = x(1) + x(2)$$

Figure 10.41 shows the synthesis of Listing 10.36.

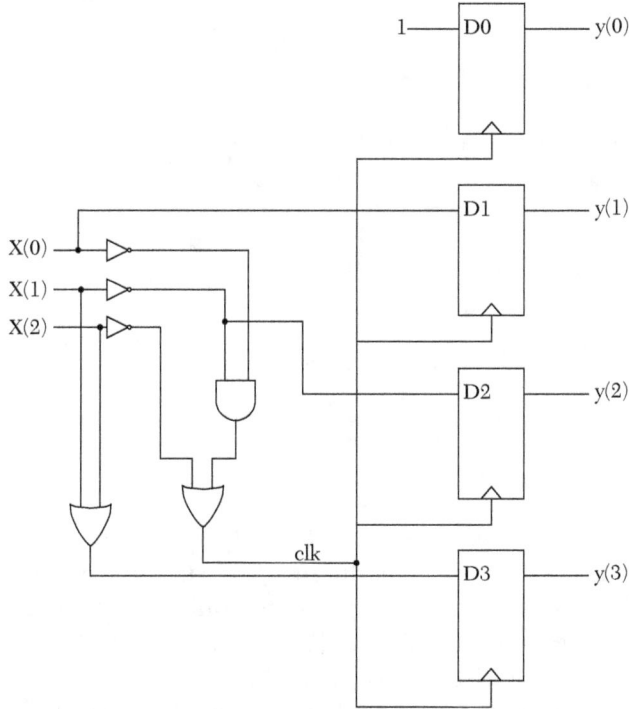

FIGURE 10.41 Synthesis of Listing 10.36.

As shown in Figure 10.41, the main components of the synthesis are latches. These latches are for storing the previous values of y. If the IF statement can be modified in Listing 10.36 to make it complete, all four latches in Figure 10.41 can be avoided (see Exercise 10.8).

10.3.9 Mapping the Verilog User-Defined Primitive

In Chapter 8, Verilog user-defined primitive (UDP) was covered. Listing 10.37 shows a copy of Listing 8.15 where UDP was implemented.

LISTING 10.37 (same as Listing 8.15) Verilog Code 2x1 Multiplexer with Active-Low Enable Using Combinational User-Defined Primitive

```
module Mux2x1Prmtv(A, B, SEL, Gbar,Y);
    input A,B,SEL,Gbar;

    output Y;

multiplexer MUX1 (Y, Gbar, SEL,A,B) ;

endmodule
primitive multiplexer (mux, enable, control, dataA, dataB) ;
output mux;
input enable, control, dataA, dataB;
table
// enable control dataA dataB mux
    1   ?   ?       ?       : 0;
    0   0   1       ?       : 1;
    0   0   0       ?       : 0;
    0   1   ?       1       : 1;
    0   1   ?       0       : 0;
    0   x   0       0       : 0;
    0   x   1       1       : 1;
endtable
endprimitive
```

To synthesize the code of Listing 10.37, we follow the same steps shown in Figure 10.1. The module can be summarized as a system with four one-bit inputs A, B, SEL, Gbar, and one one-bit output Y. The relationship between the output and the inputs of the system is shown in the statement table in Listing 10.37. To synthesize the code, a truth table is built; it is very similar to the contents of the statement table except, due to limitations of the hardware domain, the operator ? is replaced with the "don't care" operator x. Table 10.18 shows the truth table of representing the module in the hardware domain.

TABLE 10.18 Truth Table for Listing 10.37

Gbar	SEL	A	B	Y
1	x	x	x	0
0	0	1	x	1
0	0	0	x	0
0	1	x	1	1
0	1	x	0	0
0	x	0	0	0
0	x	1	1	1

Table 10.18 is the same as Table 2.4, and the logic diagram of Listing 10.37 is the same as Figure 2.9.

10.4   Summary

This chapter covered the fundamentals of hardware synthesis. We looked at synthesis as reverse engineering; HDL code was synthesized it into gates and latches. The steps of synthesizing any system can be summarized as follows:

1. Formulate the flowchart of the system.

2. Write the behavioral code of the system.

3. Simulate the behavioral code to verify the code.

4. Map the behavioral statements into hardware components and gates.

5. Write the structural code for the components and gates.

6. Simulate the structural code and compare it with the behavioral simulation to verify the mapping.

7. Download the components and gates into an electronic chip.

8. Test the chip to verify that the download represents the system.

The hardware domain is very limited in comparison to the simulation domain. For example, the hardware domain cannot distinguish between VHDL variables and signals. We learned how to map behavioral statements such as IF, case, and For-Loop. Any signal that needs to retain a value must be mapped using latches. Procedures, tasks, functions, and user-defined primitives are simulation tools; they do not have explicit hardware mappings. The operations they perform should be incorporated in the entity or in the module to be synthesized. Integers should be declared, if possible, with a range. This reduces the number of bits the synthesizer allocates for the integer. If the range is not specified, the synthesizer allocates at least 32 bits for integers.

10.5   Exercises

1. Synthesize the code in Listing 10.38, simulate it, write the structural description, and verify it.

LISTING 10.38 Code for Exercise 10.1

```
library IEEE;
use IEEE.STD_LOGIC_1164.ALL;
use ieee.numeric_std.all;

entity IF_sgned is
port (a : in signed (3 downto 0); Y : out Boolean);
end IF_sgned;

architecture IF_sgned of IF_sgned is

begin
IfB : process (a)
variable tem : Boolean;
begin
if (a < "1100") then
tem := true;
else
tem := false;
end if;
Y <= tem;
end process;
end IF_sgned;
```

2. Synthesize the code in Listing 10.39. Simulate it, write the structural description, and verify it.

LISTING 10.39 Code for Exercise 10.2

```
module elseif2 (inp, outp);
input [3:0] inp;
output [2:0] outp;
reg [2:0] outp;
always @ (inp)
begin
    if (inp[0] == 1'b1)
    outp = 3'd7;
    else if (inp[1] == 1'b1)
    outp = 3'd6;
    else if (inp[2] == 1'b1)
    outp = 3'd5;
    else
    outp = 3'd0;
end
endmodule
```

3. Verify the synthesis of Listing 10.26 by writing gate-level structural VHDL code for Figure 10.26. Simulate the code and verify that the simulation output is the same as that for Listing 10.26.

4. For the code of Listing 10.26, change the following lines:

```
else
    z_tem := 15;
    end if;
```

to just

```
    end if;
```

Synthesize the new code using multiplexers, gates, and flip-flops (if needed).

5. Simulate the VHDL behavioral code of Listing 10.30. Write the VHDL structural description of the logic diagram shown in Figure 10.33 and simulate it. Verify that the two simulations are identical.

6. Synthesize the behavioral code shown in Listing 10.40 using RTL.

LISTING 10.40 Code for Exercise 10.6

```
library IEEE;
use IEEE.STD_LOGIC_1164.ALL;

entity exercise is
```

```
port (a : in std_logic_vector (3 downto 0);
    c : in integer range 0 to 15;
    b : out std_logic_vector (3 downto 0));
    end exercise;
    architecture exercise of exercise is
    begin
    shfl : process (a, c)
    variable result, j : integer;
        variable temp : std_logic_vector (3 downto 0);
    begin

        result := 0;
        lop1 : for i in 0 to 3 loop
            if a(i) = '1' then
            result := result + 2**i;
            end if;
            end loop;
            if result > c then
            lop2 : for i in 0 to 3 loop
                j := (i + 3) mod 4;
                temp (j) := a(i);
                end loop;
            end if;
    b <= temp;
    end process shfl;
end exercise;
```

7. For Figure 10.41, write the structural code for the logic shown in the figure, simulate it, and verify that the figure is the synthesis of Listing 10.36.

8. We want to realize Figure 10.41 on a programmable device such as a FPGA. Use the synthesis tools (provided in most cases with the HDL package) to synthesize the code of Listing 10.36. Compare the outcome of the synthesizer with Figure 10.41 and report the differences. Now, use the tools provided in your HDL package to download the design into a FPGA or compatible chip. Use the same test signals to compare the software's and hardware's simulation. Report the differences and suggest how to minimize these differences.

9. In Listing 10.36, if the statement inside function fn is written as:

```
function [3:0] fn;
```

```
input [2:0] a;
begin

if (a <= 4)

fn = 2 * a + 5;
end
endfunction

endmodule
```

then it is likely that the code is intended to say that if a is greater than four, the value of fn is unimportant. If this is true, can you modify the function's code to avoid using the four latches? Redraw the synthesis of your code.

10. Synthesize the code in Listing 8.18. Hint: use RTL, use a register to synthesize the cpu address, and use a memory or group of registers to synthesize the cache. After synhesizing, write the Verilog code of the synthesis and verify it.

CREATING A *VHDL* OR *VERILOG* PROJECT USING *CAD* SOFTWARE PACKAGE

In this appendix, the necessary steps to create a new project using Xilinx ISE 13.1 or 14.1 software are covered. The steps include the source code and its test bench code. Although these steps are for ISE 13.1, the same concepts can be applied to other versions or to other vendors' products. These steps are for beginners. To find out more about these CAD packages, visit the homepage of the vendor.

Step 1: Double click on the Xilinx Project Navigator icon. From the toolbar, select File ® New Project. A dialog box will open (see Figure A.1).

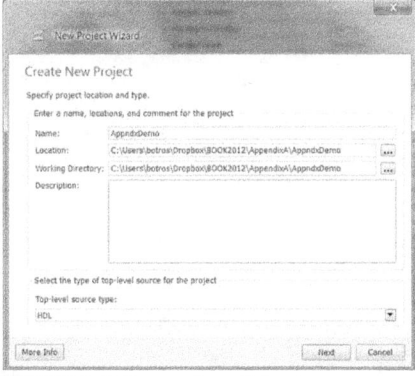

FIGURE A.1 New-project dialog box.

Step 2: In the new-project dialog box, type the desired location in the "Project Location" field or browse to the directory under which you want to create your new project directory using the browse button next to the "Project Location" field.

Step 3: Enter the name of the project. In Figure A.1, the name entered is "AppndxDemo."

Step 4: Click "Next," and Figure A.2 appears. Enter the appropriate information as shown in Figure A.2. The device is the chip where the HDL program, if desired, is downloaded after synthesis. The device in Figure A.2 is selected to be from the Spartan3E chip family.

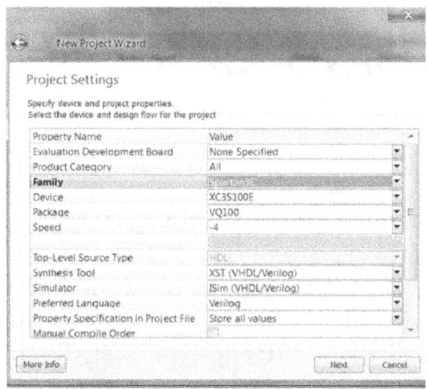

FIGURE A.2 Project dialog box.

Step 5: Click "Next" until you see the screen depicted in Figure A.3. This window summarizes the properties of the new project.

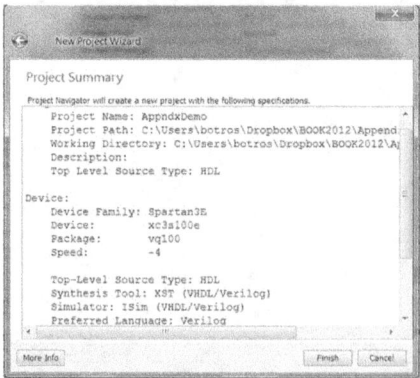

FIGURE A.3 Summary of entries to the project "AppndxDemo."

Step 6: Click "Finish" (see Figure A.4). The screen now shows the name of the project and the device.

FIGURE A.4 Simulation screen after clicking "Finish."

Step 7: Attach the HDL module to your project. Click "Project" and select "New Source" (see Figure A.5).

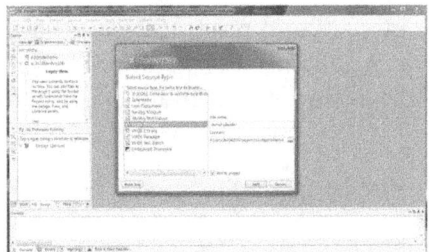

FIGURE A.5 Attaching a new source to the project.

Step 8: Because we are writing a VHDL module, select "VHDL Module" and enter the name of the entity (VHDL) or module (Verilog) as the file name in Figure A.5. The name of the file here is selected to be "DemoFulladder." It is preferable to leave the location as it is so the module and the project are stored in the same directory. If writing Verilog, select "Verilog Module" Instead of VHDL Module.

Step 9: Keep clicking "Next" until you can click "Finish." You will then have the windows shown in Figure A.6. The screen section of "Sources in Project" shows the name of the project and the VHDL module. The right-hand section of the screen shows a template for the VHDL module. Delete any comments or libraries that you do not need in your module. Referring to the left-hand side of Figure A.6, the "Processes for Source" panel shows the tools for compiling, testing, and synthesizing the VHDL module. On the bottom of the screen, the "Process View" panel accesses selected tools to display reports (logs) on various activities.

FIGURE A.6 The project and module screen.

Step 10: Enter the VHDL code for full adder. The VHDL module here is the VHDL code for the full adder that was discussed in Chapter 2 (see Figure 2.5). The copy, cut, and paste tools can be accessed via the "Edit" menu on the toolbar. After finishing writing the module, click "Save." The screen will display the code along with tools for testing and simulating the code (see Figure A.7).

FIGURE A.7 The full-adder module.

Step 11: Check the syntax of the VHDL program. This checking can be done using various tools; one of these tools is the behavioral check syntax, which appeared on the screen entitled "Processes," as shown in Figure A.7. To check the syntax, select the file "Demofulladder" and click on "Behavioral Check Syntax." The results of syntax checking, with detected errors (if any), appear on the screen entitled "Console" at the bottom of Figure A.7. If there are no errors, the VHDL code has been compiled successfully.

Step 12: After compiling the code, we need to simulate and test it. There are several ways to simulate and test. Here, we use a test bench to simulate and test the code. A test bench is a user-written code that assigns values or wave forms to the input signals of the code being tested. To build a test bench (see Figure A.8), select the file "Demofulladder" and click "Project-

New Source-VHDL Test Bench" and enter a name of the test bench. Here, we assign the name "fulladTstBnch" as the name of our test bench.

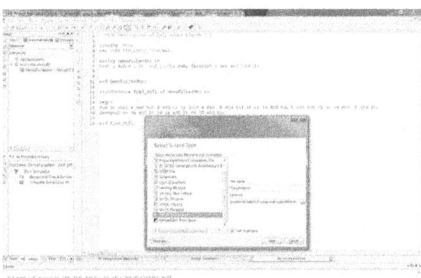

FIGURE A.8 Building a test bench.

Keep clicking "Next" and then "Finish." Figure A.9 shows the partial screen after clicking "Finish." Start cleaning up the template by deleting the comments if wanted.

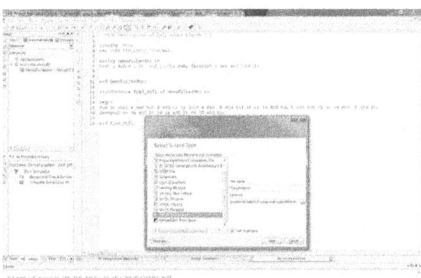

FIGURE A.9 Test-bench template.

The test-bench template declares the file "Demofulladder" (the VHDL code) as a component and declares its associated signals, here the inputs a, b, and c, and the output sum and carryout. The template instantiates the signals of the full adder with the ports of the component. The template lists several processes that can be used to instantiate values or waveforms to each input signal. To test the full adder, we insatiate each input signal with a clock waveform; the period of the clock is varying from one signal to the other, so all possible values of signals a, b, and c are generated. Figure A.10 shows the test-bench code. The statements

```
Pa: process
    begin
```

```
    wait for period;
    a <= not a;
end process;
```

describe a process that generates a clock with a period of 20 ns. The statement a <= not a; inverts signal a continuously; if a is 0, it will be inverted to 1 and vice versa. The statement wait for period; will delay the inversion of signal a for 10 ns because the period was declared to be 10 ns. This delay and the inversion generate a clock with a period of 20 ns.

FIGURE A.10 The test-bench code.

Step 13: After checking that the test-bench code has no errors, select the test-bench module and click on "Simulate Behavioral Model." On the simulation screen, adjust the scale to 10ns/division and click on "Run the Simulation." Figure A.11, which is a copy of Figure 2.7, shows the waveform that should appear on the screen.

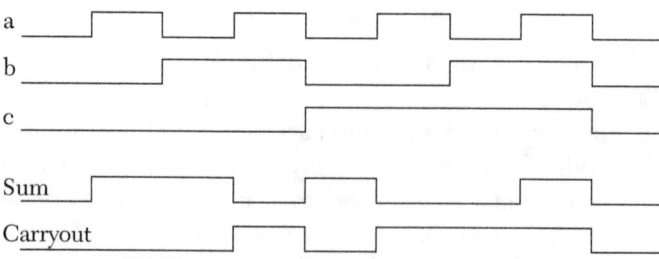

FIGURE A.11 Waveform output for full adder.

INDEX

www.ingramcontent.com/pod-product-compliance
Lightning Source LLC
Chambersburg PA
CBHW080401190526
45161CB00003B/98